Electronic Circuit and System Simulation Methods

Other McGraw-Hill Books of Interest

Handbooks

CHEN • *Computer Engineering Handbook*

COOMBS • *Printed Circuits Handbook*

COOMBS • *Electronic Instruments Handbook,* Second Edition

DI GIACOMO • *Digital Bus Handbook*

FINK AND CHRISTIANSEN • *Electronics Engineers' Handbook*

HARPER • *Electronic Packaging and Interconnection Handbook*

JURAN AND GRYNA • *Juran's Quality Control Handbook*

RORABAUGH • *Digital Filter Designer's Handbook*

TUMA • *Engineering Mathematics Handbook*

WAYNANT • *Electro-Optics Handbook*

WILLIAMS AND TAYLOR • *Electronic Filter Design Handbook*

Other

ANTOGNETTI • *Power Integrated Circuits*

BEST • *Phase-Locked Loops,* Second Edition

BUCHANAN • *CMOS / TTL Digital Systems Design*

BUCHANAN • *BiCMOS / CMOS Systems Design*

BYERS • *Printed Circuit Board Design with Microcomputers*

DAUGHERTY • *Analog to Digital Conversion*

ELLIOTT • *Integrated Circuits Fabrication Technology*

HECHT • *The Laser Guidebook*

KIELKOWSKI • *Inside SPICE*

LICARI • *Multichip Module Design*

MASSOBRIO AND ANTOGNETTI • *Semiconductor Device Modeling with SPICE,* Second Edition

SZE • *VLSI Technology*

TABAK • *Advanced Microprocessors,* Second Edition

TSUI • *LSI / VLSI Testability Design*

WATERS • *Active Filter Design*

WOBSCHALL • *Circuit Design for Electronic Instrumentation*

WALKER • *Optical Engineering Fundamentals*

Electronic Circuit and System Simulation Methods

Lawrence T. Pillage

Ronald A. Rohrer

Chandramouli Visweswariah

McGraw-Hill, Inc.

New York San Francisco Washington, D.C. Auckland Bogotá
Caracas Lisbon London Madrid Mexico City Milan
Montreal New Delhi San Juan Singapore
Sydney Tokyo Toronto

Library of Congress Cataloging-in-Publication Data

Pillage, Lawrence.
Electronic circuit and system simulation methods / Lawrence Pillage, Ronald A. Rohrer, Chandramouli Visweswariah.
p. cm.
Includes index.
ISBN 0-07-050169-6
1. Linear integrated circuits—Computer simulation. 2. SPICE (Computer file) I. Rohrer, Ronald A. II. Visweswariah, Chandramouli. III. Title.
TK7874.P52 1994
621.3815'01'1353—dc20 94-24429
CIP

1 2 3 4 5 6 7 8 9 0 DOC/DOC 9 0 9 8 7 6 5 4

ISBN 0-07-050169-6

The sponsoring editor for this book was Stephen S. Chapman.

Printed and bound by R. R. Donnelley & Sons Company.

Table of Contents

Preface

The field of circuit simulation has seen some exciting developments ever since the advent of integrated circuits. Modern integrated circuits continually challenge circuit simulation algorithms and implementations with the verification problems they pose. What makes circuit simulation unique is its multi-disciplinary nature. It is an intertwined set of concepts borrowed and adapted from mathematics, circuit theory, graph theory, physics, device modeling, electrical engineering, and software development. Although there is much active research in the subject, this book attempts to clearly explain some of the fundamentals of circuit simulation, on which most modern techniques are based. Some of the more recent advances are covered in the book, too.

This book evolved from our teaching and research activities over the years. We are indebted to all those who invented the concepts and techniques described in the book and to those who wrote earlier books on the subject.

This book would not have been possible without the collaboration, cooperation, and help of many colleagues, students, and friends. While it is not possible to mention all of them, we would particularly like to thank our spouses Leah Pillage, Casey Jones, and Patricia Buchanan for their constant support and encouragement; Catherine Rapinett for typing early drafts of the manuscript; all of the graduate students from the simulation courses at Carnegie Mellon University and the University of Texas at Austin for their critique of the notes; and David Ling, Ellen Yoffa, and Bill Joyner at the IBM T. J. Watson Research Center for their support and encouragement. We also thank Steve Chapman and Jim Halston at McGraw Hill for an outstanding job.

L.T.P., R.A.R., and C.V.

Electronic Circuit and System Simulation Methods

Chapter 1 *Introduction to Circuit Simulation*

The simulation techniques introduced in the 1950s to analyze circuits with tens of transistors are substantially the same as those used today to analyze circuits with tens of thousands of transistors. With the availability of powerful computers and the advent of workstations, circuit simulators can run much faster today than they did then. But these programs are often called upon to verify the potential performance of the next generation of workstations and parallel processors, and with each new generation of computers the underlying circuitry becomes ever more complex. Due to this increasing complexity, circuit simulation as a pre-manufacturing design verification strategy is barely keeping up with the demands that are being placed upon it. And in many areas of application it has even fallen behind, being supplemented by less precise switch- and logic-level simulation. But such reduced-precision simulation strategies may only work reliably when design styles are restricted. In the fierce competition for faster and smaller circuits, more often than not, presumed restrictions on design styles are bent or broken. In the final analysis then, only circuit simulation is trusted -- sometimes more so than it should be -- to verify the essential electrical-system behavior. As a result, huge computing resources are assigned to the circuit simulation task, which is often the bottleneck in the design process.

This chapter provides an overview of circuit simulation techniques. In particular, it is an introduction to some of the algorithms and methodologies used in the industry standard circuit simulator SPICE [Nagel71, Nagel75]. The techniques introduced in this chapter will be covered in greater detail in the chapters that follow. In addition, alternatives to SPICE and various nontraditional simulation techniques will be introduced in subsequent chapters.

1.1 Traditional Circuit Simulation

Before introducing circuit simulation techniques it may be helpful to first discuss some of the capabilities of a simulator such as SPICE. Most general purpose circuit simulation programs provide the following capabilities:

- Linear dc analysis to evaluate the dc currents and dc voltages for a lumped, linear, time-invariant circuit.
- Nonlinear dc analysis to obtain the quiescent operating point of a circuit which contains nonlinear elements, such as transistors.
- Linear ac analysis to obtain the frequency response of a lumped, linear, time invariant circuit.
- Small signal ac analysis to obtain the frequency domain response of a circuit by replacing nonlinear elements with their linearized equivalents computed from the quiescent operating point.
- Linear transient analysis to determine the time domain response of a circuit to various input waveforms starting with the initial conditions obtained from the linear dc analysis.
- Large signal transient analysis to obtain the time domain response of a circuit which contains nonlinear elements, such as transistors. The time domain responses are determined by considering the various input waveforms starting with the initial conditions obtained from the nonlinear dc analysis.

In addition to the modes listed above, circuit simulators provide a variety of other functions, such as pole/zero analysis and noise analysis. There are also special purpose simulation programs for thermal analysis, switched capacitor circuit analysis, and so on. By far the most common way in which the above modes are used is a nonlinear dc analysis to establish the quiescent point at the start of the simulation, followed by a large signal transient analysis. The analyses listed above, which form the core for most general purpose circuit simulators, also share a common analysis core in that they all rely on the linear dc analysis algorithm. Therefore, to consider how a circuit simulator provides these capabilities we start with the foundations of linear dc analysis.

1.2 Linear, Time-Invariant Circuits

Consider a simple linear time-invariant circuit comprised of the most basic two-terminal elements: resistors, ideal independent current and voltage sources. The branch relations for these circuit elements are shown in Figure 1.1. The current-voltage relationships for any of these elements can be abstracted in terms of a generic two-terminal element or a topological branch as shown in Figure 1.2. In order to treat all types of branches in a consistent and systematic manner, we use *associated reference directions*.

Associated Reference Directions: The positive (+) reference for the branch voltage (v_b) is at the tail of the branch current (i_b) reference arrow, and the negative (-) reference for the branch voltage is at the head of the branch current reference arrow.

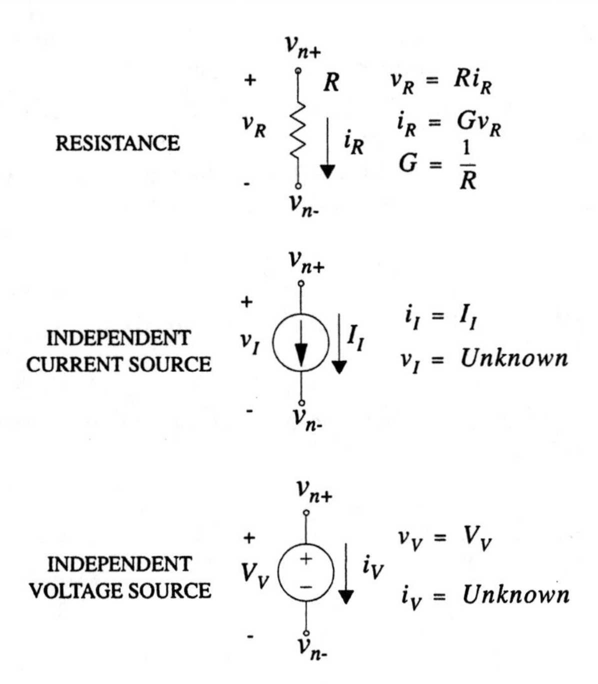

Figure 1.1 Branch relations for some two-terminal circuit elements.

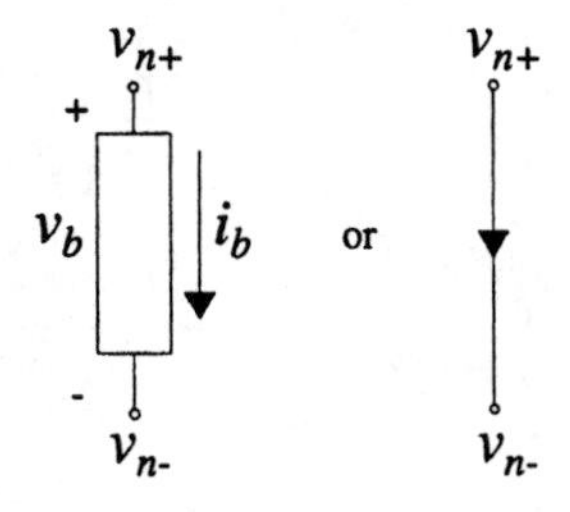

Figure 1.2 Generic two-terminal element and a topological branch.

The (positive referenced) branch voltage drop is in the same direction as the (positive referenced) branch current. Using the associated reference directions, we can compute instantaneous branch power as follows:

- $p_b = v_b i_b > 0$ implies power is being delivered to the element from the rest of the circuit.
- $p_b = v_b i_b < 0$ implies power is being delivered to the rest of the circuit from the element.

Using the topological branch model in Figure 1.2 and the associated reference direction conventions, Kirchhoff's Laws of Interconnection are easily defined:

Kirchhoff's Voltage Law (KVL): Every circuit node has a unique voltage (with respect to the ground or datum node, which is 0 volts by convention). The voltage (drop) across a branch, v_b, is equal to the difference between the (positive and negative referenced) voltages of the nodes on which it is incident.

$$v_b = v_{n+} - v_{n-} \tag{1.2.1}$$

Kirchhoff's Current Law (KCL): The (algebraic) sum of all of the currents flowing out of (or into) any circuit node is zero.

An example of the application of KCL at a node is shown in Figure 1.3. The branch relations along with the KCL expression for node N result in the following equation in terms of the node voltage variables:

$$\frac{v_N - v_M}{R_3} + \frac{v_N - v_K}{R_5} + \frac{v_N - v_R}{R_{10}} + \frac{v_N - v_S}{R_{64}} + I_6 - I_{49} = 0 \tag{1.2.2}$$

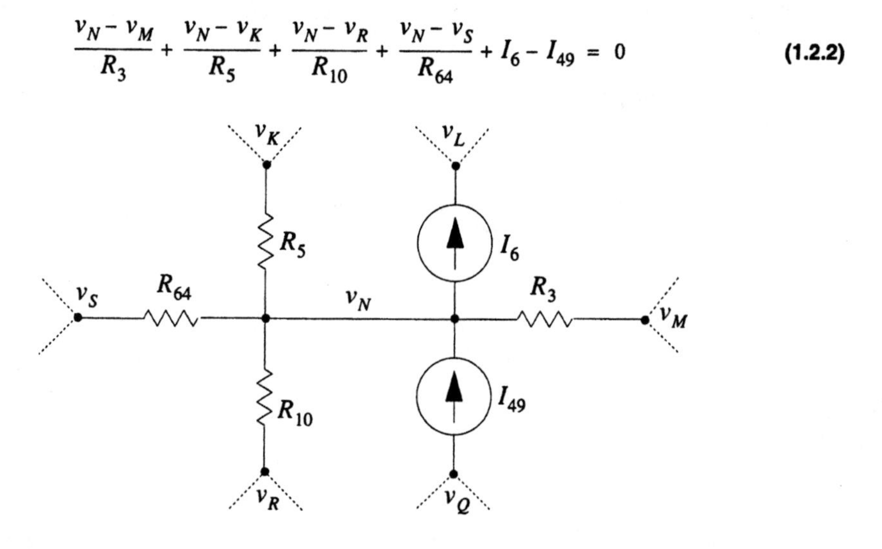

Figure 1.3 KCL at the Nth node.

1.3 Nodal Analysis

Consider a b branch, $(n+1)$ node circuit (in the rest of this book, we will consider $(n+1)$ node circuits so that the number of non-datum nodes will always be n). Writing a

KCL equation in terms of node voltages, as described above, for every non-datum node, leads to n nodal equations in terms of n non-datum node voltages. The *datum* node is also called the *ground* or *reference* node, and usually is taken to be at a potential of zero volts. We can write programs to formulate and then solve such sets of equations in general.

For example, consider the analysis of the dc circuit shown in Figure 1.4. This circuit has four nodes not counting the ground node; therefore, $n = 4$. Consider for now that all of the resistances are $1\,\Omega$ and the two current sources have values of 1A. By inspection we can write the KCL expression for the n non-datum nodes resulting in the following set of nodal equations:

$$\frac{v_1 - v_0}{R_2} + \frac{v_1 - v_2}{R_3} - I_1 = 0 \qquad \rightarrow \qquad 2v_1 - v_2 = 1$$

$$\frac{v_2 - v_1}{R_3} + \frac{v_2 - v_0}{R_4} + \frac{v_2 - v_3}{R_5} = 0 \qquad \rightarrow \qquad -v_1 + 3v_2 - v_3 = 0$$

$$\frac{v_3 - v_2}{R_5} + \frac{v_3 - v_0}{R_6} + \frac{v_3 - v_4}{R_7} = 0 \qquad \rightarrow \qquad -v_2 + 3v_3 - v_4 = 0 \tag{1.3.1}$$

$$\frac{v_4 - v_3}{R_7} + \frac{v_4 - v_0}{R_8} - I_9 = 0 \qquad \rightarrow \qquad -v_3 + 2v_4 = 1$$

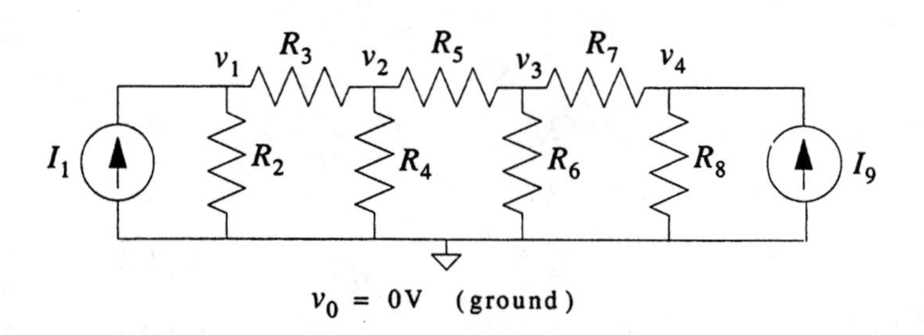

Figure 1.4 Resistor ladder circuit example.

These n nodal equations can be expressed in matrix form

$$Yv = J \tag{1.3.2}$$

where Y is the $n \times n$ nodal admittance matrix, J is the $n \times 1$ vector of current source inputs and v is the $n \times 1$ vector of node voltages which are being sought. Writing the set of nodal equations by applying KCL at each node is not, however, the most efficient means of formulation for a software program.

1.4 Nodal Admittance Equation Stamps

In general, the circuit being analyzed is described for a circuit simulation program in terms of a netlist file. Even computer-aided engineering (CAE) tools which include schematic capture will first convert the graphical representation into a netlist description. Figure 1.5 is a simple netlist format (similar to the SPICE language) which describes the circuit shown in Figure 1.4.

Branch type/ name	From node	To node	Value
I1	0	1	1.0
R2	1	0	1.0
R3	1	2	1.0
R4	2	0	1.0
R5	2	3	1.0
R6	3	0	1.0
R7	3	4	1.0
R8	4	0	1.0
I9	0	4	1.0

Figure 1.5 A netlist representation of the circuit shown in Figure 1.4.

For the nodal admittance (matrix) equations, the elements in the netlist contribute terms in a procedural manner. Branches 2 through 8, the resistors, contribute terms to the Y matrix while branches 1 and 9, the independent current sources, contribute to the J vector.

From the input list these contributions can be characterized on a branch-by-branch basis in terms of *matrix stamps* (sometimes called *element stencils* or just *stencils*). To explain how stamps work, we will assume that the circuit nodes are consecutively numbered. Of course this is not the case in general, however, unique alphanumeric node names in the user-specified netlist are easily mapped into consecutively numbered nodes internal to the circuit simulator.

The resistor matrix-stamp is shown in Figure 1.6(a). For a resistor of value R_k from node i to node j as shown in Figure 1.7(a), a *positive* conductance value $(1/R_k)$ is *added* to matrix locations (i, i) and (j, j) while a *negative* value $(-1/R_k)$ is *added* to matrix locations (i, j) and (j, i). The stamp for independent current sources is shown in Figure 1.6(b). For a current source branch of value I_k directed from node i to node j, as shown in Figure 1.7(b) a *negative* current source value is *added* to the ith entry of the $\boldsymbol{J}$ vector and a *positive* value is *added* to the jth entry. We postpone the treatment of the independent voltage source stamp until Chapter 2 where we discuss Modified Nodal Analysis.

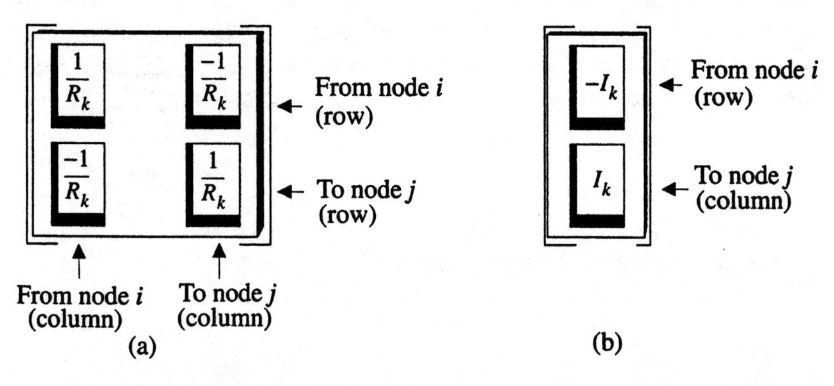

Figure 1.6 Element stamps: (a) For a resistor of value R_k connected *from* node i *to* node j; (b) For an independent current source of value I_k connected *from* node i *to* node j.

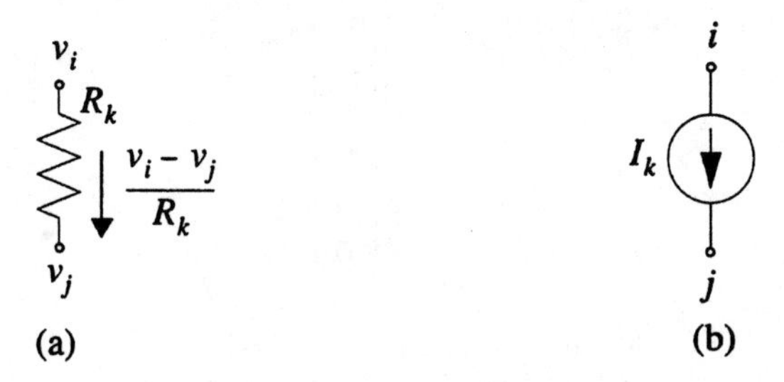

Figure 1.7 (a) A resistor connected *from* node i *to* node j, and (b) a current source connected *from* node i *to* node j.

We can obtain the overall set of nodal admittance equations by *stamping in* the branch contributions on an element-by-element basis. To understand the stamp concept, note that the current entering node i and leaving node j due to I_k adds two terms to the $\boldsymbol{J}$ vector. For example, the first line of the netlist in Figure 1.5 is stamped into the nodal matrix equations as follows:

$$\begin{bmatrix} 0 & 0 & 0 & 0 & 0 \\ 0 & 0 & 0 & 0 & 0 \\ 0 & 0 & 0 & 0 & 0 \\ 0 & 0 & 0 & 0 & 0 \\ 0 & 0 & 0 & 0 & 0 \end{bmatrix} \begin{bmatrix} v_0 \\ v_1 \\ v_2 \\ v_3 \\ v_4 \end{bmatrix} = \begin{bmatrix} -I_1 \\ I_1 \\ 0 \\ 0 \\ 0 \end{bmatrix} \tag{1.4.1}$$

Note that this section builds up the equations the way a computer program would. The intermediate forms of the equations (1.4.1), (1.4.4), and (1.4.5) are not mathematically valid equations, but the final set of equations (1.4.6) and (1.4.7) are complete. For resistors (see Figure 1.7(a)), the current is expressed in terms of the node voltages on the left hand side. For example, the current *leaving* node i due to resistor R_k is

$$\frac{1}{R_k}(v_i - v_j) \tag{1.4.2}$$

and the current *leaving* node j through resistor R_k is

$$\frac{1}{R_k}(v_j - v_i) \tag{1.4.3}$$

Therefore, stamping in the second element (R_2) from the netlist in Figure 1.5 results in

$$\begin{bmatrix} \frac{1}{R_2} & -\frac{1}{R_2} & 0 & 0 & 0 \\ -\frac{1}{R_2} & \frac{1}{R_2} & 0 & 0 & 0 \\ 0 & 0 & 0 & 0 & 0 \\ 0 & 0 & 0 & 0 & 0 \\ 0 & 0 & 0 & 0 & 0 \end{bmatrix} \begin{bmatrix} v_0 \\ v_1 \\ v_2 \\ v_3 \\ v_4 \end{bmatrix} = \begin{bmatrix} -I_1 \\ I_1 \\ 0 \\ 0 \\ 0 \end{bmatrix} \tag{1.4.4}$$

Similarly, after stamping the third element (R_3) from the netlist the partial matrix equations are

$$\begin{bmatrix} \frac{1}{R_2} & -\frac{1}{R_2} & 0 & 0 & 0 \\ -\frac{1}{R_2} & (\frac{1}{R_2}+\frac{1}{R_3}) & -\frac{1}{R_3} & 0 & 0 \\ 0 & -\frac{1}{R_3} & \frac{1}{R_3} & 0 & 0 \\ 0 & 0 & 0 & 0 & 0 \\ 0 & 0 & 0 & 0 & 0 \end{bmatrix} \begin{bmatrix} v_0 \\ v_1 \\ v_2 \\ v_3 \\ v_4 \end{bmatrix} = \begin{bmatrix} -I_1 \\ I_1 \\ 0 \\ 0 \\ 0 \end{bmatrix} \tag{1.4.5}$$

Finally, after stamping in all of the elements from the netlist we have the following set of nodal equations:

$$\begin{bmatrix} (\frac{1}{R_2}+\frac{1}{R_4}+\frac{1}{R_6}+\frac{1}{R_8}) & -\frac{1}{R_2} & -\frac{1}{R_4} & -\frac{1}{R_6} & -\frac{1}{R_8} \\ -\frac{1}{R_2} & \frac{1}{R_2}+\frac{1}{R_3} & -\frac{1}{R_3} & 0 & 0 \\ -\frac{1}{R_4} & -\frac{1}{R_3} & \frac{1}{R_3}+\frac{1}{R_4}+\frac{1}{R_5} & -\frac{1}{R_5} & 0 \\ -\frac{1}{R_6} & 0 & -\frac{1}{R_5} & (\frac{1}{R_5}+\frac{1}{R_6}+\frac{1}{R_7}) & -\frac{1}{R_7} \\ -\frac{1}{R_8} & 0 & 0 & -\frac{1}{R_7} & \frac{1}{R_7}+\frac{1}{R_8} \end{bmatrix} \begin{bmatrix} v_0 \\ v_1 \\ v_2 \\ v_3 \\ v_4 \end{bmatrix} = \begin{bmatrix} -I_1-I_9 \\ I_1 \\ 0 \\ 0 \\ I_9 \end{bmatrix} \tag{1.4.6}$$

We need not -- and usually do not -- bother to build the zeroth (ground node) row and corresponding column, but here it does show the symmetry of the nodal formulation. What we have here is the *indefinite* (or "floating") *admittance formulation:* each row and each column sums algebraically to zero, indicating that we do not have an independent set of equations. We can pick any node to be the reference and obtain the *definite admittance formulation* by crossing out its row and the corresponding column. In this case the zero row and column correspond to the designated ground node, and can be omitted:

$$\begin{bmatrix} (\frac{1}{R_2}+\frac{1}{R_3}) & -\frac{1}{R_3} & 0 & 0 \\ -\frac{1}{R_3} & (\frac{1}{R_3}+\frac{1}{R_4}+\frac{1}{R_5}) & -\frac{1}{R_5} & 0 \\ 0 & -\frac{1}{R_5} & (\frac{1}{R_5}+\frac{1}{R_6}+\frac{1}{R_7}) & -\frac{1}{R_7} \\ 0 & 0 & -\frac{1}{R_7} & (\frac{1}{R_7}+\frac{1}{R_8}) \end{bmatrix} \begin{bmatrix} v_1 \\ v_2 \\ v_3 \\ v_4 \end{bmatrix} = \begin{bmatrix} I_1 \\ 0 \\ 0 \\ I_9 \end{bmatrix} \quad (1.4.7)$$

Because all rows and columns sum algebraically to zero, we can reconstruct the indefinite admittance formulation from the definite in case we want to change reference nodes. The indefinite admittance formulation will be useful to us later when we consider multi-terminal elements. For now, we should note that the two-terminal element stamps are indefinite. They can be made definite, of course, by eliminating any matrix entries that correspond to the ground node's row or column.

The final step for dc Nodal Analysis is to solve the matrix equations in (1.4.7) to obtain the dc node voltages. Constructing the inverse of $\boldsymbol{Y}$ and multiplying it by $\boldsymbol{J}$ is one way of obtaining the vector of node voltages. However, there are more efficient means of solving circuit equations (more on this later in Chapter 3 and Chapter 7). In fact, most simulators use a direct method for solving sets of linear simultaneous equations to obtain the node voltages. Gaussian Elimination or LU factorization are the most often applied algorithms for generalized circuit simulation. In this book we will often use the matrix inverse notation to indicate solution of a set of simultaneous equations, like

$$\boldsymbol{v} = \boldsymbol{Y}^{-1}\boldsymbol{J} \quad (1.4.8)$$

In almost all such instances, a practical implementation would use Gaussian Elimination or LU factorization in the interests of efficiency. Thus any references to obtaining the inverse of a matrix can be assumed to mean factoring the equations with either of these direct methods. Further, the equations of large circuits produce sparse matrices, i.e., matrices with very few non-zero entries. The sparsity of such matrices is exploited to obtain the solution on a computer with reasonable memory and runtime requirements. The matrix in (1.4.7) is somewhat sparse since the percentage of non-zeros is only 62.5 percent. For large circuits, however, the matrices can be extremely sparse with less than 1 percent of the entries being non-zero. The special handling of *sparse matrices* will be covered in detail in Chapter 7 where it is shown that one of the reasons that we do not attempt to invert large sparse matrices is that the sparsity gets destroyed.

The importance of the linear solution cannot be overstated since the following sections will demonstrate that nonlinear dc and time domain transient circuit analysis are based upon successive applications of a linear circuit solver.

1.5 Nonlinear (dc) Circuit Analysis

Consider the problem of determining the dc bias point for the simple diode circuit shown in Figure 1.8. This single loop circuit can be most easily described in terms of a loop equation for the current i_d. Suppose that the diode in Figure 1.8 is characterized in the $i - v$ plane by the curve shown in Figure 1.9. Classically, we can obtain the dc operating point by means of the (Thevenin equivalent) load line also shown in this figure. The load line represents the voltage source and series resistor, and its intersection with the $i - v$ curve for the diode is the dc operating point for this circuit.

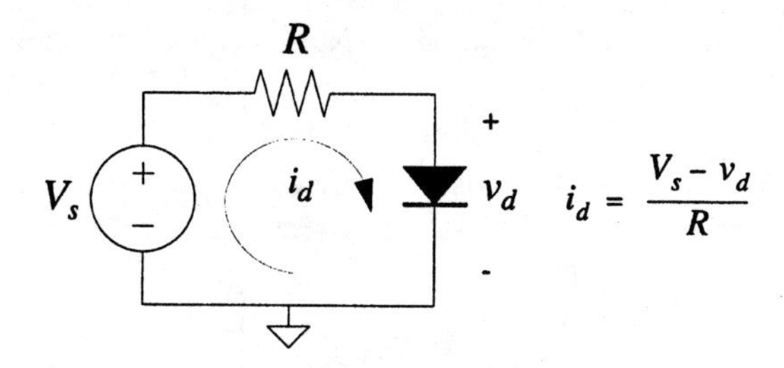

Figure 1.8 Simple diode circuit.

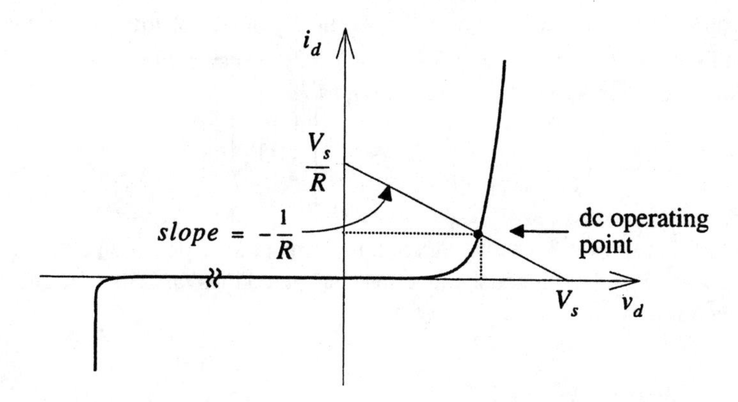

Figure 1.9 Diode i-v curve and graphical solution of dc operating point.

Computationally, we solve for this dc operating point iteratively. To describe how this circuit would be analyzed in a circuit simulator we will first convert the Thevenin equivalent in Figure 1.8 to the Norton equivalent shown in Figure 1.10 in order to simplify this discussion. The single KCL equation at the non-datum node, hence, the nodal equation is

$$\frac{v_d}{R} + i_d = \frac{V_s}{R} \tag{1.5.1}$$

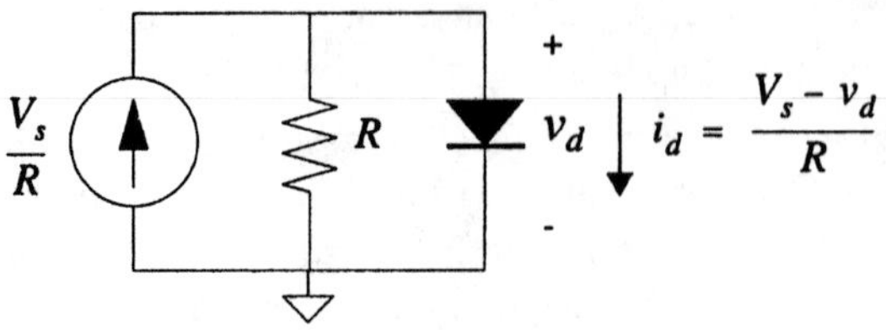

Figure 1.10 Diode circuit with a Norton equivalent source.

For completeness, however, we must also include the branch relation equation for the diode. The nonlinear expression for the diode's current could be the following:

$$i_d = I_{SAT}\left[exp\left(\frac{qv_d}{\eta kT}\right) - 1\right] \tag{1.5.2}$$

where I_{SAT} is the reverse saturation current, q is the charge on an electron, η is the non-ideality factor, k is Boltzmann's constant, and T is the absolute temperature in degrees Kelvin. Equation (1.5.2) describes a curve in the first quadrant of the i_d-v_d plane similar to the one in Figure 1.9. Combining (1.5.1) and (1.5.2) we can express the nodal equation in terms of the node voltage variable exclusively, as

$$\frac{v_d}{R} + I_{SAT}\left[exp\left(\frac{qv_d}{\eta kT}\right) - 1\right] = \frac{V_s}{R} \tag{1.5.3}$$

Equation (1.5.3) represents a nodal equation in terms of a single node voltage. The only difference between it and other nodal equations considered thus far is that the equation is a nonlinear function $f(v_d)$:

$$f(v_d) = \frac{V_s}{R} - \left(\frac{v}{R} + I_{SAT}\left[exp\left(\frac{qv_d}{\eta kT}\right) - 1\right]\right) = 0 \tag{1.5.4}$$

To solve (1.5.4) using Newton-Raphson [Ralston78] we would start with some initial guess $v_d = V_{init}$ and attempt to evaluate the node voltage as follows:

Start: $n = 0$

Guess: $v_n = V_{init}$

Linearize: $f(v_n + \Delta v_n) \approx f(v_n) + f'(v_n)\Delta v_n = 0$ (Taylor expansion)

Solve: $\Delta v_n = -\dfrac{f(v_n)}{f'(v_n)}$ where $f'(v_n) = \left.\dfrac{df}{dv}\right|_{v = v_n}$

Increment: $v_{n+1} = v_n + \Delta v_n$

Test: Is $|f(v_{n+1})| \leq \varepsilon$?

If yes, we have converged to the required solution; if no, then return to the linearization step.

Newton-Raphson iteration will converge provided that:

1. The first derivative of $f(v)$, $f'(v)$, is continuous (which presents certain restrictions on the device models).
2. The initial guess is “sufficiently close” to the final solution (SPICE uses the *node-set* concept to allow the user to force this situation).

The Newton-Raphson steps outlined above can also be explained graphically for the simple circuit in Figure 1.10. For example, consider the linear load line and the diode curve in Figure 1.11. Starting with an initial guess P1, a tangent T1 is extended from this presumed operating point. At the intersection of T1 and the load line, solution S1 is obtained. This tangent projection represents the Newton-Raphson linearization step while the solution represents the solve step. If S1 is not within ε (the error tolerance) of the actual operating point, a new guess, P2, is obtained from this first solution. The next solution attempt is then the intersection of tangent T2 with the load line. These iterations are repeated until a solution is found to be acceptably close to the exact operating point. (Since the exact solution is not known, convergence is assumed when successive solutions are within ε of each other.)

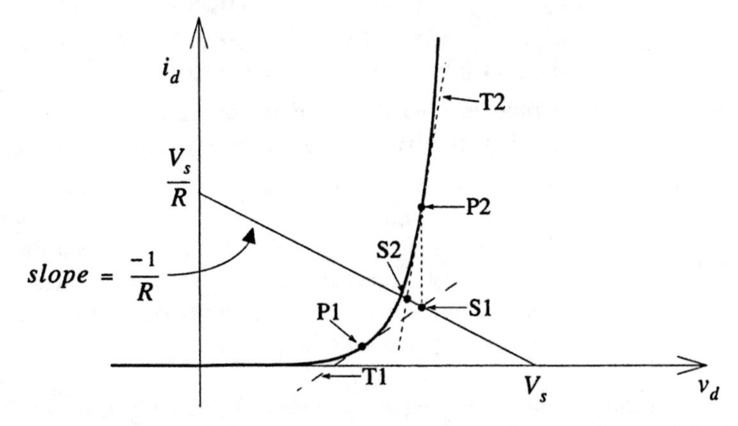

Figure 1.11 Iteration to obtain dc operating point.

The linearization about a presumed operating point can be depicted as replacing the diode by a linear Norton equivalent circuit at each iteration. The relation between the tangent linearization and a linear Norton equivalent circuit is shown in Figure 1.12. The iteration procedure amounts to successive solutions of linear(ized) dc circuits with each nonlinear device being replaced by its tangent-determined Norton equivalent circuit.

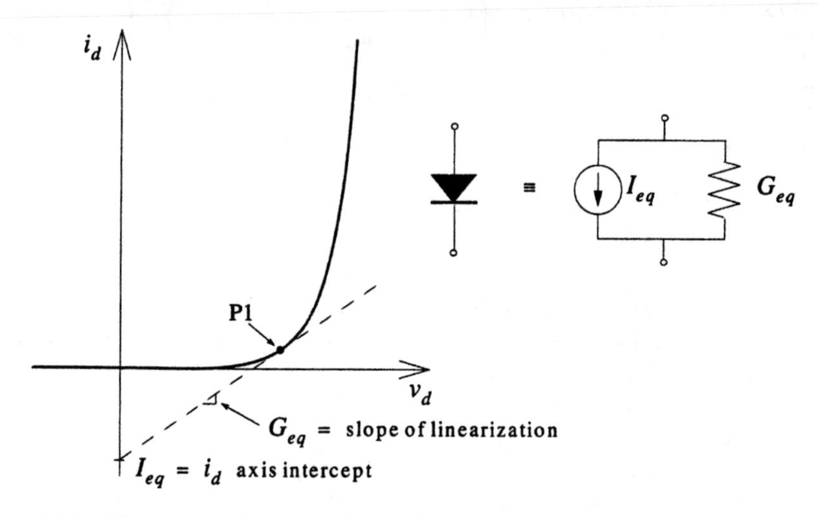

Figure 1.12 The companion model which represents the linearized approximation (tangent) about the presumed operating point.

In general, this nonlinear iteration procedure is attempted with multiple nonlinear elements, all being linearized simultaneously. The overall procedure is as follows:

1. Initialize: Guess circuit voltages and/or currents.
2. Linearize: Obtain linearized Norton equivalents for all nonlinear elements about their presumed operating points.
3. Solve: Formulate (nodal) equations that characterize the linearized circuit, and solve them for the new presumed operating points.
4. Convergence: Return to Step 1 using the new presumed operating points as the "guesses" unless the change from the last iteration has been acceptably small. If the changes are small, then the iteration terminates successfully. If the number of iterations has exceeded a pre-determined number of iterations, then the solution procedure has failed.

Since multiple elements are linearized simultaneously, each element which should "see" a nonlinear load line will actually "see" a moving linear load line from one iteration to the next. Mathematically, this is multi-dimensional Newton-Raphson iteration which will be covered in more detail in Chapter 10. Assuming that such iterations are feasible, we can extend our dc analysis of linear circuits to solve dc nonlinear circuits by adding to it *model routines* that provide linearized equivalent circuits as functions of their presumed node voltages. Note that such nonlinear elements contribute a stamp both to the Y matrix and to the J vector.

For example, we showed that each nonlinear iteration corresponded to a tangent approximation of the curve. In terms of (1.5.2) the linearized model in Figure 1.13(a) is

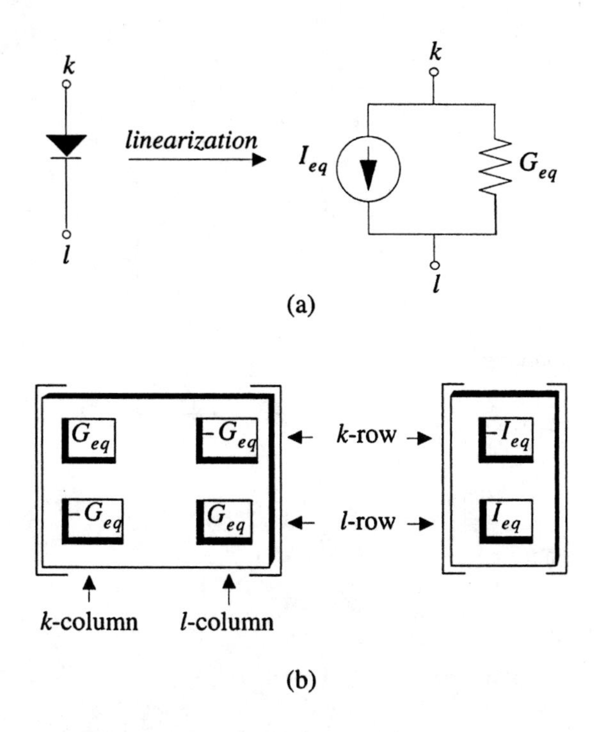

Figure 1.13 Stamp for linearized diode model.

obtained by evaluating the tangent (partial derivative of (1.5.2)) at the voltage operating point (the last nonlinear iteration value). Note that the figure shows the direction of I_{eq}, which is to be treated algebraically. I_{eq} may be negative, as shown in the figure. This linearization will be covered in more detail in Chapter 10, however, we note here that the resulting linear Norton equivalent circuit has a Y matrix and J vector stamp as shown in Figure 1.13(b). Therefore the heart of the nonlinear analysis routine is the repeated stamping and linear circuit solving as summarized by the flowchart in Figure 1.14.

We should also point out that the dc diode equation (1.5.2) does not include the capacitance effects associated with a diode junction. Capacitance is, of course, excluded from consideration simply because we are performing a dc analysis. In a dc analysis, there is no concept of time. The assumption is that time is allowed to go to infinity so that all tran-

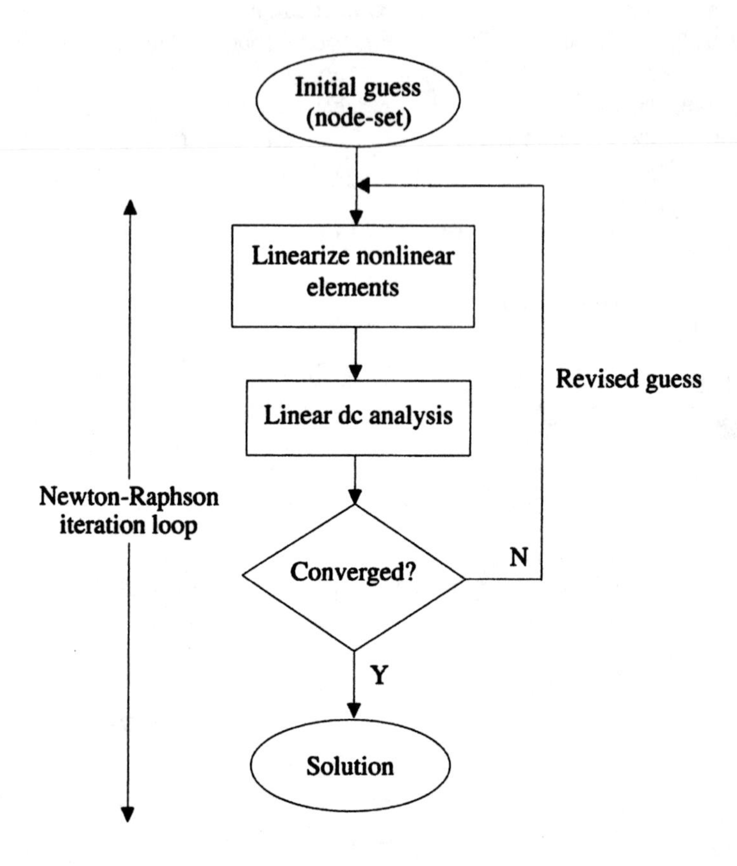

Figure 1.14 Flowchart for nonlinear dc analysis. Convergence is reached when approximately the same answer is obtained twice in a row. The rate and fact of convergence are affected by the initial guess.

sients in the circuit die out, and our only interest is in computing the final "steady state" values of the circuit. Therefore, capacitors are made into open circuits and inductors into short circuits for the purposes of a dc analysis. However, if this dc analysis is a precursor to a nonlinear transient analysis or an ac analysis, the energy storage elements are reintroduced once the dc solution has been obtained. In this case, the diffusion and depletion capacitances associated with the diode are evaluated at the computed operating point and introduced into the circuit for the transient or ac analysis.

1.6 Small Signal (ac) Analysis

Due to the way in which the nonlinear (Newton-Raphson) iterations are performed, the final Newton-Raphson linearization step produces a portion of the small signal equivalent circuit model used for ac analysis. For example, the small signal model for the diode in Figure 1.8, biased at a voltage V^*, is the tangent about the operating point (the last Newton-Raphson iteration) as shown in Figure 1.15. Of course this is not the complete small

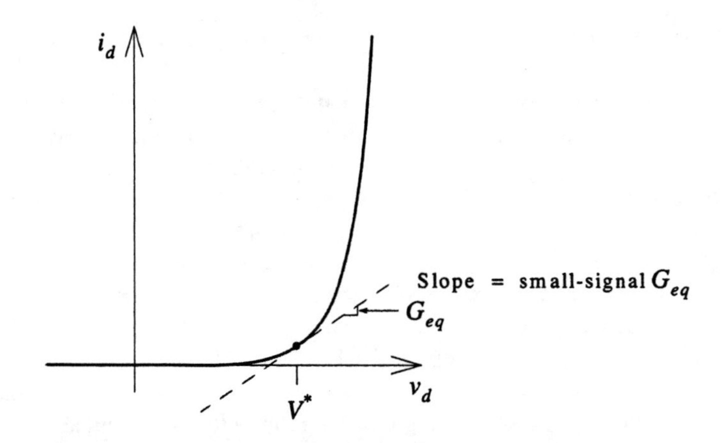

Figure 1.15 Small signal diode model at the dc bias point.

signal model since the curve in Figure 1.15 represents the dc characteristics of the diode as shown in equation (1.5.2). The model does not account for the small signal capacitance effects. The small signal capacitances (both the diffusion and depletion capacitance) are simply obtained by evaluating them at the dc operating point, V^*.

For small signal ac analysis the energy storage elements (constituting the linearized small signal capacitances and inductances) are modeled in the frequency domain, as shown in Figure 1.16. With these branch relations in terms of complex admittances, we can extend linear dc analysis to linear ac analysis simply by using complex valued voltages and currents (phasors) and stamping the complex admittances into the $\boldsymbol{Y}$ matrix. The stamps are identical to those for linear resistors in every way except that the admittances are now complex values. Therefore, we must extend the linear simultaneous equation solver to handle complex values. The analysis steps for an ac analysis are summarized in Figure 1.17.

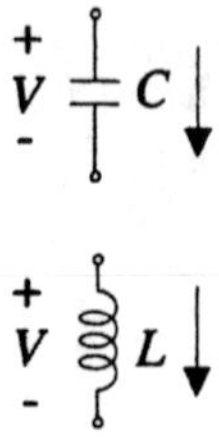

$$I = j\omega CV \Rightarrow G_{eq} = j\omega C$$

$$I = \frac{V}{j\omega L} \Rightarrow G_{eq} = \frac{1}{j\omega L}$$

Figure 1.16 Complex admittances for capacitors and inductors during ac analysis.

It is important to note that we must exercise caution when performing an ac analysis at the frequency extremes. Specifically, as $\omega \to 0$ the capacitors become open circuits and the inductors become shorts. Conversely, as $\omega \to \infty$ the inductors become open circuits and the capacitors become shorts. Nodal Analysis, which handles only admittance branch relations, may not cope well with these frequency extremes. Even values close to the extremes may cause the Y matrix entries to become excessively large or small, thus causing numerical problems due to the finite precision and range of number representation in a computer. We will discuss ways to overcome such problems later. Figure 1.17 assumes that reformulation of the nodal equations will solve the problem of extreme values. SPICE is set up to handle the extreme case of $\omega \to \infty$ only under the assumption that the frequency specified will never be sufficiently close to infinity to cause trouble. However, the limit $\omega = 0$ is a more difficult problem since this is a practical frequency (dc) and the inductor admittance becomes infinite.

1.7 Linear Transient Analysis

In contrast to small signal ac analysis, linear transient analysis is applied to evaluate the large signal behavior of a linear circuit as a function of time, t. For example, consider the lumped, linear (C is a constant and not a function of the voltage v), time-invariant capacitance shown in Figure 1.19. Suppose that we have a circuit with two such capacitors and a large signal time-varying current source as shown in Figure 1.18. The nodal equations for this circuit are:

$$\frac{v_1}{R_1} + \frac{v_1 - v_2}{R_2} + C_1 \cdot \frac{d}{dt}(v_1 - v_2) = I_{in}(t)$$

$$\frac{v_2}{R_3} + \frac{v_2 - v_1}{R_2} + C_1 \cdot \frac{d}{dt}(v_2 - v_1) + C_2 \cdot \frac{dv_2}{dt} = 0 \tag{1.7.1}$$

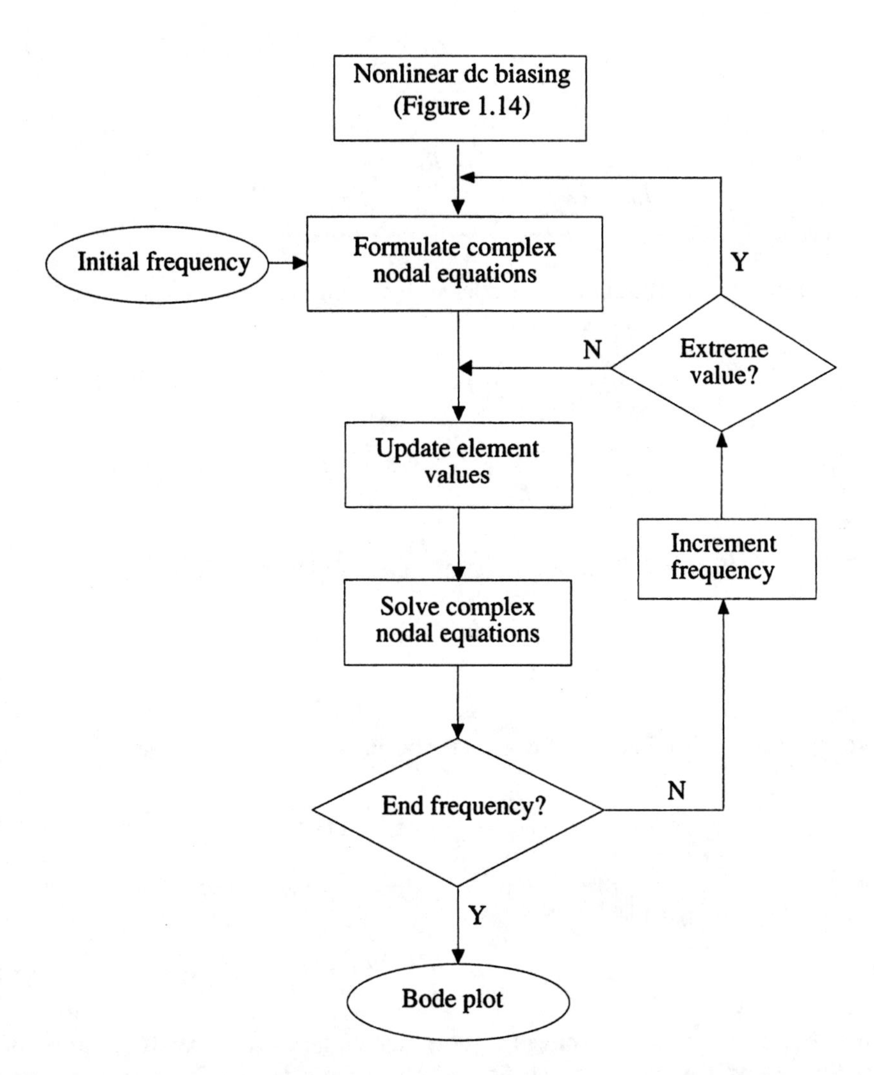

Figure 1.17 Small signal ac analysis. As the frequency is incremented it must be checked to ensure that the frequency is not at some extremal point that will cause ill-conditioning.

Notice that the equations in (1.7.1) form a set of first order linear differential equations. Suppose that we know the solution of the circuit in Figure 1.18 at time t, and we seek to find the solution at some subsequent time $t + \Delta t$. First we replace the differential equation in Figure 1.19 with an equivalent integral equation:

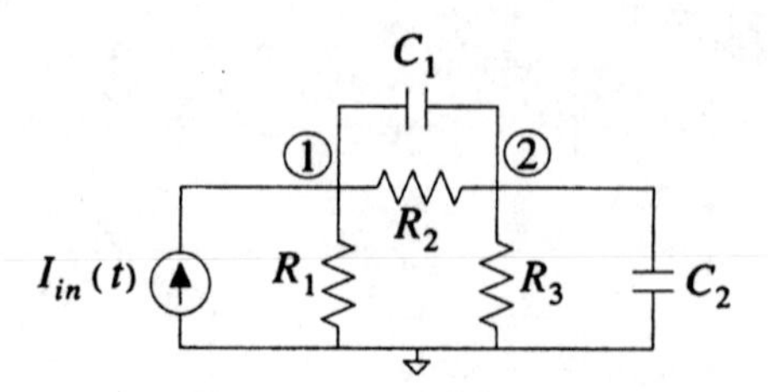

Figure 1.18 An RC circuit with a large signal time-varying current source.

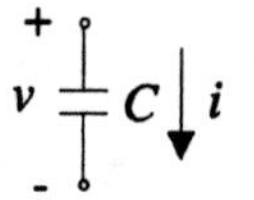

$$i = C\frac{dv}{dt}$$

Figure 1.19 Lumped, linear, time-invariant capacitor.

$$\int_t^{(t+\Delta t)} \frac{dv}{d\tau} \cdot d\tau = \frac{1}{C}\int_t^{(t+\Delta t)} i(\tau)\, d\tau \qquad \textbf{(1.7.2)}$$

Integrating over the time interval of interest results in

$$v(t+\Delta t) - v(t) = \frac{1}{C}\int_t^{(t+\Delta t)} i(\tau)\, d\tau \qquad \textbf{(1.7.3)}$$

or

$$v(t+\Delta t) = v(t) + \frac{1}{C}\int_t^{(t+\Delta t)} i(\tau)\, d\tau \qquad \textbf{(1.7.4)}$$

From (1.7.4) we see that the voltage at some future time $t + \Delta t$ depends upon the voltage at time t and the current which flows through the capacitor for the time interval $(t, t+\Delta t)$. We do not know the value of $i(t)$ for this complete interval, but only at the endpoints where its (approximate) values are computed. Therefore, we can approximate the integral of $i(t)$ in equation (1.7.4) by means of the trapezoidal approximation shown in Figure 1.20. The integral from t to $t + \Delta t$ is approximated by the shaded area, therefore,

$$\int_t^{(t+\Delta t)} i(\tau)\, d\tau \approx \frac{\Delta t}{2}\,[\,i(t) + i(t+\Delta t)\,] \qquad \textbf{(1.7.5)}$$

and (1.7.4) becomes

$$v(t+\Delta t) \approx v(t) + \frac{\Delta t}{2C}[i(t) + i(t+\Delta t)] \tag{1.7.6}$$

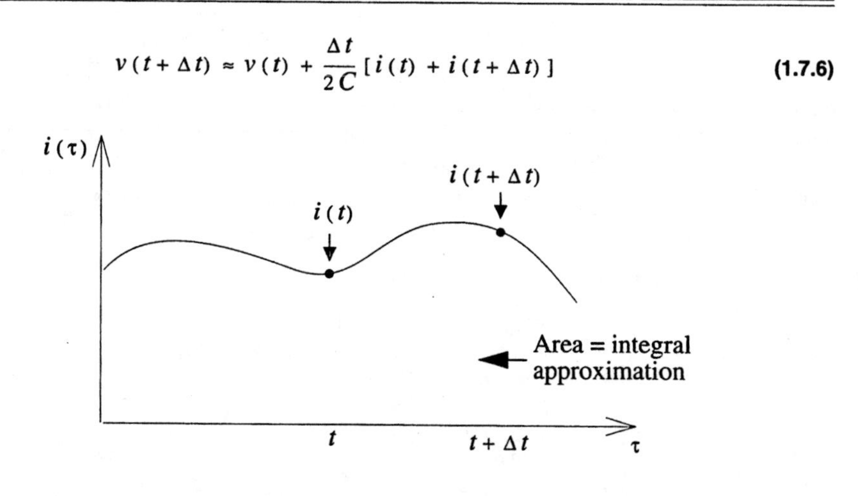

Figure 1.20 The trapezoidal approximation for the integral of the capacitor current.

We do not know $v(t+\Delta t)$ or $i(t+\Delta t)$ since we are attempting to compute them in the circuit solution at time $t+\Delta t$. Hence this is an *implicit* formula. It would be explicit if we had an expression for $v(t+\Delta t)$ or $i(t+\Delta t)$ entirely in terms of known quantities. Equation (1.7.6) has two unknowns in one equation, however, this is not a problem because it too has a simple circuit interpretation. A *companion model* for the capacitance voltage approximation in equation (1.7.6) is shown in Figure 1.21.

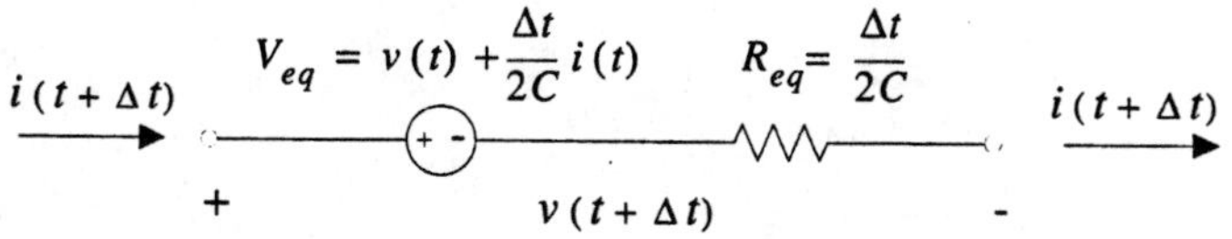

$i(t+\Delta t)$ $\quad V_{eq} = v(t) + \frac{\Delta t}{2C} i(t) \quad R_{eq} = \frac{\Delta t}{2C} \quad i(t+\Delta t)$

$+ \quad v(t+\Delta t) \quad -$

Figure 1.21 Capacitance companion model for trapezoidal integration approximation.

Notice that the unknown voltage, $v(t+\Delta t)$, appears across the Thevenin companion model which consists of a known voltage source value and a resistor which is a function of Δt. Therefore we can perform numerical integration on our differential equations in (1.7.1) by solving a linear circuit with resistors which are functions of the numerical time step.

Since we have considered only Nodal Analysis so far, we transform the Thevenin equivalent in Figure 1.21 into its Norton equivalent in Figure 1.22. (This is the approach used by SPICE with the trapezoidal integration option. We will discuss alternative integration schemes in subsequent chapters.) We note that the Norton equivalent capacitance companion model in Figure 1.22 degenerates gracefully to an open circuit for dc analysis,

$\Delta t \to \infty$. But for very small time steps, $\Delta t \to 0$, elements of the Norton equivalent capacitance model may become excessively large. This is the origin of the SPICE error message: "...Time Step Too Small." In such situations the Thevenin equivalent companion model derived originally is preferable. We should recognize, however, that the Thevenin equivalent capacitance companion model may present problems for dc analysis, $\Delta t \to \infty$. Hence the Thevenin and Norton representations of the trapezoidal companion model have their relative advantages and disadvantages.

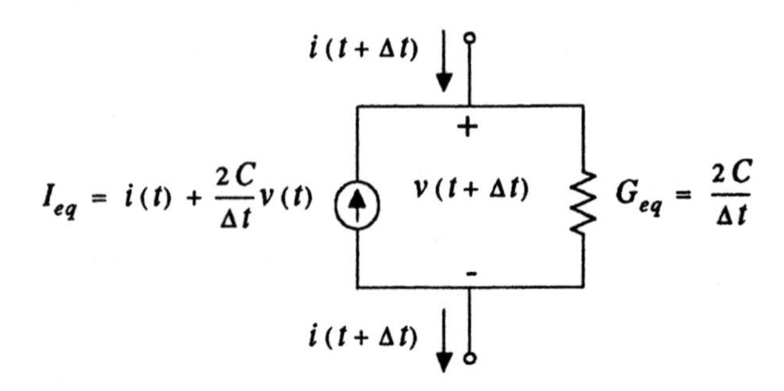

Figure 1.22 Norton equivalent circuit for capacitance trapezoidal model.

Using the Norton equivalent models in Figure 1.22 for the capacitors in Figure 1.18, the linear "dc" equivalent circuit in Figure 1.23 is solved to approximate the node voltages at time $t + \Delta t$. Then, time is moved ahead again and the solution procedure is repeated, until the required time of simulation has passed.

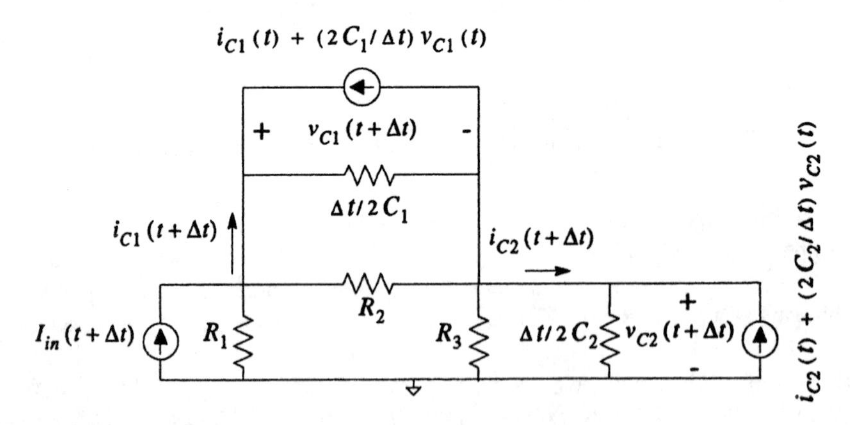

Figure 1.23 The RC equivalent circuit for a trapezoidal integration approximation.

Inductance is the dual of capacitance, so we need not repeat in detail the steps above. Figure 1.24 includes the differential equation and the corresponding trapezoidal approximation for an inductor. The trapezoidal integration approximation provides directly the Norton equivalent companion model form in Figure 1.25.

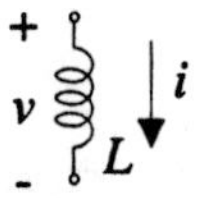

$$v = L\frac{di}{dt}$$

$$i(t+\Delta t) \approx i(t) + \frac{\Delta t}{2L}[v(t) + v(t+\Delta t)]$$

Figure 1.24 Trapezoidal integration approximation equation for an inductor.

SPICE does not use the Norton companion model for inductors; to facilitate dc analysis, it uses the Thevenin equivalent circuit inductance companion model shown in Figure 1.26. To handle the voltage source in this companion model SPICE resorts to Modified Nodal Analysis (MNA), which will be discussed in the next chapter.

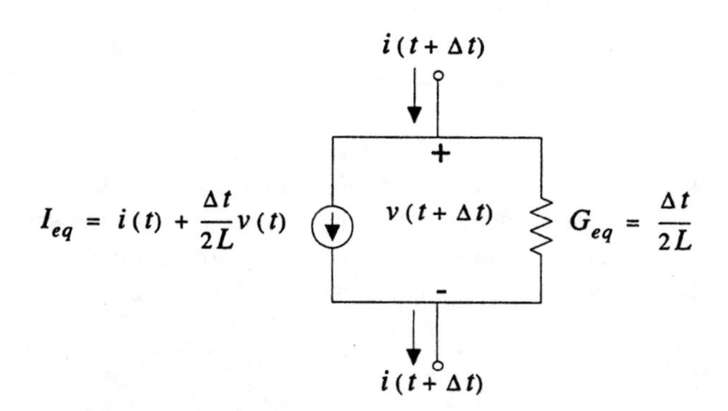

Figure 1.25 Norton equivalent companion model for an inductor.

$$V_{eq} = v(t) + \frac{2L}{\Delta t}i(t) \qquad R_{eq} = \frac{2L}{\Delta t}$$

$i(t+\Delta t)$ $\quad + \quad v(t+\Delta t) \quad - \quad$ $i(t+\Delta t)$

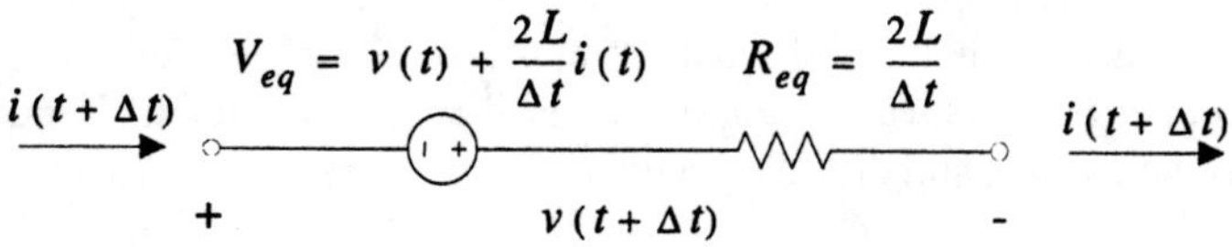

Figure 1.26 Thevenin equivalent companion model for an inductor.

We can implement linear transient analysis for RLC circuits using the companion models shown above as follows:

1. Start with a known set of capacitance voltages and inductance currents at a specified time (the initial conditions are either user-specified or obtained from a dc bias computation);
2. Increment time by a prescribed amount Δt and invoke the appropriate companion models; Δt is selected based on stability, accuracy, and efficiency considerations and can be different at different times.
3. Perform a "dc type" analysis;
4. Return to step (1) with the results and appropriately modify the companion model values until the specified end time is attained.

In this section, we have not discussed how best to choose the time step Δt. This, along with other numerical integration details, will be covered extensively in Chapter 4.

1.8 Nonlinear Transient Analysis

Finally, we consider what may be the most often applied circuit simulation analysis; namely, nonlinear transient analysis. This analysis is used to determine the large signal time domain behavior of circuits containing nonlinear elements. There are many subtleties involved, however, so we will delay the introduction of nonlinear transient analysis until later chapters. For now we will present a superficial flowchart (Figure 1.27) for nonlinear transient analysis to once again emphasize that linear dc analysis is a critical component of everything we seek to do. At each time step, companion models are used to obtain a nonlinear circuit which is solved iteratively by techniques described in Section 1.5.

1.9 Summary

The critical observation to be made is that linear circuit solution lies at the core -- the deepest inner loop -- of all of the above modes of analysis. Our linear circuit solution routine(s) must be robust, reliable, and as efficient as possible. To this end we discuss in more detail the traditional Modified Nodal Analysis approach to linear circuit solution in the next chapter. In later chapters we will discuss some alternatives to the traditional methods along with their advantages and disadvantages.

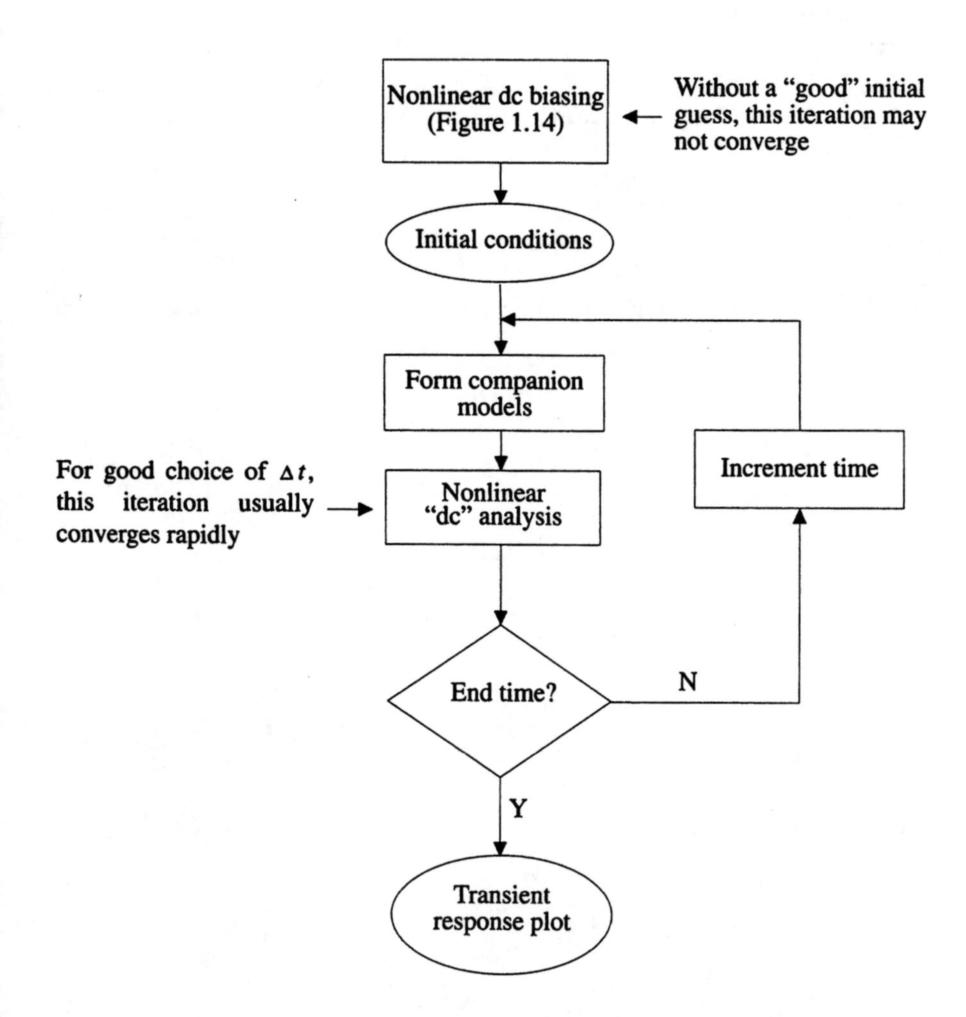

Figure 1.27 Nonlinear transient analysis. Both nonlinear dc analysis stages may require significant iteration -- especially the one for determining the dc operating point.

1.10 References

[Nagel71] L. W. Nagel and R. A. Rohrer. Computer Analysis of Nonlinear Circuits, Excluding Radiation (CANCER). *IEEE Journal of Solid State Circuits*, vol. SC-6, pp. 162-182, August 1971.

[Nagel75] L. W. Nagel. *SPICE2, A Computer Program to Simulate Semiconductor Circuits.* Technical Report ERL-M520, UC-Berkeley, May 1975.

[Ralston78] A. Ralston and P. Rabinowitz. *A First Course in Numerical Analysis.* McGraw-Hill, 1978.

Chapter 2

Linear dc Nodal Analysis

As we have seen in Chapter 1, Nodal Analysis is well suited to the formulation of equations for circuits containing only resistances (or conductances) and ideal independent current sources. In our preliminary discussion of dc, ac, transient, and nonlinear analyses, we studiously avoided voltage sources and dependent sources of any kind. In this chapter we will consider the extension of the nodal equations to handle independent voltage sources and various linear controlled sources. With these extensions, Nodal Analysis will provide the means for performing analysis in general on a wide range of electronic circuits.

2.1 Voltage-Controlled Current Sources

Of independent voltage sources and the four kinds of controlled sources, voltage-controlled current sources are the ones that most naturally fit into the nodal equation formulation. We will first discuss these controlled sources, and then generalize our discussion to the other elements listed above.

Linear controlled sources are not elements which you will find in electronic circuits explicitly. They are, however, important elements for modeling the behavior of various linear and nonlinear circuit components. For example, consider the N-MOSFET amplifier circuit shown in Figure 2.1. To analyze the frequency response of this amplifier we use a small signal model for the MOSFET and perform an ac analysis on the complete linear(ized) circuit. A small signal model for a MOSFET, such as the one shown in Figure 2.2, contains a linear voltage-controlled current source. That is, under the small signal assumption, the circuit behaves linearly such that small signal voltages applied to the gate of the MOSFET (for example, a 10 mV-peak sine wave in Figure 2.1) produce an amplified drain-to-source current sine wave. The small signal current gain is the transconductance, g_m, which will be considered in more detail in Chapter 10. For now we will consider the incorporation of such a model into our Nodal Analysis equations.

The voltage-controlled current source from Figure 2.2 is a *natural* for Nodal Analysis. For generality we model any voltage-controlled current source as a four-terminal element, such as the one shown in Figure 2.3. Note that the controlling branch in the figure is a

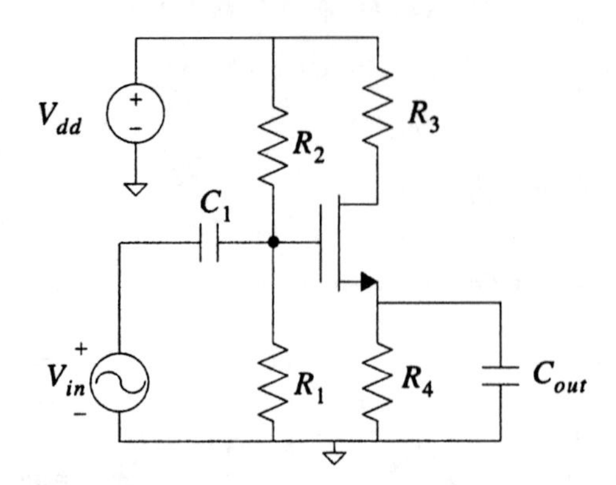

Figure 2.1 MOSFET amplifier circuit.

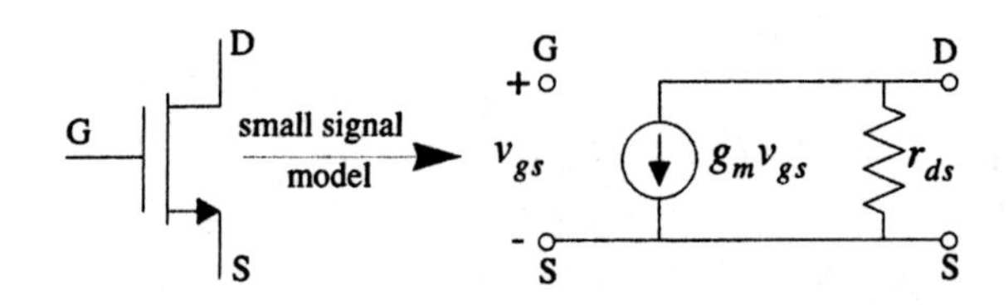

Figure 2.2 Low frequency small signal N-MOSFET model.

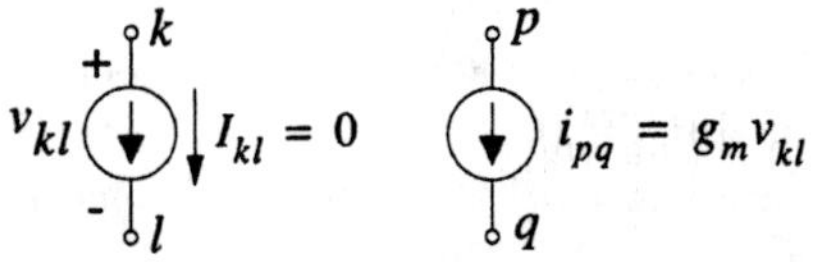

v_{kl} $I_{kl} = 0$ $\quad$ $i_{pq} = g_m v_{kl}$

Figure 2.3 Voltage-controlled current source.

zero-valued current source. This source is merely an open circuit and does not affect the operation of the circuit. It acts as an *ideal voltmeter* that helps us measure the controlling voltage in a convenient way. In a circuit, the controlling branch of a voltage-controlled current source may be anything: a resistor, a diode, a capacitor, and so on. Rather than formulate equations differently in each of these cases, we add a zero-valued current source and then build equations in a canonical fashion. Thus, representing a voltage-controlling branch by a zero-valued current source leads to notational and computational convenience. Moreover, in Chapter 9 it will be shown that this "ideal voltmeter" facilitates the calculation of the circuit sensitivities.

The voltage-controlled current source in Figure 2.3 contributes the stamp shown in Figure 2.4 to the ***Y*** matrix.

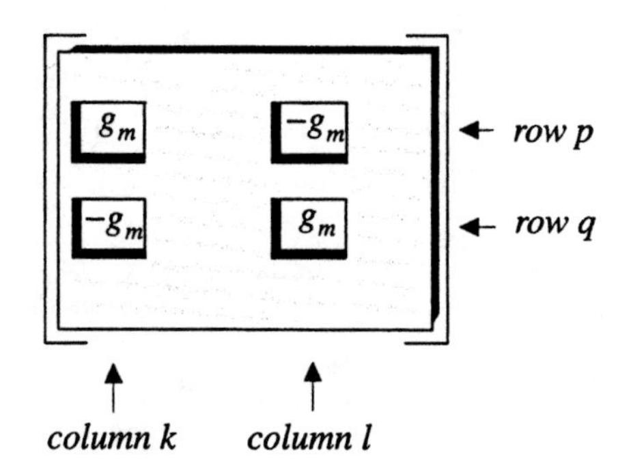

Figure 2.4 Stamp for a voltage-controlled current source.

Notice that this stamp contributes conductance terms to the ***Y*** matrix much like those for linear resistors. The distinction here is that the conductance terms do not appear symmetrically about the diagonal of the matrix. For example, the current out of node p, hence the expression appearing in row p of the matrix, is

$$i_{pq} = g_m v_{kl} \Rightarrow i_{pq} = g_m (v_k - v_l) \tag{2.1.1}$$

From (2.1.1) it is apparent that for row p of the matrix, which sums the current flowing *out* of row p, the conductance terms appear as a positive value in column k and a negative value in column l. Similarly, conductance terms appear with exactly the opposite signs for the entries in columns k and l for row q, since i_{pq} is flowing *into* this node. Note that we could show it formally, but we recognize that the zero-valued current source that represents the controlling branch contributes nothing to the ***J*** vector. In other words, we recognize that the zero-valued current source is not even required, but included only for our notational convenience.

To clarify the use of this controlled-source matrix stamp we will replace the MOSFET in Figure 2.1 with the small signal model in Figure 2.2 and construct the complete set of nodal equations. The complete small signal circuit model is shown in Figure 2.5. Note that for this small signal transistor model we have applied the low-frequency assumption. That is, we do not consider the effects in the small signal model due to the parasitic MOSFET capacitances. Of course, if the frequency of the input sine wave were high enough, these capacitance effects would have to be considered.

The stamping of all of the elements in Figure 2.5 leads to the following set of nodal equations:

$$\begin{bmatrix} 1 & 0 & 0 & 0 \\ -j\omega C_1 & (j\omega C_1 + G_1 + G_2) & 0 & 0 \\ 0 & -g_m & (G_4 + g_m + g_d + j\omega C_{out}) & -g_d \\ 0 & g_m & (-g_d - g_m) & (g_d + G_3) \end{bmatrix} \begin{bmatrix} V_1(j\omega) \\ V_2(j\omega) \\ V_3(j\omega) \\ V_4(j\omega) \end{bmatrix} = \begin{bmatrix} V_{in}(j\omega) \\ 0 \\ 0 \\ 0 \end{bmatrix} \tag{2.1.2}$$

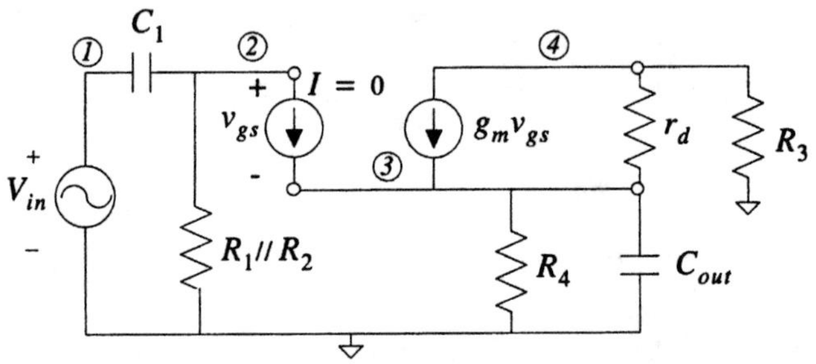

Figure 2.5 Small signal equivalent circuit for the MOSFET amplifier.

Note that the resistors in Figure 2.5 have been expressed in conductance form in (2.1.2). The voltages are expressed as complex voltages due to the complex admittances of the capacitors. As was shown in Chapter 1, capacitor stamps are identical in form to linear conductance stamps, except that the admittance is $j\omega C$ instead of simply G.

Although we haven't discussed voltage source stamps yet, equation (2.1.2) includes an independent voltage source. Since one end of the voltage source in Figure 2.5 is connected to ground, we can trivially include it with an equation $V_1 = V_{in}$, just as we have done in row one of (2.1.2). However, non-grounded, or floating voltage sources are not as easily handled. In the section which follows we seek a more generalized approach for stamping voltage sources using the *Modified Nodal Analysis* (MNA) method. Once we can handle voltage source stamps, we will return to the other three types of linear controlled sources since all of them depend upon the ability to handle voltage source elements.

2.2 Independent Voltage Sources

Since they do not have an admittance description form, it might appear that independent voltage sources complicate Nodal Analysis, but when properly handled they actually simplify it. Suppose that we have a floating voltage source connected between non-datum nodes k and l in a circuit, as shown in Figure 2.6. (The more common situation of a grounded voltage source can be easily obtained as a simpler special case of this treatment.)

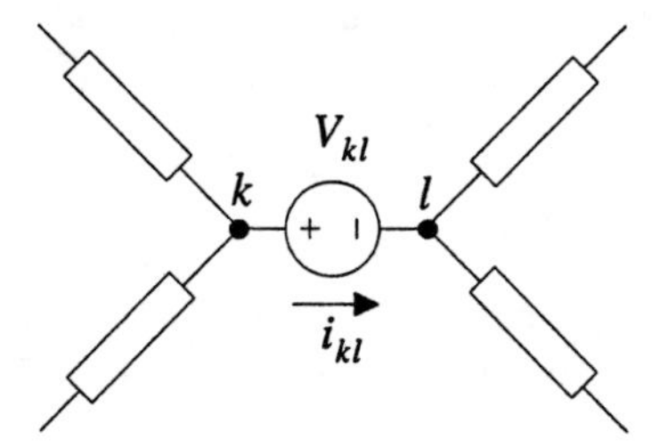

Figure 2.6 A floating voltage source between nodes k and l.

The voltage, V_{kl}, between nodes k and l is known, so we cannot declare both node voltages, v_k and v_l, to be independent variables. To resolve this problem we introduce the voltage source current, i_{kl}, as a new independent variable and add the voltage constraint,

$$v_k - v_l = V_{kl} \tag{2.2.1}$$

as a new equation. Now we have one more equation and one more unknown in addition to those for conventional Nodal Analysis. Thus, for an $(n+1)$ node (including datum) circuit with one floating voltage source V_{kl} from node k to node l, we have the following set of Modified Nodal Analysis (MNA) equations:

$$
\begin{array}{r}
\\ \\ \text{row } 1 \\ \vdots \\ \text{row } k \\ \vdots \\ \text{row } l \\ \vdots \\ \text{row } n \\ \text{row } n+1
\end{array}
\begin{array}{c}
\begin{array}{cccccc} \text{col } 1 & \text{col } k & \text{col } l & \text{col } n & \\ & & & & \text{col } n+1 \end{array} \\
\begin{bmatrix}
y_{11} & \cdots & y_{1k} & \cdots & y_{1l} & \cdots & y_{1n} & 0 \\
 & & \vdots & & & & \vdots & \\
y_{k1} & \cdots & y_{kk} & \cdots & y_{kl} & \cdots & y_{kn} & 1 \\
 & & \vdots & & & & \vdots & \\
y_{l1} & \cdots & y_{lk} & \cdots & y_{ll} & \cdots & y_{ln} & -1 \\
 & & \vdots & & & & \vdots & \\
y_{n1} & \cdots & y_{nk} & \cdots & y_{nl} & \cdots & y_{nn} & 0 \\
0 & \cdots & 1 & \cdots & -1 & \cdots & 0 & 0
\end{bmatrix}
\end{array}
\begin{bmatrix} v_1 \\ \vdots \\ v_k \\ \vdots \\ v_l \\ \vdots \\ v_n \\ i_{kl} \end{bmatrix}
=
\begin{bmatrix} I_1 \\ \vdots \\ I_k \\ \vdots \\ I_l \\ \vdots \\ I_n \\ V_{kl} \end{bmatrix}
\qquad (2.2.2)
$$

The y terms and the I terms represent the $\boldsymbol{Y}$ matrix and $\boldsymbol{J}$ vector entries without inclusion of the floating voltage source. In other words, (2.2.2) is equivalent to having the following stamp for a voltage source,

$$
\begin{array}{r} \text{row } k \\ \\ \text{row } l \\ \text{row } n+1 \end{array}
\begin{array}{c}
\left[\begin{array}{ccc:c}
 & & & 1 \\
 & \boldsymbol{Y} & & \\
 & & & -1 \\
\hdashline
1 & & -1 & 0
\end{array}\right] \\
\begin{array}{ccc} \text{col } k & \text{col } l & \text{col } n+1 \end{array}
\end{array}
\left[\begin{array}{c} \\ \boldsymbol{v} \\ \\ \hdashline i_{kl} \end{array}\right]
=
\left[\begin{array}{c} \\ \boldsymbol{J} \\ \\ \hdashline V_{kl} \end{array}\right]
\qquad (2.2.3)
$$

where the dashed lines indicate the augmentation of the original matrix $\boldsymbol{Y}$ and vector $\boldsymbol{J}$ which represent the conventional nodal admittance equations.

So long as they do not form loops, several independent voltage sources can be treated similarly, each having a simple set of stamps and adding a new row and column to the MNA equations. A loop of independent voltage sources is illegal and non-physical (more on this topic in Section 2.6) so we do not have to worry about such a situation. Note, however, that there is a zero on the diagonal in the $(n+1, n+1)$ place for equations (2.2.2) and (2.2.3). Some simultaneous equation solution schemes may expect to be able to divide the entries in a row by the corresponding diagonal element. Modified Nodal Analysis (MNA) [Ho75] overcomes this problem by exchanging row k with row n, which results in a $+1$ in the (k, k) location and a $+1$ in the $(n+1, n+1)$ location:

$$
\begin{array}{r}
\\ \\
\text{row } 1 \\ \vdots \\ \text{row } k \\ \vdots \\ \text{row } l \\ \vdots \\ \text{row } n \\ \text{row } n+1
\end{array}
\begin{array}{c}
\begin{array}{ccccc} \text{col } 1 & \text{col } k & \text{col } l & \text{col } n & \text{col } n+1 \end{array} \\
\begin{bmatrix}
y_{11} & \cdots & y_{1k} & \cdots & y_{1l} & \cdots & y_{1n} & 0 \\
 & & \vdots & & & \vdots & & \\
0 & \cdots & 1 & \cdots & -1 & \cdots & 0 & 0 \\
 & & \vdots & & & \vdots & & \\
y_{l1} & \cdots & y_{lk} & \cdots & y_{ll} & \cdots & y_{ln} & -1 \\
 & & \vdots & & & \vdots & & \\
y_{n1} & \cdots & y_{nk} & \cdots & y_{nl} & \cdots & y_{nn} & 0 \\
y_{k1} & \cdots & y_{kk} & \cdots & y_{kl} & \cdots & y_{kn} & 1
\end{bmatrix}
\end{array}
\begin{bmatrix} v_1 \\ \vdots \\ v_k \\ \vdots \\ v_l \\ \vdots \\ v_n \\ i_{kl} \end{bmatrix}
=
\begin{bmatrix} I_1 \\ \vdots \\ V_{kl} \\ \vdots \\ I_l \\ \vdots \\ I_n \\ I_k \end{bmatrix}
\qquad (2.2.4)
$$

As an example of how the nodal equations appear when there are voltage sources, consider the simple circuit in Figure 2.7. Since there are three non-datum nodes and two volt-

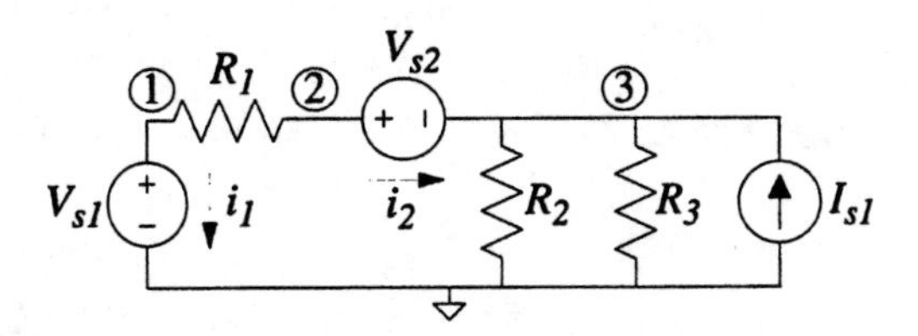

Figure 2.7 Circuit example with a floating voltage source.

age sources, the MNA equations are characterized by the following 5×5 $\boldsymbol{Y}$ matrix and 5×1 $\boldsymbol{J}$ vector:

$$
\begin{bmatrix}
G_1 & -G_1 & 0 & 1 & 0 \\
-G_1 & G_1 & 0 & 0 & 1 \\
0 & 0 & (G_2+G_3) & 0 & -1 \\
1 & 0 & 0 & 0 & 0 \\
0 & 1 & -1 & 0 & 0
\end{bmatrix}
\begin{bmatrix} v_1 \\ v_2 \\ v_3 \\ i_1 \\ i_2 \end{bmatrix}
=
\begin{bmatrix} 0 \\ 0 \\ I_{s1} \\ V_{s1} \\ V_{s2} \end{bmatrix}
\qquad (2.2.5)
$$

Of course, we could swap rows 1 and 2 with rows 4 and 5, respectively, in order to ensure that there is a nonzero term for each diagonal entry. Next, we will show that it is easier, however, just to eliminate the *excess* rows and columns.

2.3 (Conventional) Nodal Analysis

An alternative to swapping rows (k and $n+1$) as noted above is to eliminate the unknown current i_{kl} at the outset by adding together rows k and l in equation (2.2.2) [Nagel71]. (The value of i_{kl} can always be resurrected, if required, by substituting into either of these equations after solving for the vector of node voltages.) Since we have eliminated one equation on that basis, we may eliminate one more unknown. We can further simplify the set of equations by substituting in the voltage constraint equation, (2.2.1), thereby reducing the number of equations and variables by two. The overall result is as follows and is obtained by a) adding rows k and l, b) adding columns k and l and moving the V_{kl} terms to the right hand side, and c) eliminating the lth row and column.

$$\begin{array}{c} \\ \text{row } 1 \\ \vdots \\ \text{row } k \\ \vdots \\ \text{row } n\text{-}1 \end{array}
\begin{array}{c} \text{col } 1 \qquad\qquad \text{col } k \qquad\qquad \text{col } n\text{-}1 \\
\begin{bmatrix} y_{11} & \cdots & y_{1k}+y_{1l} & \cdots & y_{1n} \\ & & \vdots & & \\ y_{k1}+y_{l1} & \cdots & y_{kk}+y_{lk}+y_{kl}+y_{ll} & \cdots & y_{kn}+y_{ln} \\ & & \vdots & & \\ y_{n1} & \cdots & y_{nk}+y_{nl} & \cdots & y_{nn} \end{bmatrix} \end{array}
\begin{bmatrix} v_1 \\ \vdots \\ v_k \\ \vdots \\ v_n \end{bmatrix} = \begin{bmatrix} I_1 + y_{1l}V_{kl} \\ \vdots \\ I_k + I_l + V_{kl}(y_{kl}+y_{ll}) \\ \vdots \\ I_n + y_{n,l}V_{kl} \end{bmatrix} \qquad \textbf{(2.3.1)}$$

Note that there is one less equation now than the number of non-datum nodes since the lth row and column have been eliminated and there is no v_l unknown.

After having solved this set of equations, we can easily obtain i_{kl} from either of the original k^{th} or l^{th} equations. So long as there are no voltage loops in a circuit (more on this in Section 2.6), the above manipulations can be undertaken with any number of independent voltage sources. The result is a more compact set of nodal equations, which in our experience usually is better conditioned numerically, and can be solved more efficiently than the MNA equations.

Many elementary circuit analysis textbooks use the *supernode* concept to deal with independent voltage sources, where a supernode encompasses any subset of nodes that are voltage-source connected. For the example in Figure 2.8 we would designate *k-l* to be a supernode so that a single KCL equation can be written in terms of it.

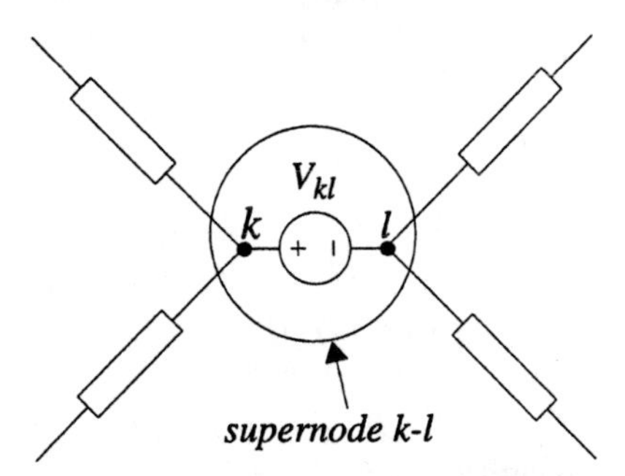

Figure 2.8 Supernode treatment of a floating voltage source.

Here, again, it must be recognized that $v_k - v_l = V_{kl}$, and that only one of these two node voltages is independent. Careful treatment of a supernode leads to the same compact nodal equations we have shown above. But it is our experience that for complicated circuits it is more straightforward and less error prone to obtain the compact nodal equations as above by first formally writing out the MNA equations and then systematically eliminating the excess currents and the dependent voltages from the unknown vector.

Another way of looking at the above formulation of compact nodal equations is in terms of voltage source transportation. With the exception of the suppressed node, l, the circuit in Figure 2.9 is equivalent to that in Figure 2.6.

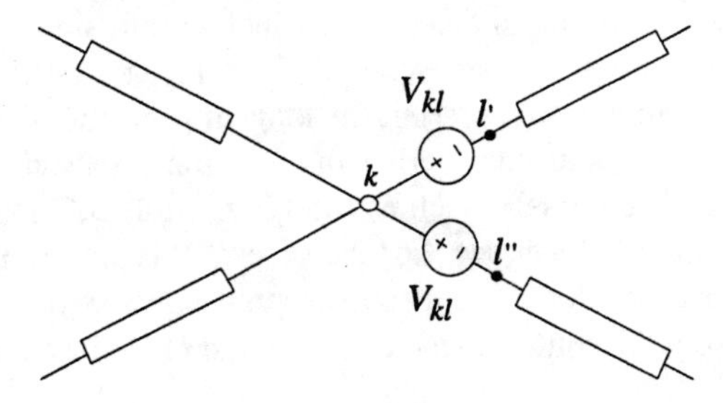

Figure 2.9 Result of voltage source transportation on the subcircuit of Figure 2.8.

A Thevenin to Norton transformation of the composite branches that include V_{kl} in Figure 2.9 yields the appropriate equivalent current source and parallel conductance. The compact nodal equations follow directly from the transformed circuit.

Applying this approach to the circuit example in Figure 2.7 would result in a single equation in terms of a single unknown as shown below:

$$(G_1 + G_2 + G_3)\, v_2 = G_1 V_{S1} + (G_2 + G_3)\, V_{S2} + I_{S1} \tag{2.3.2}$$

However, a word of caution: if there are many independent voltage sources, care must be taken with the above manipulations and the order in which the voltage sources are addressed may become important.

2.4 Controlled Sources in General

Whereas it is straightforward to deal with voltage-controlled current sources in conjunction with Nodal Analysis, manipulations similar to those for independent voltage sources are required for the other three kinds of controlled sources and for some other active elements as well. Of the remaining three types of controlled sources, a voltage source is required for either the controlled branch or the controlling branch. Similar to the ideal voltmeter in Figure 2.3, a zero-valued voltage source is used as an ammeter in series with the element whose current is the controlling variable. This method is convenient since we showed that modified Nodal Analysis maintains a current variable for every voltage source in the circuit. Thus, a zero-valued voltage source can be used as an ideal ammeter. In fact, SPICE requires the insertion of a zero-valued ideal voltage source in order to measure current if the output variable of interest is a current.

Similar to the voltage-controlled current source in Figure 2.2 used to model the small signal behavior of the MOSFET in Figure 2.1, the small signal h parameter model for the bipolar junction transistor (BJT) amplifier circuit in Figure 2.10 contains a linear *current-controlled* current source. But due to the way in which circuit simulators such as SPICE perform large signal nonlinear analysis, we will show in later chapters that linear controlled sources are also required to represent the tangent projections for certain nonlinear devices. Moreover, even without considering the Newton-Raphson linearizations, there are some large signal device models, such as the Ebers-Moll BJT model in Figure 2.11, which contain linear controlled sources. For the case of this simple BJT model, the forward and reverse linear controlled current sources are used to represent the percentage of charge which travels from the emitter to the collector and from the collector to the emitter, respectively.

Figure 2.12 is the two-branch model for a current-controlled current source, which includes a zero-valued independent voltage source (*ammeter*) as (or in series with) the controlling branch. To accommodate this element we formulate the nodal equations as usual but with i_{kl} as an *excess variable* and $v_k - v_l = 0 \Rightarrow v_k = v_l$ as an *excess equation.* The stamp for a current-controlled current source is:

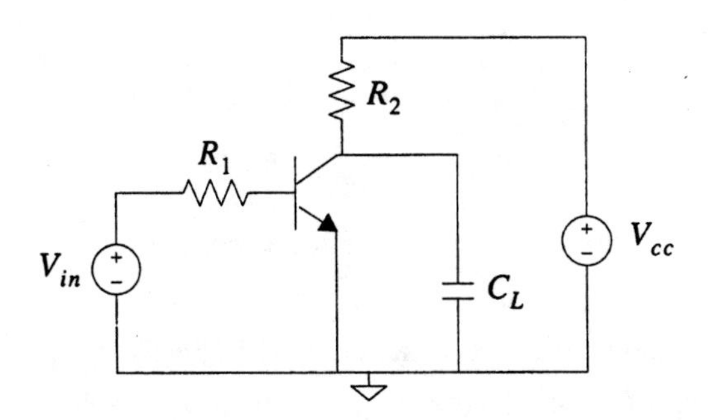

Figure 2.10 BJT inverter circuit.

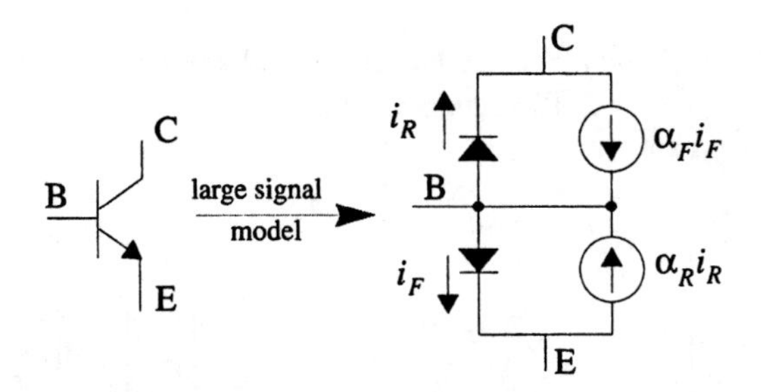

Figure 2.11 Ebers-Moll model for a BJT.

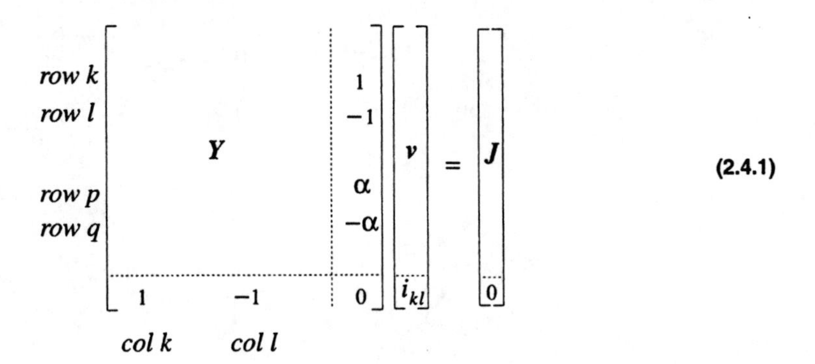

(2.4.1)

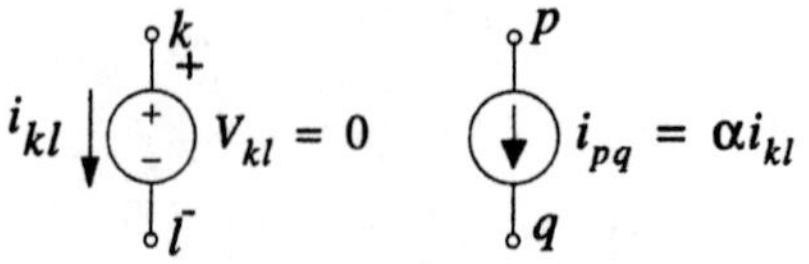

Figure 2.12 Current-controlled current source.

This set of modified nodal equations can be solved directly, or i_{kl} can be eliminated *a priori* to obtain compact nodal equations. To eliminate i_{kl} we could add rows k and l together to form a new composite (supernode) equation, and then we could use either of equations k or l to eliminate i_{kl} from equations p and q. Such *a priori* manipulations usually are not necessary since a good simultaneous equation solution program will do them for us.

The two-branch model for a voltage-controlled voltage source has only one voltage source as shown in Figure 2.13. The stamp for this controlled source, therefore, adds only one equation and one unknown:

$$
\begin{array}{r}
\\ \textit{row } p \\ \\ \textit{row } q \\ \\ \\ \\
\end{array}
\left[\begin{array}{cccc:c}
 & & & & 1 \\
 & & \boldsymbol{Y} & & \\
 & & & & -1 \\
\hdashline
-\mu & +\mu & +1 & -1 & 0
\end{array}\right]
\left[\begin{array}{c}
\\ \boldsymbol{v} \\ \\ \hdashline i_{pq}
\end{array}\right]
=
\left[\begin{array}{c}
\\ \boldsymbol{J} \\ \\ \hdashline 0
\end{array}\right]
\qquad (2.4.2)
$$

col k col l col p col q

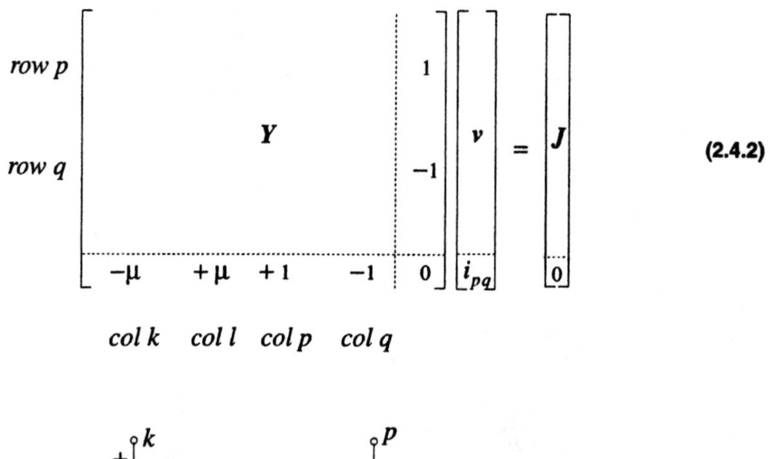

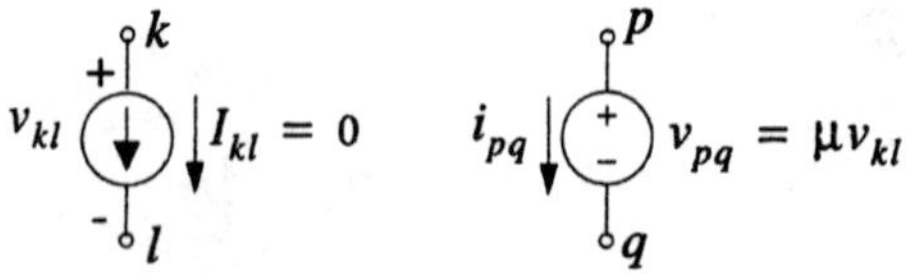

Figure 2.13 Voltage-controlled voltage source.

Because it is zero, $i_{kl} = I_{kl}$ need not be considered, and i_{pq} is the only *excess current* variable. The row equation is merely the voltage constraint

$$v_{pq} - \mu v_{kl} = 0 \Rightarrow (v_p - v_q - \mu v_k + \mu v_l = 0) \tag{2.4.3}$$

Again, we can leave these modified nodal equations as they stand and feed them directly to a simultaneous equation solver (details of that in Chapter 7), perhaps after some row and/or column swapping. Or we can easily eliminate the excess current variable i_{pq} and one of the unknown voltages v_k, v_l, v_p, or v_q, to obtain compact nodal equations.

Because it adds two excess current equations and two new equations of voltage constraints, the current-controlled voltage source is the most difficult of the dependent sources to deal with for MNA. The stamp for the two element current-controlled voltage source in Figure 2.14 is:

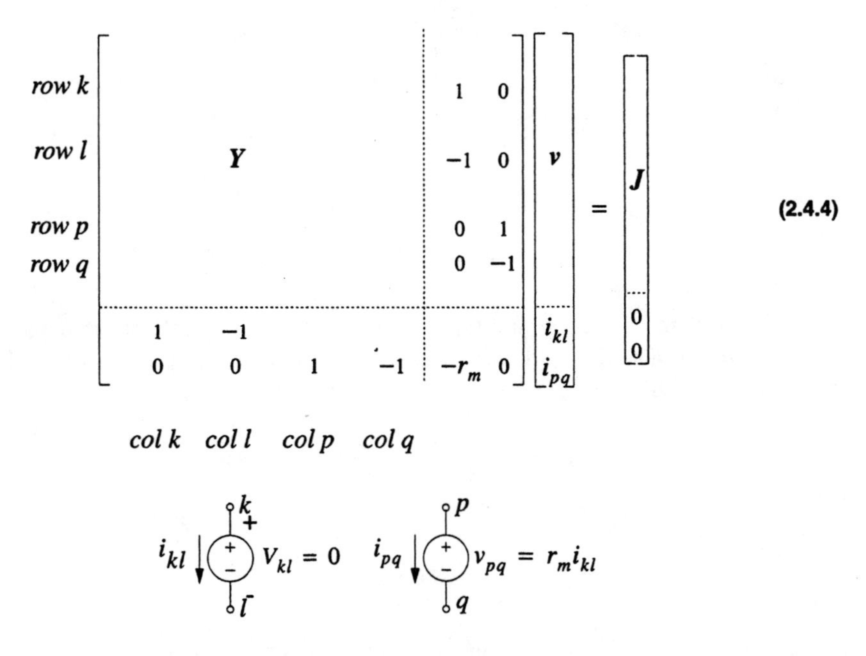

Figure 2.14 Current-controlled voltage source.

With some manipulation, both *excess currents*, i_{kl} and i_{pq}, can be eliminated, along with one of the two voltages v_k or v_l, and one of the two voltages v_p or v_q. But given the infrequency with which current-controlled voltage sources arise in active device models, we need not worry too much about how to handle this element. And a good simultaneous equation solution program will handle it for us anyway, should it ever arise.

2.5 Operational Amplifiers

A device which behaves similarly to a voltage-controlled voltage source is an operational amplifier. An often acceptable simple model for an operational amplifier such as the one in Figure 2.15 is the equation:

$$v_0 = A(v_+ - v_-) \tag{2.5.1}$$

where A is the op amp gain. Equation (2.5.1) assumes that the amplifier remains linear in its behavior, therefore it can be represented by a voltage-controlled voltage source like the one in Figure 2.13. Typically, the op amp gain, A, is very large which can lead to numerical complications in the solution of the MNA equations if the values of the other conductors in the circuit are much smaller than A.

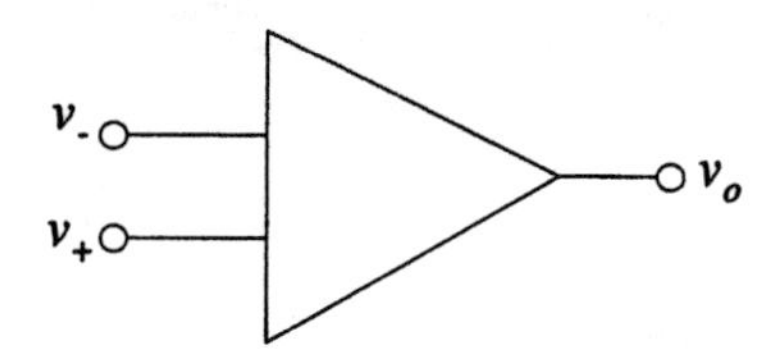

Figure 2.15 Simple operational amplifier model.

Off-diagonal matrix terms due to A are significantly larger than the diagonal terms which can result in an ill-conditioned matrix. To overcome such problems, in [Vlach83] it is recommended that we replace equation (2.5.1) with its inverse

$$v_+ - v_- = Bv_o \tag{2.5.2}$$

where

$$B = \frac{1}{A} \tag{2.5.3}$$

and B is typically a very small value. All three nodes of this op amp appear as columns and rows in Y, therefore, we are now stamping a very small value (B) in the column corresponding to v_o, instead of stamping large values (A) into the columns corresponding to v_+ and v_-. This simple artifice typically improves the numerical conditioning of Y.

In the extreme case of an *ideal* op amp, the model dictates that

$$v_+ = v_- \tag{2.5.4}$$

or A is infinite. In addition, the two separate ideal op amp currents shown in Figure 2.16 are zero:

$$i_+ = 0 \text{ and } i_- = 0 \tag{2.5.5}$$

Equation (2.5.5) stipulates that there is an independent KCL equation at each of the two input nodes for an ideal op amp, similar to the constraint on the voltage-controlled voltage source in Figure 2.13. In contrast to the controlled source, however, the op amp model also has the voltage constraint, equation (2.5.4).

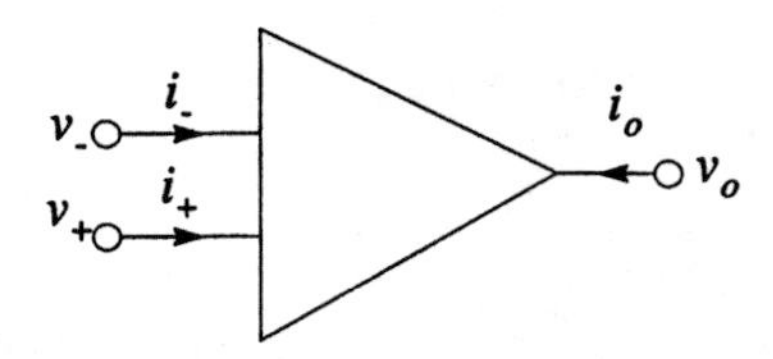

Figure 2.16 Ideal op amp model for which input currents are zero.

Because the voltage constraint equation must be incorporated into the overall set of equations which describe this element, a new unknown must be added for equation consistency. The arbitrary output current, i_o, can be added to those currents leaving the node to which the op amp output is connected. This additional unknown will make the MNA equations consistent.

Suppose, for example, that an ideal op amp were connected in an appropriate negative feedback configuration with its inputs to nodes k and l and its output to node p, as shown in Figure 2.17. Then its contributing stamp to the MNA equations would appear as follows:

$$\begin{array}{r} \\ \\ \text{row } p \\ \\ \\ \end{array} \left[\begin{array}{ccc|c} & & & \\ & Y & & \\ & & & 1 \\ \hline 1 & & -1 & 0 \end{array}\right] \left[\begin{array}{c} \\ v \\ \\ \hline i_o \end{array}\right] = \left[\begin{array}{c} \\ J \\ \\ \hline 0 \end{array}\right] \tag{2.5.6}$$

$$\text{col } k \qquad \text{col } l$$

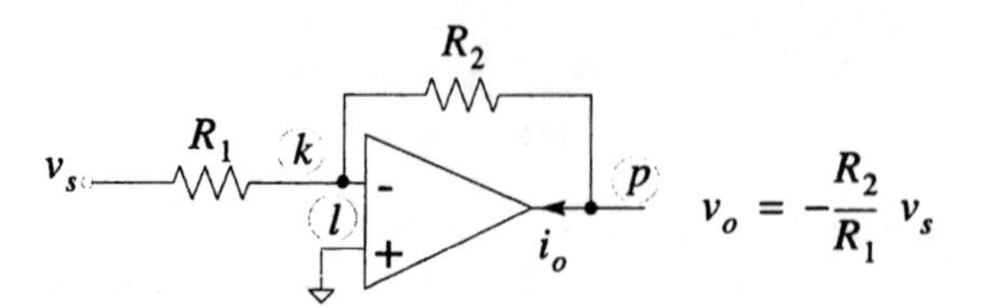

Figure 2.17 Example of an ideal op amp configured as an inverting amplifier.

2.6 *When Do Nodal Equations Fail?*

When there are several voltage sources and/or controlled sources in a circuit, we must exercise care in the formation and manipulation of the circuit equations. At the very least, we must avoid situations for which there are voltage source loops that might be in violation of KVL. Similarly, current source cutsets may be in violation of KCL (a *cutset* is a set of elements whose removal leaves two portions of the original circuit disconnected; an alternate statement of KCL is that the algebraic sum of currents through any cutset is zero). Even if these easily detectable situations do not arise, the combination of controlled sources and certain element values may render the circuit equations unsolvable.

Assuming that we have stamped in all of our circuit elements, including all of the controlled sources expressed in terms of element pairs, we must check the topology of the circuit in some way to detect the possible occurrence of voltage source loops and current source cutsets. For example, the loop of independent voltage sources in Figure 2.18 violates KVL if $V_1 + V_2 - V_3$ is not equal to zero. Moreover, even if the voltages do properly satisfy KVL, the loop current that flows through these sources is undefined. That is, the loop current can take on any value without violating the circuit equations.

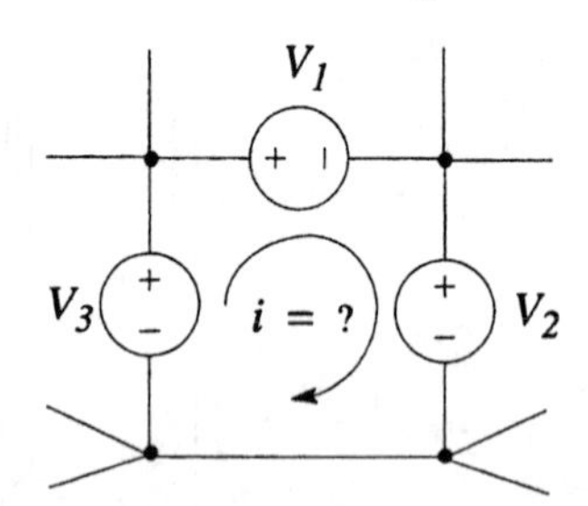

Figure 2.18 Loop of independent voltage sources.

Similarly, the current source cutset in Figure 2.19 is also non-physical. First of all, if the sum of the currents I_1, I_2, and I_3 is not zero, then KCL is violated at this node. And, even if KCL were satisfied, the node voltage value would be undefined. Therefore, instances of independent voltage source loops and independent current source cutsets should be detected after parsing the circuit netlist and before the equations are stamped.

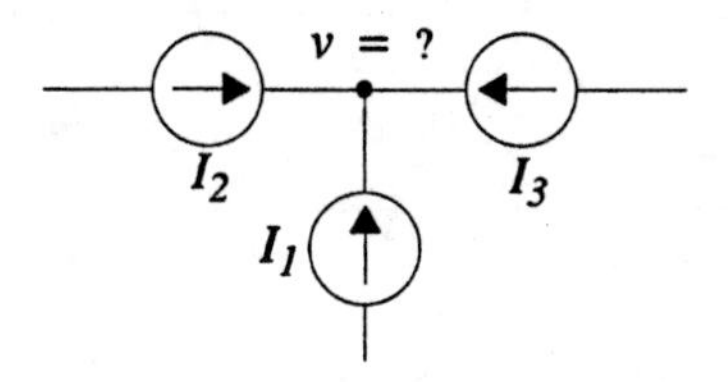

Figure 2.19 Cutset of independent current sources.

2.7 *DC Solution of Circuits with Energy Storage: C-cutsets and L-loops*

We must be careful to recognize loops and cutsets of certain energy storage elements since during initial dc analysis they are equivalent to independent sources. For example, in order to determine the initial state of a circuit for $-\infty < t < 0$, we replace all of the capacitors by open circuits and all of the inductors by short circuits. This set of equations is equivalent to requiring that in steady state, all voltage and current derivatives are zero, hence capacitor currents and inductor voltages are zero.

For the circuit in Figure 2.20, replacing all the capacitors by open circuits (current sources of value zero) results in the dc circuit shown in Figure 2.21. The opening of all the C's has resulted in an illegal cutset among I_2, I_3, and I_4, or a floating node, 4. Even though KCL is satisfied at this node, it is not obvious how we can ascertain its voltage.

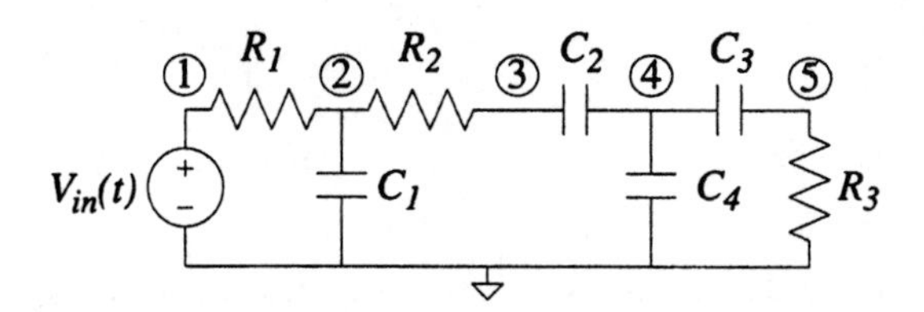

Figure 2.20 A circuit containing a cutset of capacitors.

Formulating the equations for this circuit, nodes 1, 2, 3, and 5 are easily described in terms of MNA. At node 4, with all of the current sources equal to zero, we have a *consistent, but useless* nodal equation. Since the nodal equation for node 4 is useless, a supplemental equation is required to resolve the voltage at this node.

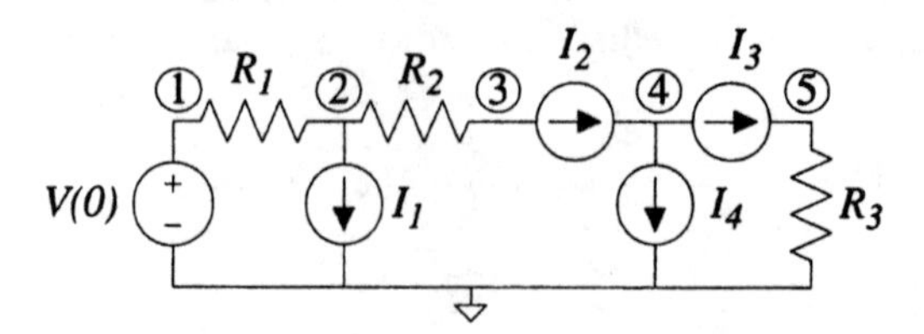

Figure 2.21 The dc equivalent for the circuit in Figure 2.20.

Since charge conservation must hold for this circuit just as it must for all dc circuits, we can use the following equation for node 4 to supplement the MNA equations:

$$C_2(v_4 - v_3) + C_3(v_4 - v_5) + C_4 v_4 = 0 \tag{2.7.1}$$

The assumption on which this equation is based is that there is no charge stored on any of the capacitors at $t = -\infty$ and that as time evolves (to $t = 0$) charge must have been conserved. Hence the complete set of dc circuit equations is

$$\begin{bmatrix} 1 & 0 & 0 & 0 & 0 \\ -G_1 & (G_1+G_2) & -G_2 & 0 & 0 \\ 0 & -G_2 & G_2 & 0 & 0 \\ 0 & 0 & -C_2 & (C_2+C_3+C_4) & -C_3 \\ 0 & 0 & 0 & 0 & G_3 \end{bmatrix} \begin{bmatrix} v_1 \\ v_2 \\ v_3 \\ v_4 \\ v_5 \end{bmatrix} = \begin{bmatrix} V_{in}(0) \\ 0 \\ 0 \\ 0 \\ 0 \end{bmatrix} \tag{2.7.2}$$

A similar procedure can be used for circuits which contain loops of inductors -- which produce a loop of voltage sources in the dc equivalent circuit -- in terms of supplemental conservation of flux equations.

2.8 Summary

The nodal and modified nodal equations are extremely general and easily implemented for any “proper” circuit. They also tend to produce the most compact set among equation formulation alternatives. For this reason the nodal equations constitute the standard formulation method for computer-based circuit analysis. In the next chapter we will consider the solution, reordering, and overall conditioning of equations that have been formulated with (Modified) Nodal Analysis.

2.9 References

[Nagel71] L. W. Nagel and R. A. Rohrer. Computer Analysis of Nonlinear Circuits, Excluding Radiation (CANCER). *IEEE Journal of Solid State Circuits*, vol. SC-6, pp. 162-182, August 1971.

[Ho75] C. W. Ho, A. E. Ruehli, and P. A. Brennan. The Modified Nodal Approach to Network Analysis. *IEEE Transactions on Circuits and Systems,* vol. CAD-25, pp. 504-509, June 1975.

[Vlach83] J. Vlach and K. Singhal. *Computer Methods for Circuit Analysis and Design.* Van Nostrand Reinhold Company, 1983.

Chapter 3

Solution of Linear Equations

In Chapter 2 we demonstrated the formulation of (nodal) equations that describe a circuit, under the assumption that there is a technique available to solve them. Because the method of formulation can affect the means of solution, in this chapter we provide a brief introduction to simultaneous equation solution. In subsequent chapters we will revisit circuit equation formulation and solution as we become more familiar with the issues involved.

3.1 Gaussian Elimination

Consider the solution of the ladder circuit from Chapter 1 shown here once again in Figure 3.1. The nodal equations which describe this circuit are

$$\begin{aligned} 2v_1 - v_2 \qquad\qquad &= 1 \\ -v_1 + 3v_2 - v_3 \qquad &= 0 \\ -v_2 + 3v_3 - v_4 &= 0 \\ -v_3 + 2v_4 &= 1 \end{aligned} \tag{3.1.1}$$

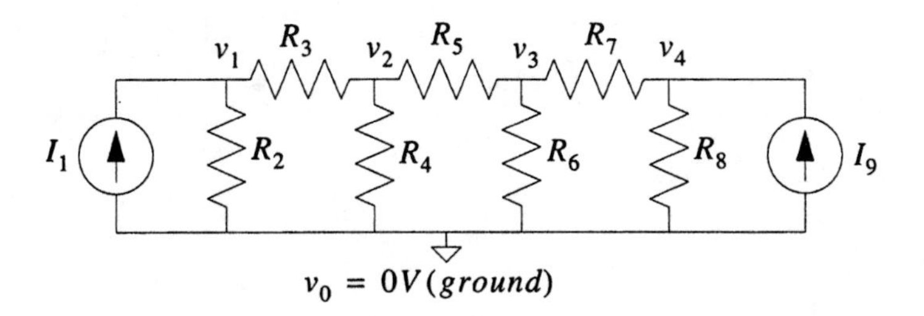

Figure 3.1 Resistor ladder circuit with 1Ω resistors and $1A$ current sources.

We can proceed to solve this set of four equations in terms of four unknowns using *Gaussian Elimination*. The elimination steps are as follows:

1. Normalize the first equation with respect to the first coefficient:

$$v_1 - \frac{1}{2}v_2 = \frac{1}{2} \tag{3.1.2}$$

2. Add the normalized first equation to the second equation to eliminate its leading coefficient:

$$\frac{5}{2}v_2 - v_3 = \frac{1}{2} \tag{3.1.3}$$

3. Normalize the new equation 2:

$$v_2 - \frac{2}{5}v_3 = \frac{1}{5} \tag{3.1.4}$$

4. Add the new, normalized equation 2 to the third equation to eliminate its leading coefficient:

$$\frac{13}{5}v_3 - v_4 = \frac{1}{5} \tag{3.1.5}$$

5. Normalize the new equation 3:

$$v_3 - \frac{5}{13}v_4 = \frac{1}{13} \tag{3.1.6}$$

6. Add the new, normalized equation 3 to the fourth equation to eliminate its leading coefficient:

$$\frac{21}{13}v_4 = \frac{14}{13} \tag{3.1.7}$$

7. Normalize the new equation 4:

$$v_4 = \frac{2}{3} \tag{3.1.8}$$

The above steps are summarized in matrix form in Figure 3.2. It is apparent that the matrix has been transformed into normalized *upper triangular form* (by upper triangular, we mean that all entries below the diagonal are zero; by normalized, we mean that each of the diagonal entries is unity):

STEPS	Y	J
	$\begin{bmatrix} 2 & -1 & 0 & 0 \\ -1 & 3 & -1 & 0 \\ 0 & -1 & 3 & -1 \\ 0 & 0 & -1 & 2 \end{bmatrix}$	$\begin{bmatrix} 1 \\ 0 \\ 0 \\ 1 \end{bmatrix}$
1) and 2)	$\begin{bmatrix} 1 & -\frac{1}{2} & 0 & 0 \\ 0 & \frac{5}{2} & -1 & 0 \\ 0 & -1 & 3 & -1 \\ 0 & 0 & -1 & 2 \end{bmatrix}$	$\begin{bmatrix} \frac{1}{2} \\ \frac{1}{2} \\ 0 \\ 1 \end{bmatrix}$
3) and 4)	$\begin{bmatrix} 1 & -\frac{1}{2} & 0 & 0 \\ 0 & 1 & -\frac{2}{5} & 0 \\ 0 & 0 & \frac{13}{5} & -1 \\ 0 & 0 & -1 & 2 \end{bmatrix}$	$\begin{bmatrix} \frac{1}{2} \\ \frac{1}{5} \\ \frac{1}{5} \\ 1 \end{bmatrix}$
5) and 6)	$\begin{bmatrix} 1 & -\frac{1}{2} & 0 & 0 \\ 0 & 1 & -\frac{2}{5} & 0 \\ 0 & 0 & 1 & -\frac{5}{13} \\ 0 & 0 & 0 & \frac{21}{13} \end{bmatrix}$	$\begin{bmatrix} \frac{1}{2} \\ \frac{1}{5} \\ \frac{1}{13} \\ \frac{14}{13} \end{bmatrix}$
7)	$\begin{bmatrix} 1 & -\frac{1}{2} & 0 & 0 \\ 0 & 1 & -\frac{2}{5} & 0 \\ 0 & 0 & 1 & -\frac{5}{13} \\ 0 & 0 & 0 & 1 \end{bmatrix}$	$\begin{bmatrix} \frac{1}{2} \\ \frac{1}{5} \\ \frac{1}{13} \\ \frac{2}{3} \end{bmatrix}$

Figure 3.2 Steps of Gaussian Elimination in matrix form.

$$\begin{aligned} v_1 - \frac{1}{2}v_2 \qquad\qquad &= \frac{1}{2} \\ v_2 - \frac{2}{5}v_3 \qquad &= \frac{1}{5} \\ v_3 - \frac{5}{13}v_4 &= \frac{1}{13} \\ v_4 &= \frac{2}{3} \end{aligned} \tag{3.1.9}$$

We can now proceed to the solution of the upper triangular equations by means of *Back Substitution*:

$$\begin{aligned} v_4 &= \frac{2}{3} \\ v_3 &= \frac{1}{13} + \frac{5}{13}v_4 = \frac{1}{3} \\ v_2 &= \frac{1}{5} + \frac{2}{5}v_3 = \frac{1}{3} \\ v_1 &= \frac{1}{2} + \frac{1}{2}v_2 = \frac{2}{3} \end{aligned} \tag{3.1.10}$$

3.2 LU Factorization

We are often interested in repeatedly solving matrix equations for which the left-hand side (LHS) is unchanged but the right-hand side (RHS) vector (called the *forcing function* because it contains the voltage and current source stimuli to the circuit) is changed. Such situations arise when performing sensitivity analysis (Section 3.6) and in some special cases of linear transient analysis. To exploit these cases where the LHS does not change, we perform *LU factorization* of the admittance matrix instead of Gaussian Elimination. During LU factorization, we operate only on the LHS matrix and the RHS input-stimulus vector is not considered. The results of LU factorization can efficiently be reused to solve sets of equations with different RHS vectors.

For example, starting with the nodal admittance equations:

$$\boldsymbol{Yv} = \boldsymbol{J} \tag{3.2.1}$$

let us assume that the nodal admittance matrix $\boldsymbol{Y}$ can be factored into a product of lower- and upper-triangular matrices:

$$Y = \begin{bmatrix} \ddots & 0 \\ L & \ddots \end{bmatrix} \begin{bmatrix} \ddots & U \\ 0 & \ddots \end{bmatrix} \tag{3.2.2}$$

A lower triangular matrix has zeros above the main diagonal and an upper triangular matrix has zeros below the main diagonal; typically the main diagonal elements of either of these matrices are not zero. Note that $\det(\boldsymbol{Y}) = \det(\boldsymbol{L})\det(\boldsymbol{U})$. Therefore, both $\boldsymbol{L}$ and $\boldsymbol{U}$ cannot have all zeros on their main diagonals for the determinant of $\boldsymbol{Y}$ to be non-zero, and hence for the equations to be solvable.

Substituting for $\boldsymbol{Y}$ in (3.2.1), we have

$$\boldsymbol{LUv} = \boldsymbol{J} \tag{3.2.3}$$

We substitute

$$\boldsymbol{x} = \boldsymbol{Uv} \tag{3.2.4}$$

to obtain

$$\boldsymbol{Lx} = \boldsymbol{J} \tag{3.2.5}$$

So we first solve the lower triangular system (3.2.5) by *Forward Substitution* to obtain $\boldsymbol{x}$ and we then solve the upper triangular system

$$\boldsymbol{Uv} = \boldsymbol{x} \tag{3.2.6}$$

by *Back Substitution* to obtain the final result $\boldsymbol{v}$.

Overall, we break the solution of a set of simultaneous equations into two steps: LU factorization (LUF) and Forward and Back Substitution (FBS). We should recognize that the intermediate vector $\boldsymbol{x}$ obtained above is the modified RHS that results from Gaussian Elimination. In other words, Gaussian Elimination provides the combination of LU factorization and Forward Substitution. By breaking down the sequence of steps we can deal more effectively with varying right hand sides.

Referring to our example in the previous section, $\boldsymbol{Y}$ was transformed into an upper triangular $\boldsymbol{U}$ matrix by Gaussian Elimination.

$$\boldsymbol{Y} = \begin{bmatrix} 2 & -1 & 0 & 0 \\ -1 & 3 & -1 & 0 \\ 0 & -1 & 3 & -1 \\ 0 & 0 & -1 & 2 \end{bmatrix} \rightarrow \begin{bmatrix} 1 & -\frac{1}{2} & 0 & 0 \\ 0 & 1 & -\frac{2}{5} & 0 \\ 0 & 0 & 1 & -\frac{5}{13} \\ 0 & 0 & 0 & 1 \end{bmatrix} = \boldsymbol{U} \tag{3.2.7}$$

Here we perform the same elimination steps, but instead of operating on the RHS vector we save the elimination steps in a lower-triangular matrix $\boldsymbol{L}$. We can obtain the $\boldsymbol{L}$ matrix from the values that appeared on and below the main diagonal immediately before their normalization (for the main diagonal elements) or annihilation (for those below the main diagonal). For the Gaussian Elimination example in Figure 3.2, the development of the $\boldsymbol{L}$ matrix is shown in Figure 3.3.

The lower triangular matrix is

$$\boldsymbol{L} = \begin{bmatrix} 2 & 0 & 0 & 0 \\ -1 & \frac{5}{2} & 0 & 0 \\ 0 & -1 & \frac{13}{5} & 0 \\ 0 & 0 & -1 & \frac{21}{13} \end{bmatrix} \tag{3.2.8}$$

and one can easily verify that

$$\boldsymbol{LU} = \boldsymbol{Y} \tag{3.2.9}$$

for this example.

Returning once again to the original problem, perform first the Forward Substitution:

$$\boldsymbol{Lx} = \boldsymbol{J} \tag{3.2.10}$$

$$\begin{bmatrix} 2 & 0 & 0 & 0 \\ -1 & \frac{5}{2} & 0 & 0 \\ 0 & -1 & \frac{13}{5} & 0 \\ 0 & 0 & -1 & \frac{21}{13} \end{bmatrix} \begin{bmatrix} x_1 \\ x_2 \\ x_3 \\ x_4 \end{bmatrix} = \begin{bmatrix} 1 \\ 0 \\ 0 \\ 1 \end{bmatrix} \tag{3.2.11}$$

$$\begin{aligned} 2x_1 &= 1 & &\Rightarrow & x_1 &= \frac{1}{2} \\ -x_1 + \frac{5}{2}x_2 &= 0 & &\Rightarrow & x_2 &= \frac{1}{5} \\ -x_2 + \frac{13}{5}x_3 &= 0 & &\Rightarrow & x_3 &= \frac{1}{13} \\ -x_3 + \frac{21}{13}x_4 &= 1 & &\Rightarrow & x_4 &= \frac{2}{3} \end{aligned} \tag{3.2.12}$$

STEPS $\quad [Y] \rightarrow [U] \qquad\qquad [L]$

$$\begin{bmatrix} 2 & -1 & 0 & 0 \\ -1 & 3 & -1 & 0 \\ 0 & -1 & 3 & -1 \\ 0 & 0 & -1 & 2 \end{bmatrix} \quad \text{then} \quad \begin{bmatrix} 2 & \bullet & \bullet & \bullet \\ -1 & \bullet & \bullet & \bullet \\ 0 & \bullet & \bullet & \bullet \\ 0 & \bullet & \bullet & \bullet \end{bmatrix}$$

1) and 2)

$$\begin{bmatrix} 1 & -\frac{1}{2} & 0 & 0 \\ 0 & \frac{5}{2} & -1 & 0 \\ 0 & -1 & 3 & -1 \\ 0 & 0 & -1 & 2 \end{bmatrix} \quad \text{then} \quad \begin{bmatrix} 2 & \bullet & \bullet & \bullet \\ -1 & \frac{5}{2} & \bullet & \bullet \\ 0 & -1 & \bullet & \bullet \\ 0 & 0 & \bullet & \bullet \end{bmatrix}$$

3) and 4)

$$\begin{bmatrix} 1 & -\frac{1}{2} & 0 & 0 \\ 0 & 1 & -\frac{2}{5} & 0 \\ 0 & 0 & \frac{13}{5} & -1 \\ 0 & 0 & -1 & 2 \end{bmatrix} \quad \text{then} \quad \begin{bmatrix} 2 & \bullet & \bullet & \bullet \\ -1 & \frac{5}{2} & \bullet & \bullet \\ 0 & -1 & \frac{13}{5} & \bullet \\ 0 & 0 & -1 & \bullet \end{bmatrix}$$

5) and 6)

$$\begin{bmatrix} 1 & -\frac{1}{2} & 0 & 0 \\ 0 & 1 & -\frac{2}{5} & 0 \\ 0 & 0 & 1 & -\frac{5}{13} \\ 0 & 0 & 0 & \frac{21}{13} \end{bmatrix} \quad \text{then} \quad \begin{bmatrix} 2 & \bullet & \bullet & \bullet \\ -1 & \frac{5}{2} & \bullet & \bullet \\ 0 & -1 & \frac{13}{5} & \bullet \\ 0 & 0 & -1 & \frac{21}{13} \end{bmatrix}$$

7)

$$\boldsymbol{U} = \begin{bmatrix} 1 & -\frac{1}{2} & 0 & 0 \\ 0 & 1 & -\frac{2}{5} & 0 \\ 0 & 0 & 1 & -\frac{5}{13} \\ 0 & 0 & 0 & 1 \end{bmatrix} \qquad \boldsymbol{L} = \begin{bmatrix} 2 & \bullet & \bullet & \bullet \\ -1 & \frac{5}{2} & \bullet & \bullet \\ 0 & -1 & \frac{13}{5} & \bullet \\ 0 & 0 & -1 & \frac{21}{13} \end{bmatrix}$$

Figure 3.3 Steps of LU factorization in matrix form.

We note that this solution, $x = L^{-1}J$, is the same as the RHS we obtained in the previous section with Gaussian Elimination, as it should be. So upon proceeding to the Back Substitution,

$$Uv = x \tag{3.2.13}$$

We thus obtain equations identical to (3.1.9), which yield upon solution

$$v_4 = \frac{2}{3},\quad v_3 = \frac{1}{3},\quad v_2 = \frac{1}{3},\quad \text{and}\quad v_1 = \frac{2}{3} \tag{3.2.14}$$

as we have seen before in equation (3.1.10).

We would tend to use LU factorization with Forward and Back Substitution instead of Gaussian Elimination because: (1) it is of equal complexity, and (2) we can change the RHS excitation vector and re-solve the equations easily and efficiently.

For example, if we wanted to solve the circuit in Figure 3.1 again, but this time with $I_1 = 2\text{A}$, we would need only to change the J vector since the LU factors still apply. Consider also the case of linear transient analysis as described briefly in Chapter 1. With a fixed time step, the equivalent conductance and resistance models for the energy storage elements would remain constant, and only the input signals and the companion model sources would change with each time step. A single LU factorization could be used throughout the entire fixed time step transient analysis. Later we will see that the LU factors prove useful for other applications as well, such as sensitivity analysis and large parameter variation analysis.

3.3 How LU Factorization Works

In the previous section, we assumed that it would be possible to factor a square matrix into lower- and upper-triangular factors. We outlined a practical procedure for computing the LU factors of a matrix. In this section, we will explain how LU factorization works, and thus validate the practical procedure described in the previous section. For clarity, we will go through the steps of factoring a 4×4 symbolic matrix; the procedure can clearly be extended to an $n \times n$ matrix.

Given the Y matrix, first label its original version with superscript zeros:

$$Y^{(0)} = \begin{bmatrix} y_{11}^{(0)} & y_{12}^{(0)} & y_{13}^{(0)} & y_{14}^{(0)} \\ y_{21}^{(0)} & y_{22}^{(0)} & y_{23}^{(0)} & y_{24}^{(0)} \\ y_{31}^{(0)} & y_{32}^{(0)} & y_{33}^{(0)} & y_{34}^{(0)} \\ y_{41}^{(0)} & y_{42}^{(0)} & y_{43}^{(0)} & y_{44}^{(0)} \end{bmatrix} \tag{3.3.1}$$

In an attempt to factor the matrix in (3.3.1) into the product of two matrices, we will start with the trivial step:

$$Y = L^{(0)}Y^{(0)} = \begin{bmatrix} 1 & 0 & 0 & 0 \\ 0 & 1 & 0 & 0 \\ 0 & 0 & 1 & 0 \\ 0 & 0 & 0 & 1 \end{bmatrix} \begin{bmatrix} y_{11}^{(0)} & y_{12}^{(0)} & y_{13}^{(0)} & y_{14}^{(0)} \\ y_{21}^{(0)} & y_{22}^{(0)} & y_{23}^{(0)} & y_{24}^{(0)} \\ y_{31}^{(0)} & y_{32}^{(0)} & y_{33}^{(0)} & y_{34}^{(0)} \\ y_{41}^{(0)} & y_{42}^{(0)} & y_{43}^{(0)} & y_{44}^{(0)} \end{bmatrix} \tag{3.3.2}$$

Of course $Y^{(0)}$ is not an upper triangular matrix yet; however, this terminology will simplify our explanation.

To begin, we normalize row 1 by dividing through by $y_{11}^{(0)}$. Then we add appropriate multiples of the first row to each of those below it to annihilate $y_{21}^{(0)}$, $y_{31}^{(0)}$, and $y_{41}^{(0)}$. Based on the previous section, the entire first column of the $Y^{(0)}$ matrix should be stored as the first column of the L matrix:

$$Y^{(0)} = \begin{bmatrix} y_{11}^{(0)} & y_{12}^{(0)} & y_{13}^{(0)} & y_{14}^{(0)} \\ y_{21}^{(0)} & y_{22}^{(0)} & y_{23}^{(0)} & y_{24}^{(0)} \\ y_{31}^{(0)} & y_{32}^{(0)} & y_{33}^{(0)} & y_{34}^{(0)} \\ y_{41}^{(0)} & y_{42}^{(0)} & y_{43}^{(0)} & y_{44}^{(0)} \end{bmatrix} \qquad L^{(1)} = \begin{bmatrix} y_{11}^{(0)} & 0 & 0 & 0 \\ y_{21}^{(0)} & 1 & 0 & 0 \\ y_{31}^{(0)} & 0 & 1 & 0 \\ y_{41}^{(0)} & 0 & 0 & 1 \end{bmatrix} \tag{3.3.3}$$

The diagonal element that is used to normalize a row and annihilate the rest of the column below it is called the *pivot*. Once we have used $y_{11}^{(0)}$ as the pivot element to normalize the first row and to annihilate the rest of the first column we have:

$$Y^{(1)} = \begin{bmatrix} 1 & y_{12}^{(1)} & y_{13}^{(1)} & y_{14}^{(1)} \\ 0 & y_{22}^{(1)} & y_{23}^{(1)} & y_{24}^{(1)} \\ 0 & y_{32}^{(1)} & y_{33}^{(1)} & y_{34}^{(1)} \\ 0 & y_{42}^{(1)} & y_{43}^{(1)} & y_{44}^{(1)} \end{bmatrix} \tag{3.3.4}$$

where the superscript (1) indicates that we have completed the first loop of LU factorization. More generally, we could write for a square matrix of dimension $n \times n$:

$$y_{1j}^{(1)} = \frac{y_{1j}^{(0)}}{y_{11}^{(0)}} \qquad j = 2, 3, 4, \ldots, n \tag{3.3.5}$$

$$y_{kj}^{(1)} = y_{kj}^{(0)} - (y_{k1}^{(0)} \cdot y_{1j}^{(1)}) \qquad \begin{matrix} j = 2, 3, 4, \ldots, n \\ k = 2, 3, 4, \ldots, n \end{matrix} \tag{3.3.6}$$

At this point, we have the partial factorization of the original $\boldsymbol{Y}$ matrix:

$$\boldsymbol{Y} = \boldsymbol{L}^{(1)}\boldsymbol{Y}^{(1)} = \begin{bmatrix} y_{11}^{(0)} & 0 & 0 & 0 \\ y_{21}^{(0)} & 1 & 0 & 0 \\ y_{31}^{(0)} & 0 & 1 & 0 \\ y_{41}^{(0)} & 0 & 0 & 1 \end{bmatrix} \begin{bmatrix} 1 & y_{12}^{(1)} & y_{13}^{(1)} & y_{14}^{(1)} \\ 0 & y_{22}^{(1)} & y_{23}^{(1)} & y_{24}^{(1)} \\ 0 & y_{32}^{(1)} & y_{33}^{(1)} & y_{34}^{(1)} \\ 0 & y_{42}^{(1)} & y_{43}^{(1)} & y_{44}^{(1)} \end{bmatrix} \tag{3.3.7}$$

Using equations (3.3.5), (3.3.6) and multiplying $\boldsymbol{L}^{(1)}$ and $\boldsymbol{Y}^{(1)}$, one can verify that this relation holds true. The first column of $\boldsymbol{L}^{(1)}$ will be the first column of the eventual lower-triangular matrix and the first row of $\boldsymbol{Y}^{(1)}$ will be the first row of its upper-triangular counterpart.

Because of the zero-valued portion of the first row of $\boldsymbol{L}^{(1)}$ and the zero-valued portion of the first column of $\boldsymbol{Y}^{(1)}$, we can focus further factorization on the lower right (3×3) portions of both matrices. These submatrices can be factored as shown below.

$$\begin{bmatrix} y_{22}^{(1)} & y_{23}^{(1)} & y_{24}^{(1)} \\ y_{32}^{(1)} & y_{33}^{(1)} & y_{34}^{(1)} \\ y_{42}^{(1)} & y_{43}^{(1)} & y_{44}^{(1)} \end{bmatrix} = \begin{bmatrix} y_{22}^{(1)} & 0 & 0 \\ y_{32}^{(1)} & 1 & 0 \\ y_{42}^{(1)} & 0 & 1 \end{bmatrix} \begin{bmatrix} 1 & y_{23}^{(2)} & y_{24}^{(2)} \\ 0 & y_{33}^{(2)} & y_{34}^{(2)} \\ 0 & y_{43}^{(2)} & y_{44}^{(2)} \end{bmatrix} \tag{3.3.8}$$

which represents another round of LU factorization as above. Once again, in general

$$y_{2j}^{(2)} = \frac{y_{2j}^{(1)}}{y_{22}^{(1)}} \qquad j = 3, 4, 5, \ldots, n \tag{3.3.9}$$

$$y_{kj}^{(2)} = y_{kj}^{(1)} - (y_{k2}^{(1)} \cdot y_{2j}^{(2)}) \qquad \begin{matrix} k = 3, 4, 5, \ldots, n \\ j = 3, 4, 5, \ldots, n \end{matrix} \tag{3.3.10}$$

The overall product, $Y = L^{(2)}Y^{(2)}$ can be verified here too, using equations (3.3.9) and (3.3.10):

$$Y = \begin{bmatrix} y_{11}^{(0)} & 0 & 0 & 0 \\ y_{21}^{(0)} & y_{22}^{(1)} & 0 & 0 \\ y_{31}^{(0)} & y_{32}^{(1)} & 1 & 0 \\ y_{41}^{(0)} & y_{42}^{(1)} & 0 & 1 \end{bmatrix} \begin{bmatrix} 1 & y_{12}^{(1)} & y_{13}^{(1)} & y_{14}^{(1)} \\ 0 & 1 & y_{23}^{(2)} & y_{24}^{(2)} \\ 0 & 0 & y_{33}^{(2)} & y_{34}^{(2)} \\ 0 & 0 & y_{43}^{(2)} & y_{44}^{(2)} \end{bmatrix} \quad (3.3.11)$$

The first two columns of $L^{(2)}$ are the actual columns of the lower-triangular result that we seek and the first two rows of $Y^{(2)}$ are the first two rows of our eventual U matrix.

Continuing now with the factorization of the lower-right-hand portions of $L^{(2)}$ and $Y^{(2)}$ we have

$$\begin{bmatrix} y_{33}^{(2)} & y_{34}^{(2)} \\ y_{43}^{(2)} & y_{44}^{(2)} \end{bmatrix} = \begin{bmatrix} y_{33}^{(2)} & 0 \\ y_{43}^{(2)} & 1 \end{bmatrix} \begin{bmatrix} 1 & y_{34}^{(3)} \\ 0 & y_{44}^{(3)} \end{bmatrix} \quad (3.3.12)$$

And finally

$$\begin{bmatrix} y_{44}^{(3)} \end{bmatrix} = \begin{bmatrix} y_{44}^{(3)} \end{bmatrix} \begin{bmatrix} 1 \end{bmatrix} \quad (3.3.13)$$

The $Y^{(4)}$ matrix is now the upper-triangular matrix U that we determined previously by Gaussian Elimination. The $L^{(4)}$ matrix is the lower-triangular counterpart L. Therefore,

$$Y = LU = \begin{bmatrix} y_{11}^{(0)} & 0 & 0 & 0 \\ y_{21}^{(0)} & y_{22}^{(1)} & 0 & 0 \\ y_{31}^{(0)} & y_{32}^{(1)} & y_{33}^{(2)} & 0 \\ y_{41}^{(0)} & y_{42}^{(1)} & y_{43}^{(2)} & y_{44}^{(3)} \end{bmatrix} \begin{bmatrix} 1 & y_{12}^{(1)} & y_{13}^{(1)} & y_{14}^{(1)} \\ 0 & 1 & y_{23}^{(2)} & y_{24}^{(2)} \\ 0 & 0 & 1 & y_{34}^{(3)} \\ 0 & 0 & 0 & 1 \end{bmatrix} \quad (3.3.14)$$

Note that since the diagonal terms of U are always one, we need not store them and the relevant terms of U and the terms of L could be stored in one matrix. In practice, we would start with Y and begin elimination on column 1. After terms are annihilated, the first column of L could be stored there. Similar manipulations would be used for column 2, column 3, and so on. Let us assume that we are dealing with *dense matrices*, where every element of the matrix is stored, whether it is a zero or a non-zero. In this case, the naive implementation of LU factorization would require double the storage required for a single $n \times n$ matrix, with the original matrix being massaged into U, and a new matrix being created to store L. However, by carrying out the factorization in-place, we can get by with just as much storage as required for the original matrix. If we are using *sparse matrices*, where only the non-zeros of a matrix are stored, then the memory requirements situation is totally different. For a more detailed discussion of sparse matrices, see Chapter 7.

In the above discussion we performed the LU factorization on a row basis. We should mention that LU factorization can be equally well performed on a column basis. Then the L matrix would have ones on the main diagonal and the U matrix would not. An easy way to think of this is that we could LU factor the transpose of the Y matrix as above, and then transpose the result:

$$Y^T = \hat{L}\hat{U} \Rightarrow Y = \hat{U}^T\hat{L}^T = LU \tag{3.3.15}$$

where $L = \hat{U}^T$ is a lower-triangular matrix with ones on the diagonal and $U = \hat{L}^T$ is an upper-triangular matrix.

3.4 Pivot Conditioning

Floating point representation of variables on computers with finite machine precision results in *rounding (or round-off) errors* during arithmetic operations in general and LU factorization in particular. Especially when the matrix condition is poor -- which is when the determinant of Y is very small -- we are particularly susceptible to the dangers of round-off error. We must recognize though that poor matrix conditioning and round-off errors are separate problems, but we are generally worst affected when the two combine to create numerical problems. While we are stuck with the conditioning of the matrix we can use some clever techniques to minimize round-off error.

Round-off errors usually occur when the difference of two large but roughly equal numbers is computed or when a large number and a tiny one are added or subtracted. Multiplication does not affect round-off error significantly. The rationale behind the above statements is most easily explained by way of a simple example.

Consider the linear set of equations:

$$\begin{bmatrix} a_{11} & a_{12} \\ a_{21} & a_{22} \end{bmatrix} \begin{bmatrix} x_1 \\ x_2 \end{bmatrix} = \begin{bmatrix} b_1 \\ b_2 \end{bmatrix} \tag{3.4.1}$$

which we might solve by Gaussian Elimination or LU factorization. First we choose a_{11} to be the pivot element to obtain

$$\begin{bmatrix} 1 & \frac{a_{12}}{a_{11}} \\ a_{21} & a_{22} \end{bmatrix} \begin{bmatrix} x_1 \\ x_2 \end{bmatrix} = \begin{bmatrix} \frac{b_1}{a_{11}} \\ b_2 \end{bmatrix} \tag{3.4.2}$$

Then we annihilate the a_{21} term:

$$\begin{bmatrix} 1 & \frac{a_{12}}{a_{11}} \\ 0 & \left(a_{22} - \frac{a_{21}a_{12}}{a_{11}}\right) \end{bmatrix} \begin{bmatrix} x_1 \\ x_2 \end{bmatrix} = \begin{bmatrix} \frac{b_1}{a_{11}} \\ \left(b_2 - \frac{a_{21}b_1}{a_{11}}\right) \end{bmatrix} \tag{3.4.3}$$

At this point we note that round-off error would be small if the following inequalities were true:

$$|a_{22}| \gg \left|\frac{a_{21} \cdot a_{12}}{a_{11}}\right| \tag{3.4.4}$$

and

$$|b_2| \gg \left|\frac{a_{21} \cdot b_1}{a_{11}}\right| \tag{3.4.5}$$

If, however, these inequalities were

$$|a_{22}| \ll \left|\frac{a_{21} \cdot a_{12}}{a_{11}}\right| \tag{3.4.6}$$

and

$$|b_2| \ll \left|\frac{a_{21} \cdot b_1}{a_{11}}\right| \tag{3.4.7}$$

then significant information from the original a_{22} or b_{22} values would be lost.

The reason we are concerned with the round-off error when the inequality in (3.4.6) is true, and not for the condition in (3.4.4), is because the absolute round-off error is related to the size of matrix entries. The conditions in (3.4.6) promote the growth of absolute values of matrix elements, while those in (3.4.4) do not.

For example, the new a_{22} entry after one factorization step, $a_{22}^{(1)}$ is

$$a_{22}^{(1)} = \left(a_{22}^{(0)} - \frac{a_{21}^{(0)} a_{12}^{(0)}}{a_{11}^{(0)}}\right) \tag{3.4.8}$$

where $a_{22}^{(0)}$ is the original a_{22} entry. The actual $a_{22}^{(1)}$ value that we calculate, $\tilde{a}_{22}^{(1)}$, has a certain amount of round-off error, δ_{22}:

$$\tilde{a}_{22}^{(1)} = a_{22}^{(1)} + \delta_{22} \tag{3.4.9}$$

Since the matrix entries are stored as floating point numbers, this absolute round-off error is related to the magnitudes of the element values. When the inequality in (3.4.6) is true, the magnitude of $a_{22}^{(1)}$ is much greater than the magnitude of the original value, $a_{22}^{(0)}$. We would expect, therefore, that the absolute round-off error would be largest when (3.4.6) is true.

We can demonstrate this change in round-off error due to growth of the matrix entry values with the following example. Consider the following numerical values for the a_{ij} terms and the b_i terms in (3.4.1):

$$\begin{bmatrix} 1.00e-4 & 1.00 \\ 1.00 & 1.00 \end{bmatrix} \begin{bmatrix} x_1 \\ x_2 \end{bmatrix} = \begin{bmatrix} 1.00 \\ 2.00 \end{bmatrix} \tag{3.4.10}$$

Also, to demonstrate our point, we'll assume that the computer used to factorize this matrix has only three decimal digits of precision. For such an exaggerated case, after normalization of the first row and annihilation of the a_{21} term we would obtain

$$\begin{bmatrix} 1.00 & 1.00e4 \\ 0.00 & -1.00e4 \end{bmatrix} \begin{bmatrix} x_1 \\ x_2 \end{bmatrix} = \begin{bmatrix} 1.00e4 \\ -1.00e4 \end{bmatrix} \tag{3.4.11}$$

For this example, with only three digits of machine precision, the round-off error for $\tilde{a}_{22}^{(1)}$ is 1.00. This large round-off error is a consequence of the large magnitude term, or the growth of the a_{22} term. Back Substitution of the equations in (3.4.11) yields $x_2 = 1.00$ and $x_1 = 0.00$, clearly an incorrect solution.

From (3.4.4) we see that we would like the diagonal entries, a_{11} and a_{22}, to be much larger than the off-diagonal terms, a_{12} and a_{21}. These conditions also deter the growth of the element values, hence, the absolute round-off error.

For example, rearranging the order of the rows in (3.4.10),

$$\begin{bmatrix} 1.00 & 1.00 \\ 1.00e-4 & 1.00 \end{bmatrix} \begin{bmatrix} x_1 \\ x_2 \end{bmatrix} = \begin{bmatrix} 2.00 \\ 1.00 \end{bmatrix} \tag{3.4.12}$$

Then normalizing and annihilating as was done previously yields

$$\begin{bmatrix} 1.00 & 1.00 \\ 0.00 & 1.00 \end{bmatrix} \begin{bmatrix} x_1 \\ x_2 \end{bmatrix} = \begin{bmatrix} 2.00 \\ 1.00 \end{bmatrix} \tag{3.4.13}$$

The absolute round-off error for a_{22} is now $1e-4$. Moreover, upon Back Substitution we find that $x_2 = 1.00$ and $x_1 = 1.00$, which are the correct answers for the given machine precision.

Thus far we have considered only the round-off error as it affects the matrix terms. We should point out that the round-off errors in the right hand side terms can also be significant.

To proceed to the solution of (3.4.3) in terms of Back Substitution

$$x_2 = \frac{b_2 - \dfrac{a_{21}b_1}{a_{11}}}{a_{22} - \dfrac{a_{21}a_{12}}{a_{11}}} \tag{3.4.14}$$

and

$$x_1 = \frac{b_1}{a_{11}} - \frac{a_{12}}{a_{11}}\left(\frac{b_2 - \dfrac{a_{21}b_1}{a_{11}}}{a_{22} - \dfrac{a_{21}a_{12}}{a_{11}}}\right) \tag{3.4.15}$$

Here, again, we would not want to lose significant information from the original value of b_1, so we would ask that the following inequality be true:

$$|b_1| \gg \left| a_{12} \left(\frac{b_2 - \frac{a_{21} b_1}{a_{11}}}{a_{22} - \frac{a_{21} a_{12}}{a_{11}}} \right) \right| \qquad \textbf{(3.4.16)}$$

In order to avoid numerical problems, the general strategy is to exchange rows or columns to have as large a pivot on the diagonal as possible. There are two approaches to adjusting the pivot element, or *pivoting*. The first approach is *full pivoting*, where we swap rows and/or columns to get the largest possible pivot from among all the rows and columns of the submatrix being factored. The second approach is *partial pivoting*, where we swap rows or columns to get the largest available pivot in the *first row* or *first column* of the submatrix being factored. Note that neither of these pivoting strategies takes the right hand side vector into account. But we see from inequalities (3.4.5) and (3.4.16) that the right hand side magnitudes $|b_1|$ and $|b_2|$ are important as well. Since LU factorization ignores the right hand side of a matrix equation, problems may arise later during Forward and Back Substitution. In particular, those rows which may have small or zero valued right hand sides should be given higher pivoting priority than others. Such prioritization can only be invoked for Gaussian Elimination, where the right hand side of the matrix equation is considered along with the left, and only a single solution is sought. In such a situation, we can consider normalizing the equations in terms of the right hand sides as shown below:

$$\begin{bmatrix} \frac{a_{11}}{b_1} & \frac{a_{12}}{b_1} \\ \frac{a_{21}}{b_2} & \frac{a_{22}}{b_2} \end{bmatrix} \begin{bmatrix} x_1 \\ x_2 \end{bmatrix} = \begin{bmatrix} 1 \\ 1 \end{bmatrix} \qquad \textbf{(3.4.17)}$$

Then we can proceed with pivot selection for Gaussian Elimination as usual. We note finally that a row with a small magnitude or zero-valued right hand side should be given a very high pivot priority. Such a prioritization would not be possible with LU factorization which ignores the right hand side vector entirely.

To reduce round-off error, partial pivoting is often applied, whereby the pivot element for the (i, i) location is selected as the largest element in the ith column or row, but not both. For example, consider the LU factorization of the following matrix equations with partial pivoting:

$$\begin{bmatrix} 1 & 2 & 1 \\ 1 & 1 & 3 \\ 4 & 1 & 1 \end{bmatrix} \begin{bmatrix} x_1 \\ x_2 \\ x_3 \end{bmatrix} = \begin{bmatrix} 5 \\ 8 \\ 7 \end{bmatrix} \tag{3.4.18}$$

To begin, the largest element in column 1, the 4 in position $(3, 3)$, would be selected as the first pivot element, as shown in Figure 3.4. This requires swapping rows 1 and 3, including the corresponding elements in the RHS vector. The values saved in the $\boldsymbol{L}$ matrix are the values *prior* to annihilation and normalization of column 1, but *after* the reordering of the rows for partial pivoting. The remaining steps of factorization (with partial pivoting) are outlined in Figure 3.4. One can easily verify that the product of the final $\boldsymbol{L}$ and $\boldsymbol{U}$ is equal to the original matrix with the new row ordering 3, 1, 2.

Although it is rarely warranted in practice, full pivoting is the selection of the pivot element as the largest element in the submatrix which remains to be factored. The LU factorization and solution of equation (3.4.18) with full pivoting is shown in Figure 3.5. We should note that the swapping of columns required during full pivoting does involve the reordering of variables, as demonstrated by this example. The swapping of columns 2 and 3 permuted the order of the variables as shown in the results.

For all of these factorization examples

$$\boldsymbol{Y} = \boldsymbol{L}\boldsymbol{U} \tag{3.4.19}$$

and

$$det(\boldsymbol{Y}) = det(\boldsymbol{L})\, det(\boldsymbol{U}) \tag{3.4.20}$$

The determinant of $\boldsymbol{U}$ is simply 1 since the $\boldsymbol{U}$ matrix is triangular with only ones along the main diagonal. Therefore

$$det(\boldsymbol{Y}) = det(\boldsymbol{L}) \tag{3.4.21}$$

and the determinant of $\boldsymbol{L}$ is simply the product of its diagonal terms, which are the pivot values prior to normalization. So, in the course of LU factorization we can monitor the determinant of $\boldsymbol{Y}$, hence the condition of the matrix, according to the values of the pivot elements. Even with full pivoting we may be ultimately forced to use a very small pivot value, since no matter what pivoting scheme we use, the determinant of $\boldsymbol{Y}$ is invariant! So a nearly singular $\boldsymbol{Y}$ matrix will ultimately result in a small pivot.

$$Y \rightarrow U \qquad\qquad L$$

$$\begin{bmatrix} 1 & 2 & 1 \\ 1 & 1 & 3 \\ 4 & 1 & 1 \end{bmatrix} \xRightarrow{\text{swap rows 1 and 3}} \begin{bmatrix} 4 & 1 & 1 \\ 1 & 1 & 3 \\ 1 & 2 & 1 \end{bmatrix} \qquad \begin{bmatrix} 4 & \bullet & \bullet \\ 1 & \bullet & \bullet \\ 1 & \bullet & \bullet \end{bmatrix}$$

$$\begin{bmatrix} 1 & \frac{1}{4} & \frac{1}{4} \\ 0 & \frac{3}{4} & \frac{11}{4} \\ 0 & \frac{7}{4} & \frac{3}{4} \end{bmatrix} \xRightarrow{\text{swap rows 2 and 3}} \begin{bmatrix} 1 & \frac{1}{4} & \frac{1}{4} \\ 0 & \frac{7}{4} & \frac{3}{4} \\ 0 & \frac{3}{4} & \frac{11}{4} \end{bmatrix} \qquad \begin{bmatrix} 4 & 0 & \bullet \\ 1 & \frac{7}{4} & \bullet \\ 1 & \frac{3}{4} & \bullet \end{bmatrix}$$

$$\begin{bmatrix} 1 & \frac{1}{4} & \frac{1}{4} \\ 0 & 1 & \frac{3}{7} \\ 0 & 0 & \frac{17}{7} \end{bmatrix} \qquad L = \begin{bmatrix} 4 & 0 & 0 \\ 1 & \frac{7}{4} & 0 \\ 1 & \frac{3}{4} & \frac{17}{7} \end{bmatrix}$$

$$U = \begin{bmatrix} 1 & \frac{1}{4} & \frac{1}{4} \\ 0 & 1 & \frac{3}{7} \\ 0 & 0 & 1 \end{bmatrix} \qquad LU = \begin{bmatrix} 4 & 1 & 1 \\ 1 & 2 & 1 \\ 1 & 1 & 3 \end{bmatrix} \qquad (new)\, RHS = \begin{bmatrix} 7 \\ 5 \\ 8 \end{bmatrix}$$

Figure 3.4 LU factorization with partial pivoting.

3.5 *Iterative Refinement*

Even with the best pivoting schemes, some round-off error is incurred during factorization. Hence we obtain a solution that does not quite satisfy the original set of equations. Let us say v_{exact} is the exact solution and v is our computed solution. Then, we can write

$Y \rightarrow U$ $\qquad\qquad$ L

$$\begin{bmatrix} 1 & 2 & 1 \\ 1 & 1 & 3 \\ 4 & 1 & 1 \end{bmatrix} \overset{\text{swap rows 1 and 3}}{\Rightarrow} \begin{bmatrix} 4 & 1 & 1 \\ 1 & 1 & 3 \\ 1 & 2 & 1 \end{bmatrix} \qquad \begin{bmatrix} 4 & \bullet & \bullet \\ 1 & \bullet & \bullet \\ 1 & \bullet & \bullet \end{bmatrix}$$

$$\begin{bmatrix} 1 & \frac{1}{4} & \frac{1}{4} \\ 0 & \frac{3}{4} & \frac{11}{4} \\ 0 & \frac{7}{4} & \frac{3}{4} \end{bmatrix} \overset{\text{swap columns 2 and 3}}{\Rightarrow} \begin{bmatrix} 1 & \frac{1}{4} & \frac{1}{4} \\ 0 & \frac{11}{4} & \frac{3}{4} \\ 0 & \frac{3}{4} & \frac{7}{4} \end{bmatrix} \qquad \begin{bmatrix} 4 & 0 & \bullet \\ 1 & \frac{11}{4} & \bullet \\ 1 & \frac{3}{4} & \bullet \end{bmatrix}$$

$$\begin{bmatrix} 1 & \frac{1}{4} & \frac{1}{4} \\ 0 & 1 & \frac{3}{11} \\ 0 & 0 & \frac{68}{44} \end{bmatrix} \qquad L = \begin{bmatrix} 4 & 0 & 0 \\ 1 & \frac{11}{4} & 0 \\ 1 & \frac{3}{4} & \frac{68}{44} \end{bmatrix}$$

$$U = \begin{bmatrix} 1 & \frac{1}{4} & \frac{1}{4} \\ 0 & 1 & \frac{3}{11} \\ 0 & 0 & 1 \end{bmatrix} \qquad x = \begin{bmatrix} x_1 \\ x_3 \\ x_2 \end{bmatrix} \qquad RHS = \begin{bmatrix} 7 \\ 8 \\ 5 \end{bmatrix}$$

$$LU = \begin{bmatrix} 4 & 1 & 1 \\ 1 & 3 & 1 \\ 1 & 1 & 2 \end{bmatrix}$$

Figure 3.5 LU factorization with full pivoting.

$$J - Yv_{exact} = 0 \tag{3.5.1}$$

and

$$J - Yv = r \tag{3.5.2}$$

where r is the vector of residues due to the inaccuracy in our solution process. Each residue component is the algebraic sum of the currents entering the corresponding node when our vector of computed node voltages is applied to the circuit. If the ith component of r is zero, then our solution correctly satisfies KCL at the ith node. If not, the ith component of r is the erroneous sum of currents at the ith node.

Intuitively, if the sum of the computed currents into node i were greater than zero $(r_i > 0)$, we could consider increasing the computed node voltage v_i in our next guess to decrease that net current. Similarly, if the sum of the computed currents flowing into node i were less than zero $(r_i < 0)$, we could attempt to increase that net current by decreasing the computed node voltage at node i in our next guess. Iterating on a solution to improve it by successively guessing better solutions is called *iterative refinement*. In theory, we could continue iterating on the node voltages until all of the residue components were acceptably small. But this procedure is *ad hoc*, therefore we consider a more formal iterative improvement scheme described below.

Subtracting (3.5.1) from (3.5.2), we get

$$Yv_{exact} - Yv = r \tag{3.5.3}$$

Define a vector e of node voltage errors in our solution.

$$e = v_{exact} - v \tag{3.5.4}$$

Then

$$Yv_{exact} - Yv = Y(\mathrm{v}_{exact} - \mathrm{v}) = Ye = r \tag{3.5.5}$$

To improve the computed solution, we add these errors computed by means of (3.5.5) to the factorized solution. But since r is not exact and the computed error, e, also has some round-off error, we repeat this improvement as an iterative process:

1. Treat the solution from LU factorization as the initial guess $v^{(0)}$.
2. Calculate the residue vector at the jth iteration from

$$r^{(j)} = J - Yv^{(j)} \tag{3.5.6}$$

and compute the error at the jth iteration using the same LU factors

$$Ye^{(j)} = r^{(j)} \Rightarrow e^{(j)} = Y^{-1}r^{(j)} \tag{3.5.7}$$

3. *If* the jth error term is sufficiently small for every node i

$$\left|e_i^{(j)}\right| < \varepsilon_{relative}\left|v_i^{(j)}\right| + \varepsilon_{absolute} \tag{3.5.8}$$

   we accept the solution $v^{(j)}$, where $\varepsilon_{relative}$ is a relative error tolerance and $\varepsilon_{absolute}$ an absolute error tolerance. This type of tolerance equation is often used to test convergence of any iterative scheme. For answers that are large in absolute value, iteration stops when the solution is sufficiently close. For small numbers, iteration stops when we are within the absolute tolerance of the right answer.

4. *Else* improve the solution for the node voltages by adding the jth error terms to them

$$v^{(j+1)} = v^{(j)} + e^{(j)} \tag{3.5.9}$$

   then increment j and go back to step 2.

Without any formal proof, we will make the statement that this *iterative improvement* will always work (*converge*) unless the factorization round-off error is nearly as large as the initial solution $v^{(0)}$ [Ralston78]. If the iterations fail to converge, (or, in other words, *diverge*) and equation (3.5.8) is never satisfied, then the LU decomposition must be carried out with greater machine precision.

In practice, most circuit simulation programs use double precision (8 bytes) to represent floating point numbers. They use pivoting schemes aimed at maintaining numerical accuracy while taking into account some other considerations (see Chapter 7). They do not check residues at the end of each LU factorization, thus assuming that their circuit equations were well-conditioned enough and their pivoting schemes were good enough to obviate any need to "check" the answer.

3.6 Sensitivity Analysis

Given a (nominal) circuit, the previous section discussed numerical problems that can lead to errors in the solution of the circuit. However, it is possible that the element values in the circuit are themselves not correct. In other words, an element that is nominally 1Ω may not actually be 1Ω but a little less or more due to manufacturing tolerance. Particularly in integrated circuit design, manufacturing variations must be taken into account. In addition, a circuit designer may want to conduct a "what-if" analysis. For example, a designer may ask the question, "If I increase a resistor R_1 by a small amount, by how much will its current change?" All of these situations call for a *sensitivity analysis*. A sensitivity analysis can help us understand the variation of circuit response with respect to variation in circuit parameters like element values.

There are two methods of sensitivity analysis: the *direct method* and the *adjoint method.* These two methods are discussed in detail in Chapter 9. This section provides an introduction to the adjoint method by differentiating the circuit equations in matrix form. A more formal derivation of the adjoint method (based on Tellegen's theorem) can be found in Chapter 9.

In Chapter 1 we showed that the core analysis of circuit simulation is the solution of a matrix equation:

$$\boldsymbol{Mx} = \boldsymbol{b} \Rightarrow \boldsymbol{x} = \boldsymbol{M}^{-1}\boldsymbol{b} \tag{3.6.1}$$

where $\boldsymbol{M}$ is an $n \times n$ matrix, $\boldsymbol{b}$ is an $n \times 1$ vector, and $\boldsymbol{x}$ is the $n \times 1$ vector of unknowns. As usual, we reiterate that we would never actually invert $\boldsymbol{M}$, rather we would LU factor it and then perform Forward and Back Substitution; we use $\boldsymbol{M}^{-1}$ to indicate that process. Now, suppose that we perform a slight variation on the elements that have been stamped into $\boldsymbol{M}$ and $\boldsymbol{b}$, and we wish to ascertain the effects that these variations might have on the solution $\boldsymbol{x}$. Let us assume that $\boldsymbol{M}$ changes to $\boldsymbol{M} + \delta\boldsymbol{M}$, where $\delta\boldsymbol{M}$ is an $n \times n$ matrix of $\boldsymbol{M}$ parameter variations and $\boldsymbol{b}$ changes to $\boldsymbol{b} + \delta\boldsymbol{b}$ where $\delta\boldsymbol{b}$ is an $n \times 1$ vector of RHS parameter variations. Further, let us assume that the changes to $\boldsymbol{M}$ and $\boldsymbol{b}$ are small. Then

$$(\boldsymbol{M} + \delta\boldsymbol{M})\ (\boldsymbol{x} + \delta\boldsymbol{x}) = \boldsymbol{b} + \delta\boldsymbol{b} \tag{3.6.2}$$

where $\delta\boldsymbol{x}$ is the $n \times 1$ vector of changes in the circuit unknowns. Upon expanding the left hand side we obtain

$$\boldsymbol{Mx} + \delta\boldsymbol{Mx} + \boldsymbol{M}\delta\boldsymbol{x} + \delta\boldsymbol{M}\delta\boldsymbol{x} = \boldsymbol{b} + \delta\boldsymbol{b} \tag{3.6.3}$$

Subtracting the nominal solution (3.6.1) and neglecting the second-order variation,

$$\cancel{\boldsymbol{Mx}} + \delta\boldsymbol{Mx} + \boldsymbol{M}\delta\boldsymbol{x} + \cancel{\delta\boldsymbol{M}\delta\boldsymbol{x}} = \cancel{\boldsymbol{b}} + \delta\boldsymbol{b} \tag{3.6.4}$$

we obtain

$$\boldsymbol{M}\delta\boldsymbol{x} = -\delta\boldsymbol{Mx} + \delta\boldsymbol{b} \tag{3.6.5}$$

Because we have already LU factored $\boldsymbol{M}$ to obtain the nominal solution $\boldsymbol{x}$, we could easily perform a perturbation analysis by solving this expression for $\delta\boldsymbol{x}$ in terms of a specific $\delta\boldsymbol{M}$ and $\delta\boldsymbol{b}$:

$$\delta x = M^{-1}(-\delta M x + \delta b) \tag{3.6.6}$$

Thus, with the expense of one additional matrix multiplication, one additional vector addition and one additional Forward and Back Substitution, we can obtain δx, the vector of circuit response variations. But suppose that we want to perform a "what if" sensitivity analysis. In such a case it would be useful to be able to find the change δx with respect to specific parameter changes δM_{ij} and δb_i. If the components of M or b depend in a complicated way on circuit or physical parameters, we could contemplate the use of the chain rule to express the sensitivities in terms of those parameters.

Under the assumption that we may be interested in the sensitivity of a particular response variable (at the circuit output, for example), and that we are not concerned with any others, then we have

$$\delta x_i = \underbrace{[i^{th} \text{ row of } M^{-1}]}_{\text{row vector}} (-\delta M x + \delta b) \tag{3.6.7}$$

for the ith variable in δx. Note that

$$\begin{aligned} i^{th} \text{ row of } M^{-1} &= i^{th} \text{ column of } (M^{-1})^T \\ &= i^{th} \text{ column of } (M^T)^{-1} \end{aligned} \tag{3.6.8}$$

But

$$(M^T)(M^T)^{-1} = \mathbf{1} \tag{3.6.9}$$

so

$$(M^T)(i^{th} \text{ column of } (M^T)^{-1}) = i^{th} \text{ column of } \mathbf{1} \tag{3.6.10}$$

Equation (3.6.10) can be restated as

$$M^T \xi_i = e_i \tag{3.6.11}$$

where ξ_i is a column vector representing the ith row of $(M)^{-1}$ and e_i is a column vector with all zeros except for a one in its ith row (the ith unit vector). Then the ith component sensitivity is

$$\delta x_i = \xi_i^T(-\delta M x + \delta b) \tag{3.6.12}$$

But we do not have to go so far as to compute

$$\xi_i = (M^T)^{-1} e_i = (M^{-1})^T e_i \tag{3.6.13}$$

In the solution of the nominal circuit we have already found

$$M = LU \tag{3.6.14}$$

and

$$M^T = U^T L^T \tag{3.6.15}$$

which provides the lower (U^T) and upper (L^T) triangular factors of M^T. Therefore, we can solve

$$U^T g_i = e_i \tag{3.6.16}$$

by Forward Substitution to obtain the intermediate variable g_i, and then solve

$$L^T \xi_i = g_i \tag{3.6.17}$$

by Back Substitution to obtain the column vector ξ_i. Returning to the sensitivity expression (3.6.12), we can recognize the following individual sensitivities:

$$\frac{\partial x_i}{\partial b_j} = \xi_{ij} \tag{3.6.18}$$

which is the jth component (row) of the column vector ξ_i, and

$$\frac{\partial x_i}{\partial M_{kl}} = -\xi_{ik} x_l \tag{3.6.19}$$

which is the negative of the kth component (row) of the column vector ξ_i multiplied by the lth component (row) of the nominal solution x.

To validate the above we can consider in symbolic form the following 2×2 matrix example.

$$\begin{bmatrix} M_{11} & M_{12} \\ M_{21} & M_{22} \end{bmatrix} \begin{bmatrix} x_1 \\ x_2 \end{bmatrix} = \begin{bmatrix} b_1 \\ b_2 \end{bmatrix} \tag{3.6.20}$$

The LU factors of the matrix M are given by

$$\begin{bmatrix} M_{11} & M_{12} \\ M_{21} & M_{22} \end{bmatrix} = \begin{bmatrix} M_{11} & 0 \\ M_{21} & (M_{22} - M_{21}M_{11}^{-1}M_{12}) \end{bmatrix} \begin{bmatrix} 1 & (M_{11}^{-1}M_{12}) \\ 0 & 1 \end{bmatrix} \quad (3.6.21)$$

Suppose that we seek the sensitivity of x_1 with respect to the components of $\boldsymbol{M}$ and $\boldsymbol{b}$, then we must solve the following transposed system:

$$\begin{bmatrix} 1 & 0 \\ (M_{11}^{-1}M_{12}) & 1 \end{bmatrix} \begin{bmatrix} M_{11} & M_{21} \\ 0 & (M_{22} - M_{21}M_{11}^{-1}M_{12}) \end{bmatrix} \begin{bmatrix} \xi_{11} \\ \xi_{12} \end{bmatrix} = \begin{bmatrix} 1 \\ 0 \end{bmatrix} \quad (3.6.22)$$

From Forward Substitution

$$\begin{bmatrix} 1 & 0 \\ (M_{11}^{-1}M_{12}) & 1 \end{bmatrix} \begin{bmatrix} g_{11} \\ g_{12} \end{bmatrix} = \begin{bmatrix} 1 \\ 0 \end{bmatrix} \quad (3.6.23)$$

we obtain

$$g_{11} = 1 \qquad \text{and} \qquad g_{12} = -M_{11}^{-1}M_{12} \quad (3.6.24)$$

Then from Back Substitution

$$\begin{bmatrix} M_{11} & M_{21} \\ 0 & M_{22} - M_{21}M_{11}^{-1}M_{12} \end{bmatrix} \begin{bmatrix} \xi_{11} \\ \xi_{12} \end{bmatrix} = \begin{bmatrix} 1 \\ -M_{11}^{-1}M_{12} \end{bmatrix} \quad (3.6.25)$$

we obtain

$$\xi_{12} = \frac{-M_{12}}{M_{11}M_{22} - M_{21}M_{12}} \qquad \text{and} \qquad \xi_{11} = \frac{M_{22}}{M_{11}M_{22} - M_{21}M_{12}} \quad (3.6.26)$$

From (3.6.18) and (3.6.19) and the result in (3.6.26) we have the following sensitivities:

$$\frac{\partial x_1}{\partial b_1} = \xi_{11} = \frac{M_{22}}{M_{11}M_{22} - M_{21}M_{12}} \quad (3.6.27)$$

$$\frac{\partial x_1}{\partial b_2} = \xi_{12} = \frac{-M_{12}}{M_{11}M_{22} - M_{21}M_{12}} \quad (3.6.28)$$

$$\frac{\partial x_1}{\partial M_{11}} = -\xi_{11}x_1 = \frac{-M_{22}}{M_{11}M_{22}-M_{21}M_{12}}x_1 \tag{3.6.29}$$

$$\frac{\partial x_1}{\partial M_{12}} = -\xi_{11}x_2 = \frac{-M_{22}}{M_{11}M_{22}-M_{21}M_{12}}x_2 \tag{3.6.30}$$

$$\frac{\partial x_1}{\partial M_{21}} = -\xi_{12}x_1 = \frac{M_{12}}{M_{11}M_{22}-M_{21}M_{12}}x_1 \tag{3.6.31}$$

$$\frac{\partial x_1}{\partial M_{22}} = -\xi_{12}x_2 = \frac{M_{12}}{M_{11}M_{22}-M_{21}M_{12}}x_2 \tag{3.6.32}$$

We can check these results in terms of the explicit solution for this 2×2 matrix problem:

$$\begin{bmatrix} x_1 \\ x_2 \end{bmatrix} = \frac{1}{M_{11}M_{22}-M_{21}M_{12}} \begin{bmatrix} M_{22} & -M_{12} \\ -M_{21} & M_{11} \end{bmatrix} \begin{bmatrix} b_1 \\ b_2 \end{bmatrix} \tag{3.6.33}$$

which yields

$$x_1 = \frac{M_{22}}{M_{11}M_{22}-M_{21}M_{12}}b_1 + \frac{-M_{12}}{M_{11}M_{22}-M_{21}M_{12}}b_2 \tag{3.6.34}$$

It is easy to see that the $\frac{\partial x_1}{\partial b_1}$ and $\frac{\partial x_1}{\partial b_2}$ for (3.6.34) match those in (3.6.27) and (3.6.28) respectively. With a little more effort we can show that $\frac{\partial x_1}{\partial M_{22}}$ of equation (3.6.34) yields

$$\begin{aligned} &= \frac{1}{M_{11}M_{22}-M_{21}M_{12}}b_1 + \frac{-M_{11}M_{22}}{(M_{11}M_{22}-M_{21}M_{12})^2}b_1 + \frac{M_{11}M_{12}}{(M_{11}M_{22}-M_{21}M_{12})^2}b_2 \\ &= \frac{1}{(M_{11}M_{22}-M_{21}M_{12})^2}(-M_{21}M_{12}b_1 + M_{11}M_{12}b_2) \\ &= \frac{M_{12}}{(M_{11}M_{22}-M_{21}M_{12})}\left[\frac{-M_{21}}{M_{11}M_{22}-M_{21}M_{12}}b_1 + \frac{M_{11}}{M_{11}M_{22}-M_{21}M_{12}}b_2\right] \\ &= \frac{M_{12}}{(M_{11}M_{22}-M_{21}M_{12})}x_2 \end{aligned} \tag{3.6.35}$$

which matches the result in (3.6.32). Similar results can be obtained from the direct symbolic differentiations $\frac{\partial x_1}{\partial M_{11}}, \frac{\partial x_1}{\partial M_{12}}, \frac{\partial x_1}{\partial M_{21}}$, etc.

It is important to note that the technique described above is *not* a perturbation method. It provides exact sensitivities to within machine accuracy. In a perturbation method, two simulations are performed with the sensitivity parameter being perturbed by a small amount. Then a difference method is used to obtain sensitivity. The first disadvantage of perturbation methods is that an extra simulation is required in addition to the nominal circuit solution. Further, the amount of the perturbation is hard to determine. Accuracy and numerical reasons make it hard to predict the right amount of perturbation in order to get good answers.

In general, more than one entry in the $\boldsymbol{M}$ matrix and/or the $\boldsymbol{b}$ vector may depend on the same physical parameter p. Consider, for example, the simple circuit in Figure 3.6. The nodal equations are:

$$\boldsymbol{M} = \boldsymbol{Y} = \begin{bmatrix} (\frac{1}{R_2} + \frac{1}{R_5}) & (-\frac{1}{R_5}) \\ (-\frac{1}{R_5}) & (\frac{1}{R_3} + \frac{1}{R_5}) \end{bmatrix} \qquad \boldsymbol{b} = \boldsymbol{J} = \begin{bmatrix} I_1 \\ I_4 \end{bmatrix} \qquad \boldsymbol{x} = \boldsymbol{v} = \begin{bmatrix} v_1 \\ v_2 \end{bmatrix} \tag{3.6.36}$$

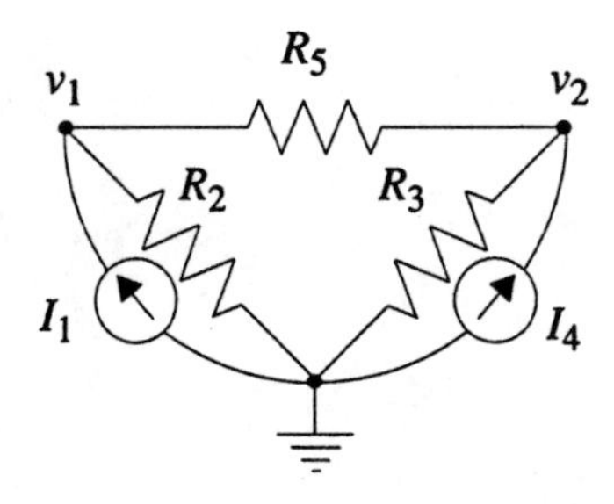

Figure 3.6 A simple two-node circuit for which sensitivities are calculated.

We see, for instance, that all four entries of $\boldsymbol{M}_{kl}$ depend on R_5. Therefore, to determine the sensitivity of the ith node voltage to changes in p we must apply the chain rule:

$$\delta x_i = \left(\sum_{k,l} \frac{\partial x_i}{\partial M_{kl}} \frac{\partial M_{kl}}{\partial p} + \sum_k \frac{\partial x_i}{\partial b_k} \frac{\partial b_k}{\partial p} \right) \delta p \tag{3.6.37}$$

For such computations to be possible, the partial derivatives $\frac{\partial M_{kl}}{\partial p}$ and $\frac{\partial b_k}{\partial p}$ must be obtained analytically and stored *a priori* as part of each element's model description.

3.7 Summary

In this chapter we have introduced several approaches for *directly* solving the Nodal Analysis matrix equations. We have also considered the errors introduced in doing so, as well as the means to perform a sensitivity analysis after a nominal solution of the circuit. This is not the last word on any of these topics. We will consider alternatives to Gaussian Elimination and LU factorization when we consider *sparse matrices* in Chapter 7. We will also describe sensitivity analysis in much greater detail in Chapter 9 and apply sensitivity to noise analysis and performance optimization.

However, we will close our discussion of dc equation formulation, analysis, and solution for now in order to begin time domain, or transient analysis, which will be the topic of the next two chapters. Of course, as was shown in Chapter 1, even the time domain analysis will employ the dc analysis techniques which we have covered so far.

3.8 References

[Ralston78] A. Ralston and P. Rabinowitz. *A First Course in Numerical Analysis*. McGraw-Hill, 1978.

Chapter 4

Linear Transient Analysis I

Transient analysis (or time domain analysis) involves computing the waveforms of a circuit as a function of time. If we were blindly to set up Modified Nodal Analysis (MNA) equations for a circuit that contained energy storage elements, we would find that we have terms like dv/dt (for the currents of capacitors) and di/dt (for the voltage of inductors). The challenge in transient analysis is to move time forward by integrating these terms to get sets of equations at each time step that we know how to solve. In particular, we will integrate the current through capacitances and the voltage across inductances.

In this chapter we will elaborate on the introduction to time domain analysis provided in Chapter 1. Chapter 1 discussed only Trapezoidal integration. In this chapter, we will also discuss other integration algorithms such as Forward Euler and Backward Euler. We will study the accuracy and stability properties of these integration methods in terms of a single exponential transient response for a simple series RC circuit.

4.1 The One-Step Integration Approximations

Consider once again a lumped, linear, time-invariant capacitor, shown in Figure 4.1. In equation (1.7.4) we expressed the capacitor voltage in terms of the integral of the capacitor current

$$v(t+\Delta t) = v(t) + \frac{1}{C}\int_{t}^{t+\Delta t} i(\tau)\,d\tau \tag{4.1.1}$$

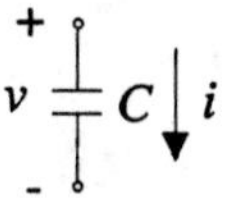

v + − C i

$$i = C\frac{dv}{dt}$$

Figure 4.1 Lumped, linear, time-invariant capacitor.

Here we consider approximating the integral in equation (4.1.1) in three possible ways:

$$\int_{t}^{t+\Delta t} i(\tau)\,d\tau \approx \begin{cases} \Delta t \cdot i(t) & \text{Forward Euler (FE)} \\ \Delta t \cdot i(t+\Delta t) & \text{Backward Euler (BE)} \\ \dfrac{\Delta t}{2}\,[i(t) + i(t+\Delta t)] & \text{Trapezoidal (TR)} \end{cases} \tag{4.1.2}$$

The graphical interpretation of each of these integration approximations is shown in Figures 4.2, 4.3, and 4.4, respectively. All of these are *one-step integration* approximations in that they rely only on the value of the integrand at the single time point immediately preceding that being computed. These methods are in contrast to multistep approximations which rely on other prior values of the integrand further back in time as well. From Figures 4.2, 4.3, and 4.4 one could make the intuitive (correct) conclusion that the Trapezoidal approximation is the most accurate among the three, although it may be more difficult to implement than the others.

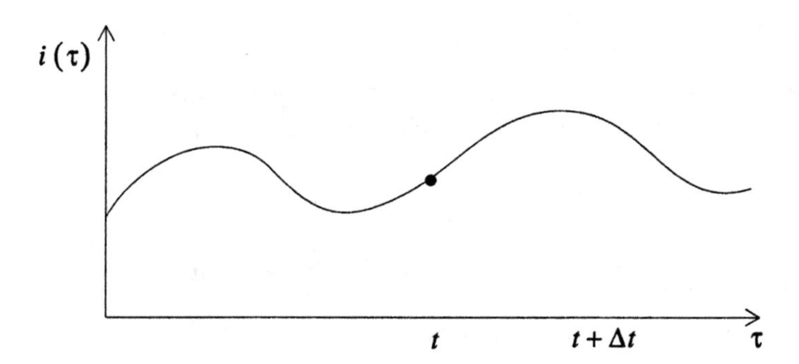

Figure 4.2 Forward Euler integration.

Inserting the approximations in (4.1.2) into (4.1.1) yields the following

$$v(t+\Delta t) \approx v(t) + \frac{1}{C} \cdot \begin{cases} \Delta t \cdot i(t) & \text{(FE)} \\ \Delta t \cdot i(t+\Delta t) & \text{(BE)} \\ \dfrac{\Delta t}{2}\,[i(t) + i(t+\Delta t)] & \text{(TR)} \end{cases} \tag{4.1.3}$$

The approximations in (4.1.3) can also be arrived at using difference approximations:

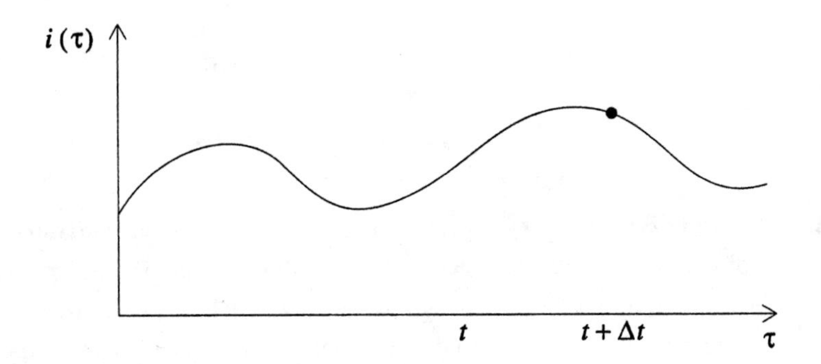

Figure 4.3 Backward Euler integration.

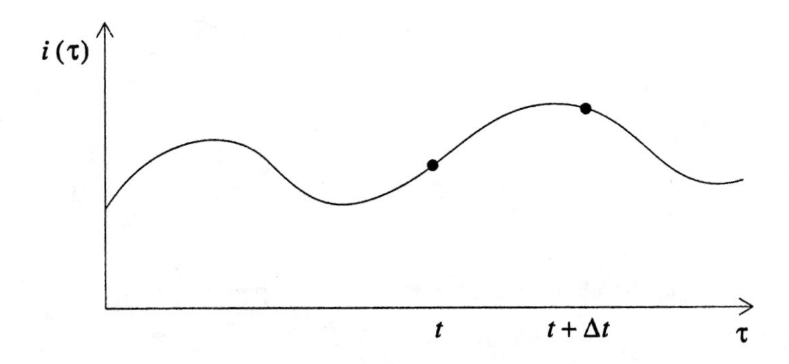

Figure 4.4 Trapezoidal integration.

$$\frac{i}{C} = \frac{dv}{dt} \approx \frac{v(t+\Delta t) - v(t)}{\Delta t} \tag{4.1.4}$$

where we are approximating the voltage derivative, hence $\frac{i}{C}$, in the time range t to $t+\Delta t$ by

$$\dot{v}(t) \approx \begin{cases} \dfrac{i(t)}{C} & \text{(FE)} \\ \dfrac{i(t+\Delta t)}{C} & \text{(BE)} \\ \dfrac{1}{2C}[i(t) + i(t+\Delta t)] & \text{(TR)} \end{cases} \tag{4.1.5}$$

For example, with FE we approximate $\dot{v}(\tau)$ for $t \le \tau \le t + \Delta t$ by the derivative at time t in order to compute the voltage at time $t + \Delta t$ via extrapolation. This FE projection is shown graphically in Figure 4.5. In other words, since $i = C\dot{v}$, FE is approximating the current over the interval $(t, t + \Delta t)$ as a constant equal to $i(t)$, as we would expect from the development of equation (4.1.2).

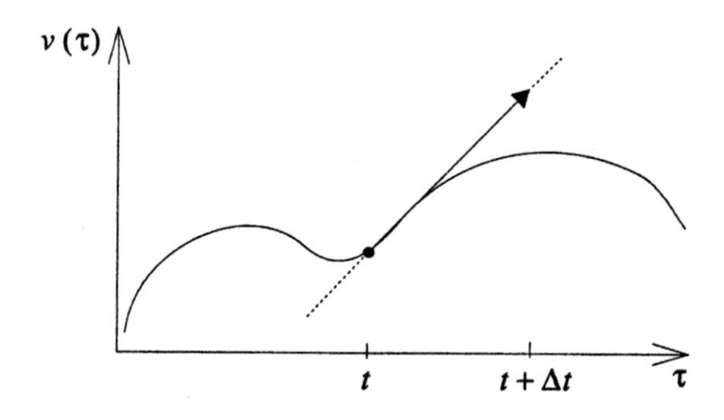

Figure 4.5 Forward Euler difference approximation.

The BE and TR difference approximations are formulated in a similar fashion, but with $\dot{v}$ approximated by $\dot{v}(t + \Delta t)$ for BE and the average of $\dot{v}(t)$ and $\dot{v}(t + \Delta t)$ for TR integration. The graphical depictions of the BE and TR difference approximations are shown in Figure 4.6 and Figure 4.7, where the end of the arrowhead indicates the next voltage value that will be computed at time $t + \Delta t$ as opposed to the actual value of $v(t + \Delta t)$. It may be apparent from Figure 4.6 and Figure 4.7 that since we do not know the value of the voltage or current at $t + \Delta t$ (it is what we are trying to find), the derivative used to project to $t + \Delta t$ is based upon the *approximate* or *guessed* voltage at $t + \Delta t$. Moreover, Figure 4.3 and Figure 4.4 seem to imply that the BE and TR integration approximations comprise areas for which the exact current $i(t + \Delta t)$ is known, which is not the case. The area is a function of the approximate current at $(t + \Delta t)$. This is why BE and TR integration are referred to as *implicit* integration algorithms in contrast to Forward Euler integration

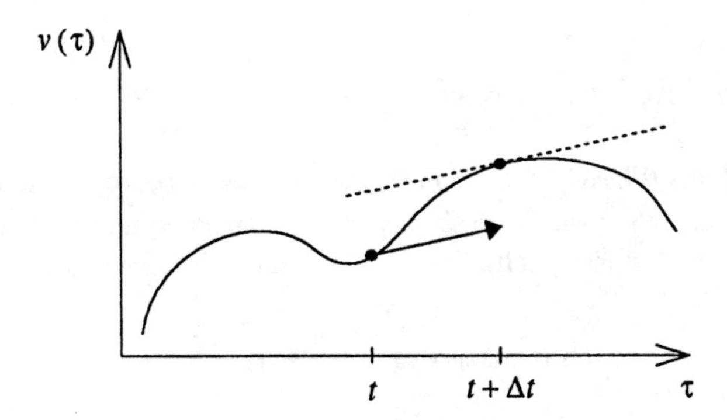

Figure 4.6 Backward Euler difference approximation.

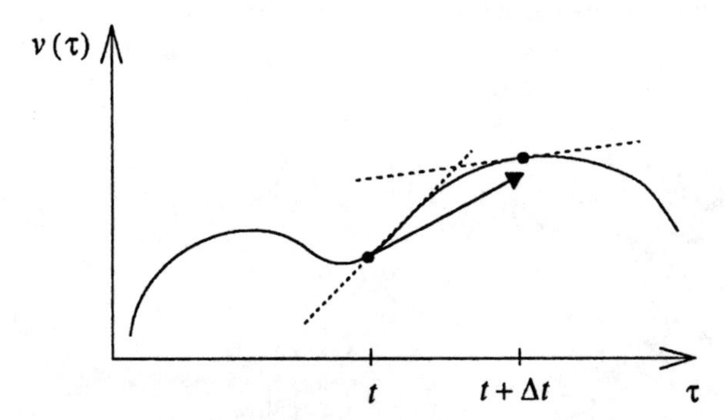

Figure 4.7 Trapezoidal difference approximation.

which is *explicit*.

Inserting the appropriate $\dot{v}$ approximation into (4.1.4) and solving for $v(t+\Delta t)$, yields the following FE, BE, and TR *difference equations*:

$$v(t+\Delta t) \approx v(t) + \Delta t \cdot \begin{cases} \dfrac{i(t)}{C} & \text{(FE)} \\ \dfrac{i(t+\Delta t)}{C} & \text{(BE)} \\ \dfrac{i(t) + i(t+\Delta t)}{2C} & \text{(TR)} \end{cases} \qquad \textbf{(4.1.6)}$$

which are identical to the numerical integration approximations in (4.1.3).

4.2 Forward Euler Approximation

For a Forward Euler (FE) approximation we compute the response at time $t + \Delta t$ in terms of $v(t)$ and $i(t)$, both of which are already known from the solution at the previous time point (hence the method is explicit). The Forward Euler approximation

$$v(t+\Delta t) \approx v(t) + \frac{\Delta t}{C} i(t) \tag{4.2.1}$$

can be represented electrically by a *companion model*, which in this case is an independent voltage source as shown in Figure 4.8. This approximation is appealing in its simplicity. Every independent voltage source simplifies dc circuit analysis, so circuits which typically have many capacitors can become very simple with the FE companion model. Integrated circuits typically have many capacitors due to transistor parasitics and wiring and interconnect models.

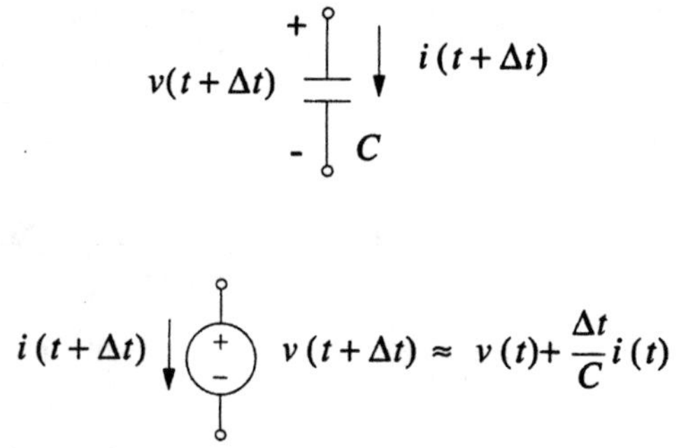

Figure 4.8 Forward Euler companion model for a capacitor.

Some simulators make an assumption that they can model all the capacitors in integrated circuits with a capacitor to ground from each node. This assumption leads to a dc equivalent circuit that is trivial to analyze with the FE approximation. Of course, if there were any floating capacitors (a floating capacitor has neither end grounded), the dc equivalent circuit would contain loops of independent voltage sources that are seemingly impossible to accommodate. But we will soon demonstrate simple ways to overcome such limitations.

Before we become overly enamored with the simplicity of Forward Euler integration, we should mention that Forward Euler carries with it the undesirable problems of reduced accuracy and instability, which are to be discussed later. Implicit integration algorithms -- those which depend on as yet unknown circuit values at time $t + \Delta t$ -- are more commonly applied in practice due to these accuracy and stability issues. However, in Chapter 11 we will describe some simulation algorithms that in part use Forward Euler integration and exploit its simplicity and efficiency.

4.3 Backward Euler Approximation

The Backward Euler (BE) approximation is an *implicit* formula since the voltage at time $t + \Delta t$ is a function of the current at time $t + \Delta t$.

$$v(t+\Delta t) \approx v(t) + \frac{\Delta t}{C} i(t+\Delta t) \tag{4.3.1}$$

In Figure 4.9, the companion model for Backward Euler integration includes a resistor to account for this implicit relation. This Thevenin equivalent companion model can be transformed to the Norton form shown in Figure 4.10, which is more convenient for Nodal Analysis.

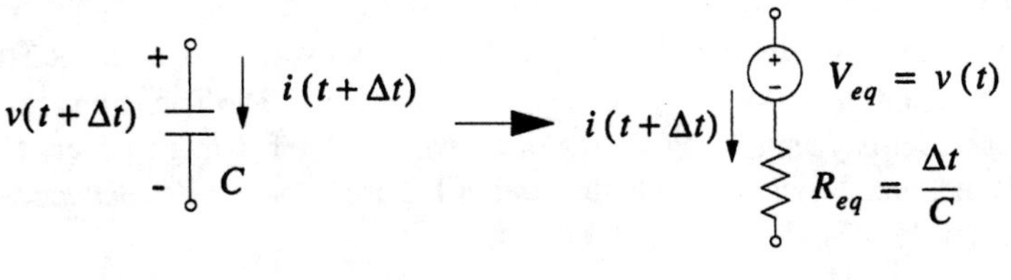

Figure 4.9 Backward Euler companion model for a capacitor.

As was the case for the Trapezoidal integration approximation that we discussed briefly in Chapter 1, the Norton equivalent model in Figure 4.10 may be numerically better suited to a dc steady state analysis, where $\Delta t \rightarrow \infty$ and a capacitor is modeled by an open circuit. On the other hand, the Thevenin equivalent companion model may be better computationally for situations with specified initial conditions or for very small time steps. The corresponding arguments for inductors will be made later in this chapter. We will also show that Backward Euler integration has accuracy comparable to that of Forward Euler but is more stable.

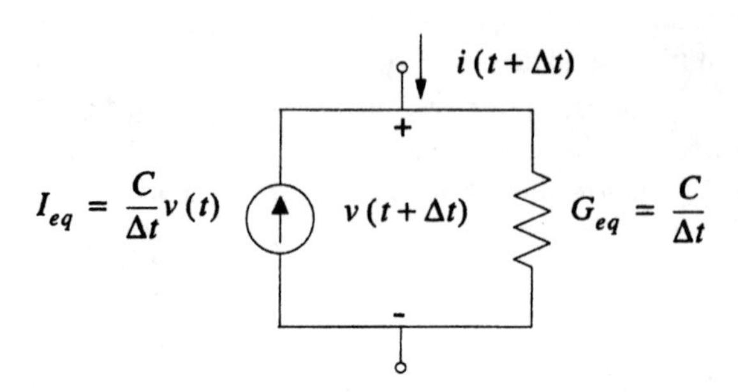

Figure 4.10 Norton representation of BE model for a capacitor.

4.4 Trapezoidal Approximation

For completeness, we review the Trapezoidal (TR) integration approximation:

$$v(t+\Delta t) \approx v(t) + \frac{\Delta t}{2C}i(t) + \frac{\Delta t}{2C}i(t+\Delta t) \tag{4.4.1}$$

As with the BE approximation, the TR algorithm is implicit since it is dependent on unknown quantities. The TR companion models for a capacitor in Thevenin and Norton form are shown in Figure 4.11(b) and Figure 4.11(c), respectively. Again, the Norton equivalent companion model may be preferable for large values of Δt or dc steady state computation, when a capacitor is replaced by an open circuit. The Thevenin model may be preferable for extremely small time steps. As we will see later in this chapter, Trapezoidal integration is in general more accurate than either FE or BE, and it lies between the two in terms of its stability behavior.

4.5 Companion Models for Inductors

In the previous three sections, we discussed applying FE, BE, and TR to capacitors. In this section, we will extend those same arguments to inductors. The basic equation for a lumped, linear, time-invariant inductor is

$$v(t) = L\frac{di(t)}{dt} \tag{4.5.1}$$

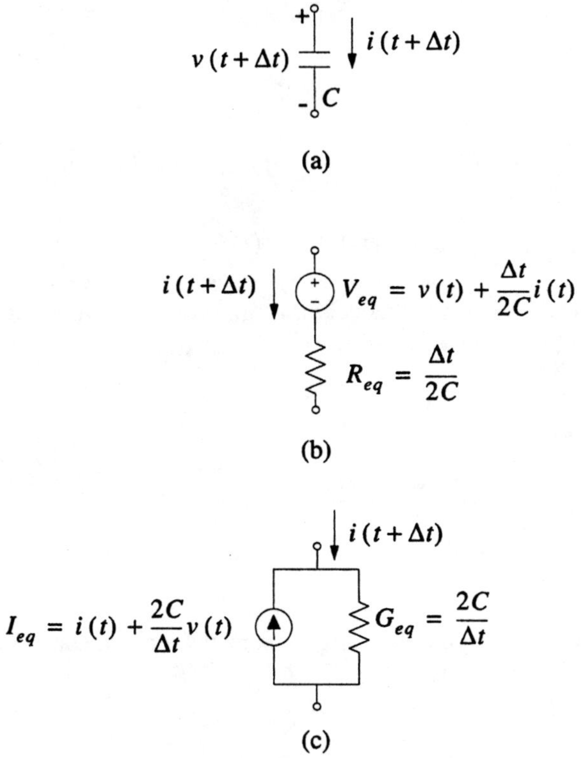

$v(t+\Delta t)$ $i(t+\Delta t)$ C

(a)

$i(t+\Delta t)$ $V_{eq} = v(t) + \frac{\Delta t}{2C} i(t)$

$R_{eq} = \frac{\Delta t}{2C}$

(b)

$i(t+\Delta t)$

$I_{eq} = i(t) + \frac{2C}{\Delta t} v(t)$ $G_{eq} = \frac{2C}{\Delta t}$

(c)

Figure 4.11 (a) A capacitor and its Trapezoidal companion models in (b) Thevenin and (c) Norton forms.

and the circuit element is shown in Figure 4.12. Given the current at time t, we can write the current at time $(t+\Delta t)$ as

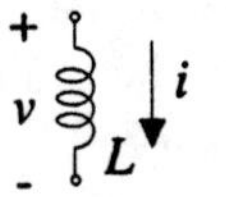

v i L $v = L\frac{di}{dt}$

Figure 4.12 Lumped, linear, time-invariant inductor and its basic equation.

$$i(t+\Delta t) = i(t) + \frac{1}{L}\int_{t}^{t+\Delta t} v(t)\,dt \qquad (4.5.2)$$

Forward Euler

Applying the FE approximation to the integral in equation (4.5.2), we get

$$i(t+\Delta t) \approx i(t) + \frac{\Delta t}{L} v(t) \qquad (4.5.3)$$

The Norton form of the FE companion model for an inductor is shown in Figure 4.13 (the FE companion model has no Thevenin form).

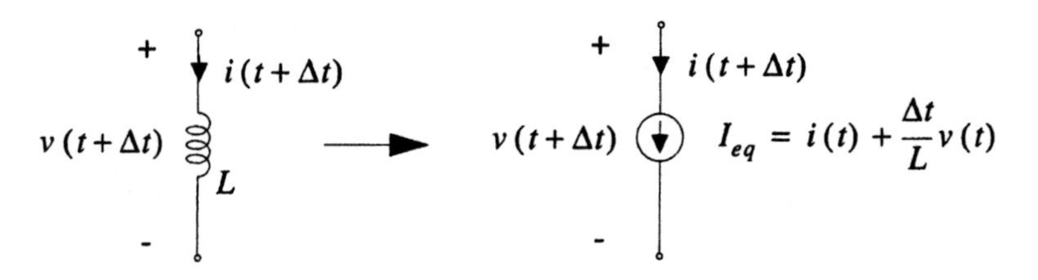

Figure 4.13 Forward Euler companion model for an inductor.

Backward Euler

Applying the BE approximation to the integral in equation (4.5.2), we obtain

$$i(t+\Delta t) = i(t) + \frac{\Delta t}{L} v(t+\Delta t) \qquad (4.5.4)$$

The Norton and Thevenin forms of the BE companion model for the inductor are shown in Figure 4.14(b) and Figure 4.14(c), respectively. The Thevenin form is good for dc steady state analysis in which $\Delta t \rightarrow \infty$. The Norton form is good for situations involving initial conditions and when Δt is very small.

Trapezoidal

Applying the TR approximation to the integral in (4.5.2), we obtain

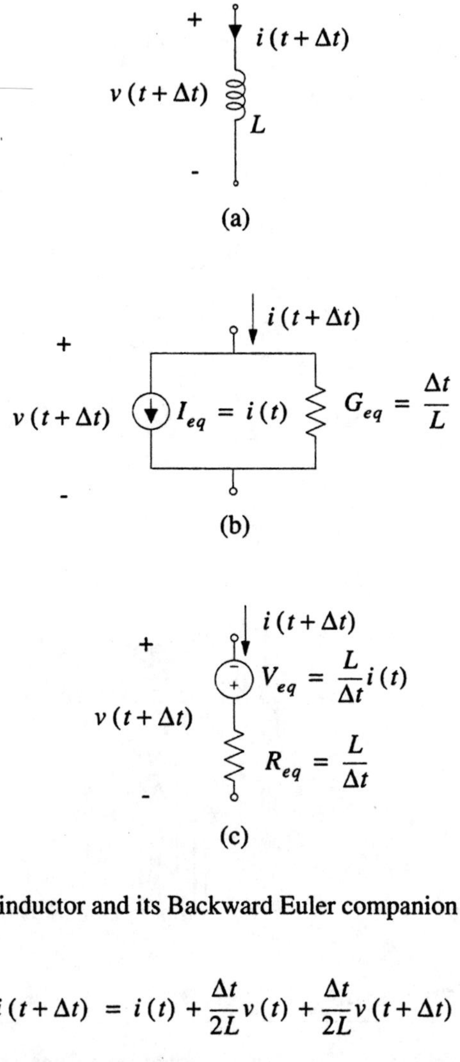

(a)

(b)

(c)

Figure 4.14 (a) An inductor and its Backward Euler companion models in (b) Norton and (c) Thevenin forms.

$$i(t+\Delta t) = i(t) + \frac{\Delta t}{2L} v(t) + \frac{\Delta t}{2L} v(t+\Delta t) \tag{4.5.5}$$

The Norton and Thevenin forms of the TR companion model of an inductor are shown in Figure 4.15(b) and Figure 4.15(c), respectively. As in the case of Backward Euler, the Thevenin form is good for dc steady state analysis in which $\Delta t \rightarrow \infty$. The Norton form is good for situations involving initial conditions and when Δt is very small.

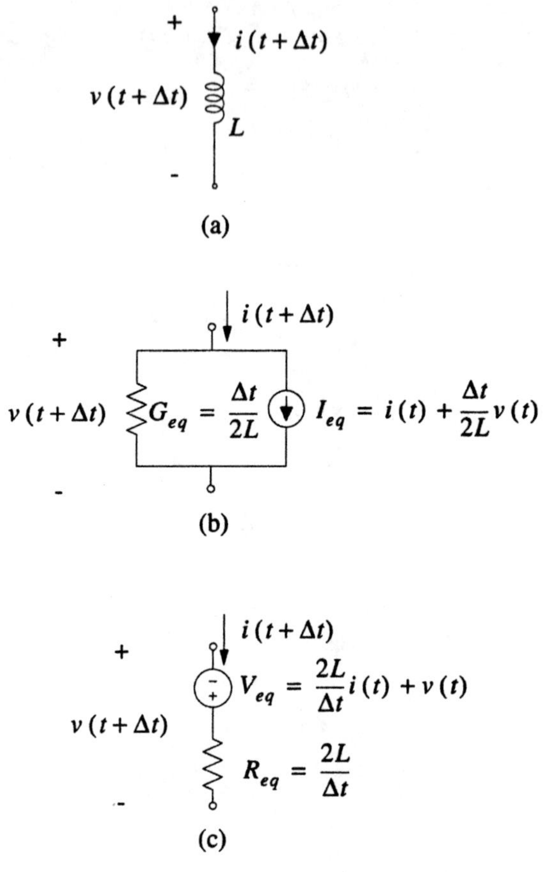

$+$ $i(t+\Delta t)$
$v(t+\Delta t)$ L
$-$

(a)

$i(t+\Delta t)$
$+$
$v(t+\Delta t)$ $G_{eq} = \dfrac{\Delta t}{2L}$ $I_{eq} = i(t) + \dfrac{\Delta t}{2L} v(t)$
$-$

(b)

$i(t+\Delta t)$
$+$
$V_{eq} = \dfrac{2L}{\Delta t} i(t) + v(t)$
$v(t+\Delta t)$
$R_{eq} = \dfrac{2L}{\Delta t}$
$-$

(c)

Figure 4.15 (a) An inductor and its Trapezoidal companion models in (b) Norton and (c) Thevenin form.

4.6 Preliminary Comments on Accuracy

All of the integration algorithms that we have discussed involve some degree of approximation. In the case of the implicit algorithms (BE and TR), we should restate that the current through the capacitance companion model, $i(t+\Delta t)$, is an *approximation* to the actual current at $t+\Delta t$. In the case of an inductor, $v(t+\Delta t)$ is an *approximation* to the actual voltage at $t+\Delta t$. The FE formula would appear at first to be an approximation in terms of an exact quantity, but after the initial time step, all voltage points are approximate (which is true for all methods of integration). It is because these values are approximate that we can study only the *local error* associated with these methods. The local error includes only the error incurred in the present time step, unlike the *cumulative error*, which includes the error due to the starting point of the present time step being inexact. The tacit assumption here is that if we can control the local error to be sufficiently small, then the cumulative error after n steps is at most n times the local error involved in a single step. The local error tolerance is chosen such that the cumulative error is still acceptable. As a practical matter, we are helped along by the local truncation error being a signed quantity, and hence there is the possibility of some negative and positive errors canceling out over time.

In general, the integration error is evaluated in terms of the *local truncation error* for these one-step integration methods. If the required integration is expressed as a Taylor expansion, the one-step integration methods can be shown to be truncated versions of those expansions, hence the term *truncation error.* The values at time t are treated as exact and the *local error* incurred in stepping from t to $t + \Delta t$ is considered. Higher order integration can also be considered, which employs several preceding time points in an attempt to better predict the value at time $t + \Delta t$. In such cases a polynomial of appropriate order is fitted to these time points and then extrapolated (for explicit integration) or interpolated (for implicit integration) to the present time point [Gear71].

These multi-step integration approaches lead once again to Thevenin or Norton companion models as before, but with the independent source components dependent on values from more than one past time point. With careful planning, the companion models can be updated efficiently from one time point to the next [Brayton72].

The higher order integration algorithms are usually applied when the time domain waveform accuracy is critical and the device nonlinearities are slight. For example, for small signal transient analysis of analog circuits, the one-step methods may yield numerical noise which might be perceived as unusual response behavior. Multi-step methods can reduce this numerical noise. However, this scheme works for small signal analog circuits since the nonlinearities are slight. In general, if the nonlinear elements change dramatically from one time point to the next, the use of previous time points to predict future behavior could have an adverse affect. In particular, in digital circuits, devices such as transistors and diodes and their associated parasitic capacitances often switch rapidly over a very small time period. When that occurs, using more history than one time point to perform a prediction does not serve a useful purpose. Therefore, only the one-step algorithms will be discussed in further detail in this book. The accuracy and stability of the one-step algorithms will be studied in terms of an example in the following sections.

4.7 The Exact Solution of a Simple Series RC Circuit

To study the behaviors of the one-step integration approximations, we apply them to a simple problem, the solution to which can be computed exactly. The series RC circuit shown in Figure 4.16 provides such a problem that will allow us to gain some insight regarding these numerical approximations.

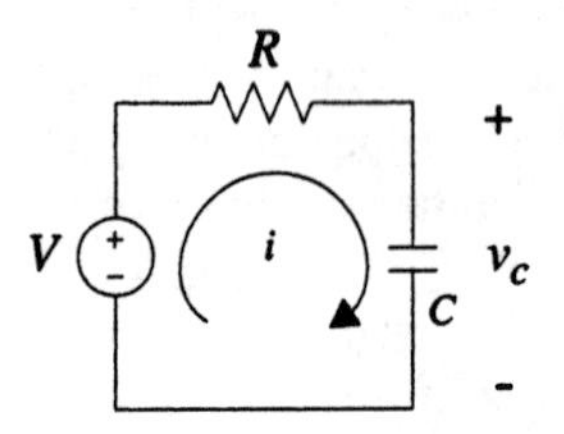

Figure 4.16 Series RC circuit.

We can write a single equation to describe this single-loop circuit

$$i = C\frac{dv_c}{dt} = \frac{V - v_c}{R} \tag{4.7.1}$$

and then reformulate it as a first order differential equation

$$\frac{dv_c}{dt} = -\frac{1}{RC}v_c + \frac{1}{RC}V \tag{4.7.2}$$

To establish some formalism which we will use again in the next chapter, we first consider the homogeneous equation for the capacitor voltage, which arises when the excitation V is zero and there is a nonzero initial condition on the capacitor, v_{c0}:

$$\frac{dv_{ch}}{dt} = -\frac{1}{RC}v_{ch} \qquad v_{ch}(t_0) = v_{c0} \tag{4.7.3}$$

Separating the terms in (4.7.3),

$$\frac{dv_{ch}}{v_{ch}} = -\frac{1}{RC}dt \tag{4.7.4}$$

and integrating over the appropriate limits

$$\int_{v_{ch}(t_0)}^{v_{ch}(t)} \frac{dv_{ch}}{v_{ch}} = -\frac{1}{RC}\int_{t_0}^{t} d\tau$$

$$ln(v_{ch}) - ln(v_{c0}) = -\frac{1}{RC}(t - t_0) \tag{4.7.5}$$

$$ln\left(\frac{v_{ch}}{v_{c0}}\right) = -\frac{1}{RC}(t - t_0)$$

$$\frac{v_{ch}}{v_{c0}} = e^{-\frac{1}{RC}(t - t_0)}$$

results in

$$v_{ch}(t) = v_{c0}e^{-\frac{1}{RC}(t - t_0)} \tag{4.7.6}$$

This solution satisfies the specified initial condition, and is what would prevail in the absence of an excitation, $V(t) \equiv 0$. For this reason it is also referred to as the zero-input response.

Next we turn our attention to the overall solution which we attempt to obtain in terms of an unknown function multiplying the homogeneous solution:

$$v_c(t) = z(t)\, v_{ch}(t) \tag{4.7.7}$$

Note that

$$z(t_0) = 1 \tag{4.7.8}$$

in order to satisfy the specified initial condition

$$v_{ch}(t_0) = v_{c0} \tag{4.7.9}$$

Substituting equation (4.7.7) into equation (4.7.2) we obtain

$$\frac{dz}{dt}v_{ch} + z\frac{dv_{ch}}{dt} = -\frac{1}{RC}zv_{ch} + \frac{1}{RC}V \tag{4.7.10}$$

Because of (4.7.3) the terms immediately to the left and right of the equals sign cancel, leaving

$$\frac{dz}{dt}v_{ch} = \frac{1}{RC}V \tag{4.7.11}$$

Therefore

$$\frac{dz}{dt} = \frac{1}{RC}\frac{V}{v_{ch}} \tag{4.7.12}$$

Inserting (4.7.6),

$$\frac{dz}{dt} = \frac{1}{RC}\frac{V}{v_{c0}}e^{\frac{1}{RC}(t-t_0)} \tag{4.7.13}$$

and integrating

$$\int_{z(t_0)}^{z(t)} dz = \int_{t_0}^{t} \frac{1}{RCv_{c0}}e^{\frac{1}{RC}(\tau-t_0)}V(\tau)\,d\tau \tag{4.7.14}$$

results in

$$z(t) - z(t_0) = \frac{1}{RCv_{c0}}\int_{t_0}^{t} e^{\frac{1}{RC}(\tau-t_0)}V(\tau)\,d\tau \tag{4.7.15}$$

Substituting (4.7.7) and (4.7.8) for $z(t_0)$ and $z(t)$ into (4.7.15) and using the homogeneous solution (4.7.5) we have finally

$$v_c(t) = v_{c0}e^{-\frac{1}{RC}(t-t_0)} + \frac{1}{RC}\int_{t_0}^{t} e^{-\frac{1}{RC}(t-\tau)}V(\tau)\,d\tau \tag{4.7.16}$$

We see here the classic separation between the zero-input response, which in this case is the homogeneous solution

$$v_{c0_i}(t) = v_{c0}e^{-\frac{1}{RC}(t-t_0)} \tag{4.7.17}$$

and the zero-state response, which is the convolution integral

$$v_{c0_s}(t) = \frac{1}{RC}\int_{t_0}^{t} e^{-\frac{1}{RC}(t-\tau)} V(\tau)\, d\tau \qquad \textbf{(4.7.18)}$$

For example, with a $1\,\mu F$ capacitor, a $10\ \text{k}\Omega$ resistor, a voltage input $V(t) = 5u(t)$, and an initial capacitor voltage $v_{c0} = 1$ volt, the zero-input, zero-state, and complete response are shown in Figures 4.17, 4.18, and 4.19, respectively.

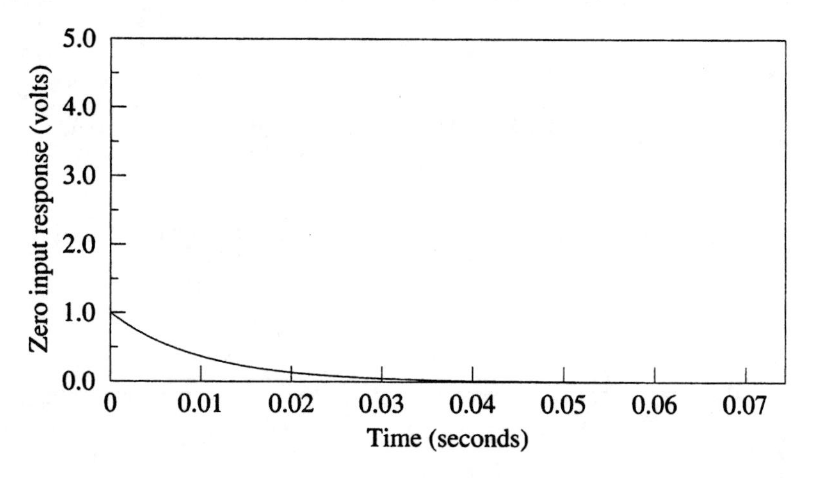

Figure 4.17 Zero-input response for series RC circuit.

Given any initial condition and driving voltage function, we can easily obtain the capacitance voltage from equations (4.7.17) and (4.7.18). To be able to compare such solutions with those we may obtain using integration approximations, we confine the input voltage function to be a sequence of steps and ramps in time as shown in Figure 4.20. As long as the time points are sufficiently close together any input voltage (or current) of interest can be approximated in this way. It may seem that we need not consider step changes since they cannot actually occur in reality and we could consider approximating them with very fast ramps. It turns out though that such an approach may be more computationally expensive than the explicit treatment of steps, so we will retain them in our discussion without further explanation at this time. To proceed, we focus on the nth open input segment for which $t_n^+ \le t \le t_{n+1}^-$, as shown in Figure 4.23. We establish some nomenclature for this open interval, $t \in (t_n, t_{n+1})$, as follows:

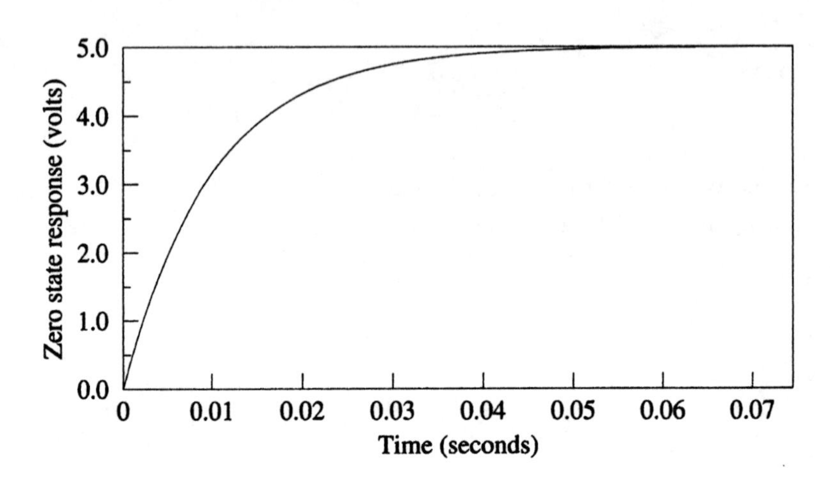

Figure 4.18 Zero-state response for series RC circuit.

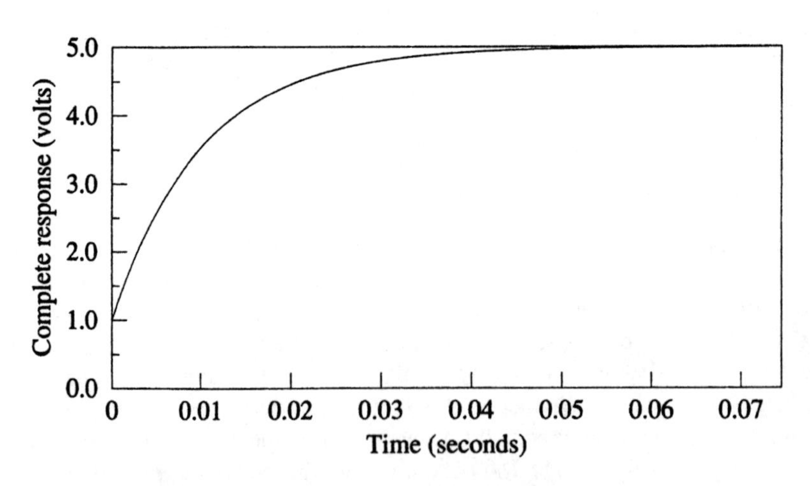

Figure 4.19 Complete response for series RC circuit.

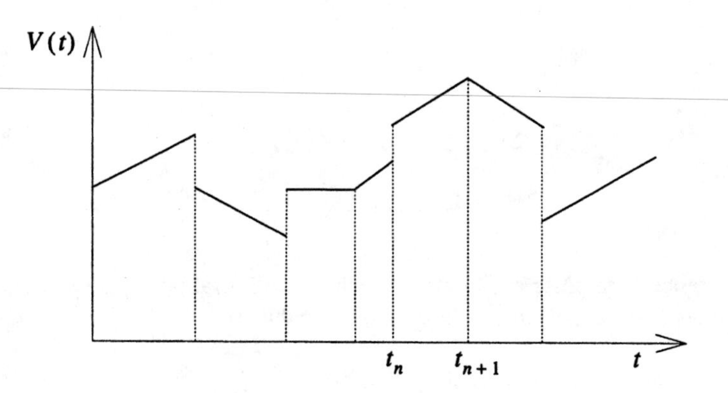

Figure 4.20 Piecewise input segments.

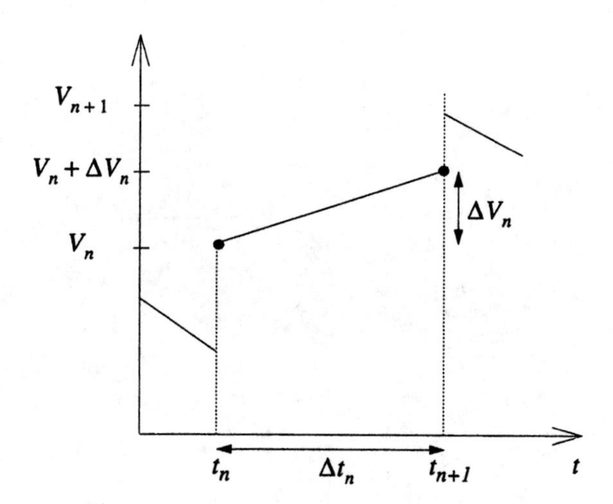

Figure 4.21 The input segment (t_n, t_{n+1}).

$$
\begin{aligned}
V(t) &= V_n + (t - t_n)\frac{\Delta V_n}{\Delta t_n} \\
V_n &\equiv V(t_n^+) \\
\Delta V_n &\equiv V(t_{n+1}^-) - V(t_n^+) \\
\Delta t_n &\equiv t_{n+1} - t_n
\end{aligned}
\tag{4.7.19}
$$

In the absence of impulse excitations, which we will disallow for now, there can be no discontinuities in the capacitance voltage state variable, so

$$
v_c(t_n^-) = v_c(t_n^+) = v_c(t_n) \equiv v_{c,n} \tag{4.7.20}
$$

Referring back to the general solution, equation (4.7.16), we can write the following expression for the state at time t_{n+1} in terms of the state at time t_n and the behavior of the input on the open interval $t \in (t_n, t_{n+1})$:

$$
\begin{aligned}
v_{c,n+1} &= v_{c,n} e^{-\frac{1}{RC}(t_{n+1}-t_n)} + \frac{1}{RC}\int_{t_n}^{t_{n+1}} e^{-\frac{1}{RC}(t_{n+1}-\tau)} \left[V_n + (\tau - t_n)\frac{\Delta V_n}{\Delta t_n}\right] d\tau \\
&= v_{c,n} e^{-\frac{\Delta t_n}{RC}} + \frac{1}{RC}\int_{t_n}^{t_{n+1}} e^{-\frac{1}{RC}(t_{n+1}-\tau)} \left(V_n - t_n\frac{\Delta V_n}{\Delta t_n}\right) d\tau \\
&\quad + \frac{1}{RC}\int_{t_n}^{t_{n+1}} e^{-\frac{1}{RC}(t_{n+1}-\tau)} \tau\frac{\Delta V_n}{\Delta t_n} d\tau \\
&= v_{c,n} e^{-\frac{\Delta t_n}{RC}} + \left(1 - e^{-\frac{\Delta t_n}{RC}}\right)\left(V_n - t_n\frac{\Delta V_n}{\Delta t_n}\right) \\
&\quad + (t_{n+1} - RC)\frac{\Delta V_n}{\Delta t_n} - (t_n - RC)\frac{\Delta V_n}{\Delta t_n} e^{-\frac{\Delta t_n}{RC}} \\
&= \left(v_{c,n} - V_n + RC\frac{\Delta V_n}{\Delta t_n}\right) e^{-\frac{\Delta t_n}{RC}} + \left(V_n + \Delta V_n - RC\frac{\Delta V_n}{\Delta t_n}\right)
\end{aligned}
\tag{4.7.21}
$$

The last three terms in (4.7.21) account for the ramp portion of the overall response.

Note, if the slope of the excitation were zero ($\Delta V_n = 0$), we would have the familiar step response augmented by a transient term due to the initial condition $v_{c,n}$:

$$v_{c,n+1} = v_{c,n} e^{-\frac{\Delta t_n}{RC}} + V_n \left(1 - e^{-\frac{\Delta t_n}{RC}}\right) \tag{4.7.22}$$

4.8 *Comparison of One-Step Integration Approximations with the Exact Solution*

Next, we consider the various integration approximations and compare them with the exact solution as shown in (4.7.21).

Forward Euler

For a Forward Euler approximation we replace the original circuit with the sequence of dc equivalents shown in Figure 4.22. So we see that

$$i_c(t_n) = \frac{V_n - v_{c,n}}{R} \tag{4.8.1}$$

and

$$\begin{aligned} v_{c,n+1} &= v_{c,n} + \frac{\Delta t_n}{C} i_c(t_n) \\ &= v_{c,n} + \frac{\Delta t_n}{RC} [V_n - v_{c,n}] \end{aligned} \tag{4.8.2}$$

is the recursive relation with which we calculate v_c.

Now, we go back to the exact expression (equation (4.7.21))

$$v_{c,n+1} = \left(v_{c,n} - V_n + RC\frac{\Delta V_n}{\Delta t_n}\right) e^{-\frac{\Delta t_n}{RC}} + \left(V_n + \Delta V_n - RC\frac{\Delta V_n}{\Delta t_n}\right) \tag{4.8.3}$$

and expand the exponential in a Taylor series

$$e^{-\frac{\Delta t_n}{RC}} = 1 - \frac{\Delta t_n}{RC} + \frac{1}{2!}\left(\frac{\Delta t_n}{RC}\right)^2 - \frac{1}{3!}\left(\frac{\Delta t_n}{RC}\right)^3 + \ldots - \ldots \tag{4.8.4}$$

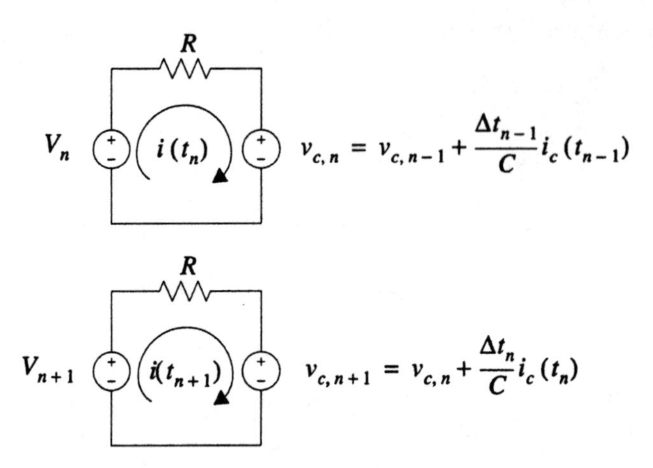

Figure 4.22 The dc equivalent circuits for FE analysis of a series RC circuit.

Retaining only the first two terms in this series we obtain, after considerable cancellation,

$$v_{c,n+1} \cong v_{c,n} + \frac{\Delta t_n}{RC}(V_n - v_{c,n}) \tag{4.8.5}$$

This expression is identical to the Forward Euler recursion in (4.8.2), thus implying that the Forward Euler approximation amounts to taking into account only the first two terms in the Taylor expansion of $e^{-\frac{\Delta t_n}{RC}}$.

When using the Forward Euler approximation in the presence of step function excitations, however, as shown in Figure 4.23, we must compute the capacitance current that would prevail immediately after the onset of the step, and use that in the subsequent time point computation, as shown in Figure 4.24.

When the step change in the input occurs, the voltage of the capacitor cannot change instantaneously. Hence the additional voltage drop occurs across the resistor, thus instantaneously changing the current through the capacitor. During the step, no time passes and no integration steps are necessary. Once we move on to the ramp, we need to apply the integration approximation. Hence the FE approximation uses the capacitance current that prevails immediately after the onset of the step. It is therefore necessary to perform a computation at every time point that any driving function has a step function. The post-step capacitance currents are used for the subsequent time point computation. This distinction between pre- and post-step capacitance currents is a subtle point that is often missed in the actual implementation of the Forward Euler integration approximation. We could alterna-

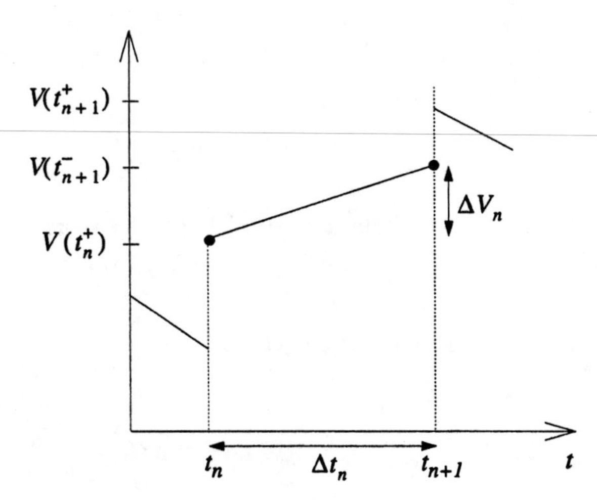

Figure 4.23 The input segment (t_n, t_{n+1}).

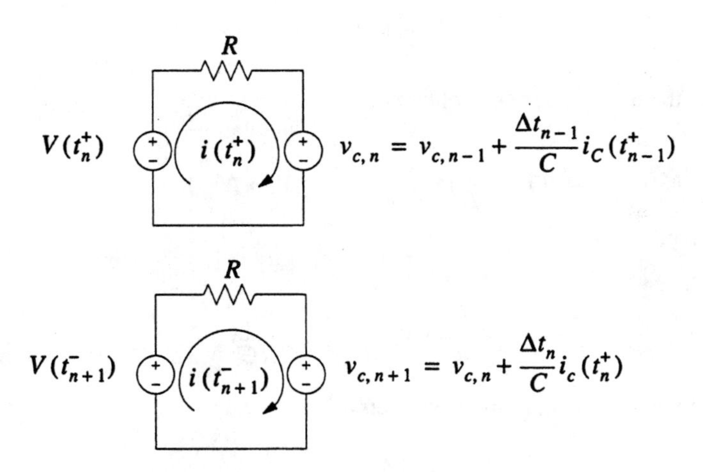

Figure 4.24 Modified FE circuit analysis.

tively prohibit step function changes in input, but then we would be forced to approximate step inputs with very steep ramps. Such an approach often necessitates using very small time steps. In practice we are better off allowing steps and even approximating very steep ramps by steps. As a final note, we should mention that not only should we recompute with Forward Euler (or any one-step integration) at step function changes in the input, but also at any change in input slope even if the function is continuous. If the input were a sinusoid, we would take appropriately "small" time steps so as not to lose accuracy due to the change of the input during the time step. Accuracy and stability considerations may also influence the time step and force us to recompute at intermediate time points (between t_n and t_{n+1} in Figure 4.23).

Backward Euler

Next we turn our attention to the Backward Euler integration approximation. We first rewrite the exact solution (equation (4.7.21)) by expressing the voltage at t_n as a function of the voltage at t_{n+1}, and then multiply through by $e^{\frac{\Delta t_n}{RC}}$:

$$v_{c,n} = V_n - RC\frac{\Delta V_n}{\Delta t_n} + \left[v_{c,n+1} - (V_n + \Delta V_n) + RC\frac{\Delta V_n}{\Delta t_n}\right]e^{\frac{\Delta t_n}{RC}} \tag{4.8.6}$$

Now, we expand the exponential in equation (4.8.6) in a Taylor series

$$e^{\frac{\Delta t_n}{RC}} = 1 + \frac{\Delta t_n}{RC} + \frac{1}{2!}\left(\frac{\Delta t_n}{RC}\right)^2 + \frac{1}{3!}\left(\frac{\Delta t_n}{RC}\right)^3 + \ldots + \tag{4.8.7}$$

and retain only the first two terms to obtain

$$v_{c,n} = \left(1 + \frac{\Delta t_n}{RC}\right)v_{c,n+1} - \frac{\Delta t_n}{RC}(V_n + \Delta V_n) \tag{4.8.8}$$

or

$$v_{c,n+1} = v_{c,n} + \frac{\Delta t_n}{RC}(V_n + \Delta V_n) - \frac{\Delta t_n}{RC}v_{c,n+1} \tag{4.8.9}$$

Equation (4.8.9) is equivalent to the BE expression in (4.3.1)

$$v_{c,n+1} \approx v_{c,n} + \frac{\Delta t_n}{C}i(t_{n+1}) \tag{4.8.10}$$

This BE expression matches the dc equivalent circuit in Figure 4.25.

Again, we would always want to recompute at every occurrence of a step change in the input, but such a change is accounted for naturally in Backward Euler integration which focuses on the t_{n+1} side of the computational interval. Note that we are computing the voltage $v_{c,n+1}$ (which is the same on either side of the step), and $i(t_{n+1}^-)$ (before the

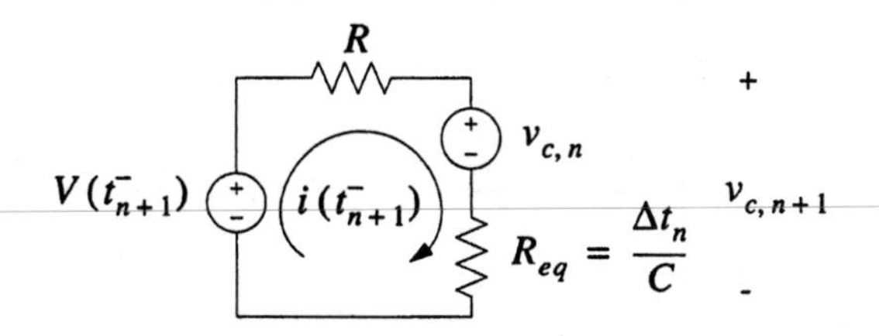

Figure 4.25 The dc equivalent circuit for BE analysis.

step) by means of the integration.

Trapezoidal Analysis

To compare Trapezoidal integration with the exact solution, we rewrite equation (4.7.21) as

$$\left[v_{c,n+1} - (V_n + \Delta V_n) + RC\frac{\Delta V_n}{\Delta t_n}\right]e^{\frac{\Delta t_n}{2RC}} = \left[v_{c,n} - V_n + RC\frac{\Delta V_n}{\Delta t_n}\right]e^{-\frac{\Delta t_n}{2RC}} \quad (4.8.11)$$

Next, we expand each of the exponentials in a Taylor series and retain the first two terms of both. After a great deal of cancellation, we are then left with the Trapezoidal integration approximation:

$$(1 + \frac{\Delta t_n}{2RC})\, v_{c,n+1} = (1 - \frac{\Delta t_n}{2RC})\, v_{c,n} + \frac{\Delta t_n}{2RC}\,[V_n + (V_n + \Delta V_n)] \quad (4.8.12)$$

or

$$v_{c,n+1} = v_{c,n} + \frac{\Delta t_n}{2RC}\,[V_n + (V_n + \Delta V_n)] - \frac{\Delta t_n}{2RC}v_{c,n} - \frac{\Delta t_n}{2RC}v_{c,n+1} \quad (4.8.13)$$

which is equivalent to the TR expression in (4.4.1)

$$v_{c,n+1} \approx v_{c,n} + \frac{\Delta t_n}{2C}\,[i(t_n) + i(t_{n+1})] \quad (4.8.14)$$

As we should expect from our experience with the above, the dc equivalent circuit in Figure 4.26 that matches expression (4.8.12) follows from the handling of the input voltage discontinuity in the *Forward Euler* companion model. Once we have obtained $v_{c,n+1}$

we see the need to compute $i_c(t^+_{n+1})$, as shown in Figure 4.27, for use in the subsequent time point computation, given by

$$i_c(t^+_{n+1}) = \frac{V(t^+_{n+1}) - v_{c,n}}{R} \tag{4.8.15}$$

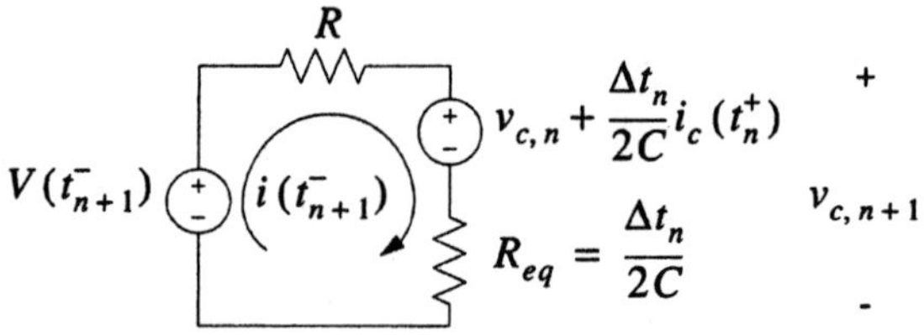

Figure 4.26 The dc equivalent circuit for TR analysis.

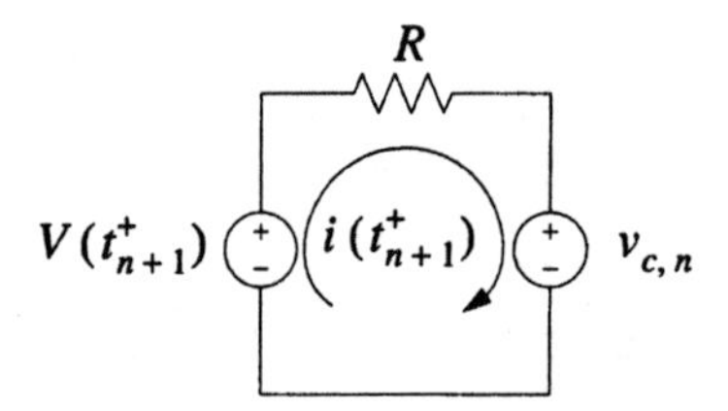

Figure 4.27 TR integration requires modification similar to FE when the input has steps.

4.9 *Accuracy of One-Step Approximations*

We can estimate the accuracy of one-step integration approximations in terms of the local truncation error that follows from the largest of the neglected terms in the Taylor series expansion for the exponentials. We are aided in our analysis by the fact that in a convergent series, each term dominates the sum of all of the remaining terms.

Forward Euler

For the Forward Euler approximation we (in effect) approximated the exponential by

$$e^{-\frac{\Delta t_n}{RC}} \approx 1 - \frac{\Delta t_n}{RC} \tag{4.9.1}$$

and neglected the remaining terms. Assuming a convergent series, the dominant term that we neglected is

$$\frac{1}{2}\left(\frac{\Delta t_n}{RC}\right)^2 \tag{4.9.2}$$

So from the exact solution

$$v_{c,n+1} = \left(v_{c,n} - V_n + RC\frac{\Delta V_n}{\Delta t_n}\right)e^{-\frac{\Delta t_n}{RC}} + \left(V_n + \Delta V_n - RC\frac{\Delta V_n}{\Delta t_n}\right) \tag{4.9.3}$$

we have an upper bound on the local truncation error ε:

$$\varepsilon \le \frac{1}{2}\left(\frac{\Delta t_n}{RC}\right)^2 \left(v_{c,n} - V_n + RC\frac{\Delta V_n}{\Delta t_n}\right) \tag{4.9.4}$$

Returning to the Forward Euler approximation (4.8.5) itself,

$$v_{c,n+1} \approx v_{c,n} + \frac{\Delta t_n}{RC}(V_n - v_{c,n}) \tag{4.9.5}$$

With some rearranging we have

$$v_{c,n} - V_n \approx \frac{RC}{\Delta t_n}(v_{c,n} - v_{c,n+1}) \tag{4.9.6}$$

Substituting (4.9.6) into (4.9.4) we have

$$\begin{aligned} \varepsilon &\le \frac{1}{2}\left(\frac{\Delta t_n}{RC}\right)^2 \frac{RC}{\Delta t_n}(v_{c,n} - v_{c,n+1} + \Delta V_n) \\ &= \frac{1}{2}\left(\frac{\Delta t_n}{RC}\right)(v_{c,n} - v_{c,n+1} + \Delta V_n) \end{aligned} \tag{4.9.7}$$

We could leave it at equation (4.9.7) and say that the local truncation error is proportional to $\frac{1}{2}(\frac{\Delta t_n}{RC})$ and then have two components, one proportional to the change in state voltage and the other proportional to the change in input voltage. Or we can go further:

$$\varepsilon \leq \frac{1}{2}(\frac{\Delta t_n}{RC})\ \{[(V_n + \Delta V_n) - v_{c,n+1}] - [V_n - v_{c,n}]\} \tag{4.9.8}$$

and since

$$\frac{(V_n + \Delta V_n) - v_{c,n+1}}{R} = i_c(t_{n+1}^-) \tag{4.9.9}$$

and

$$\frac{V_n - v_{c,n}}{R} = i_c(t_n^+) \tag{4.9.10}$$

then

$$\varepsilon \leq \frac{1}{2}\frac{\Delta t_n}{C}[i_c(t_{n+1}^-) - i_c(t_n^+)] \tag{4.9.11}$$

This error approximation makes good intuitive sense. It suggests that the error will be small if the difference in the capacitor current from one end of the time interval to the other is small, and of course if the time step Δt_n is small. In fact, if the exact current through the capacitor during the time step is constant (i.e., the voltage slope is constant), then Forward Euler integration does not incur any local truncation error. Forward Euler approximates the voltage slope for all of the time step as being the voltage slope at the start of the time step. Hence, if the voltage slope is constant (or the voltage is a linear function of time), no error is incurred. If, however, the voltage as a function of time is very convex or concave, Forward Euler integration may incur a relatively large error.

The local truncation error as expressed in (4.9.11) can now be used to adjust the time step to stay within a pre-set error bound. At the outset, we can try to traverse an entire time interval, throughout which no steps or ramp slope changes occur. If we subsequently discover the (voltage) truncation error to be too great we can limit the time step and settle for the computation of an intermediate solution. Such a solution need not be computationally expensive, since adjustment of the Forward Euler time step only entails a change in the companion model equivalent source value. So the LU factorization obtained for the discarded solution can be reused with a new Forward and Back Substitution. Note finally that although the Forward Euler error appears to be proportional to Δt_n, it is also dependent on the change of current in the capacitor. The larger the time step, the larger is the worst-case change in capacitor current and the larger the worst-case truncation error.

Backward Euler

From the exact expression (4.8.6)

$$v_{c,n} = V_n - RC\frac{\Delta V_n}{\Delta t_n} + \left[v_{c,n+1} - (V_n + \Delta V_n) + RC\frac{\Delta V_n}{\Delta t_n}\right]e^{\frac{\Delta t_n}{RC}} \quad (4.9.12)$$

we can expand the exponential terms as

$$\left[v_{c,n+1} - (V_n + \Delta V_n) + RC\frac{\Delta V_n}{\Delta t_n}\right]\left[1 + \frac{\Delta t_n}{RC} + \frac{1}{2!}\left(\frac{\Delta t_n}{RC}\right)^2 + \ldots\right] \quad (4.9.13)$$

Since Backward Euler corresponds to approximating this Taylor series by the first two terms, we can estimate the error from (4.9.13) as we did for the Forward Euler case:

$$\varepsilon \le -\frac{1}{2}\left(\frac{\Delta t_n}{RC}\right)^2\left[v_{c,n+1} - (V_n + \Delta V_n) + RC\frac{\Delta V_n}{\Delta t_n}\right] \quad (4.9.14)$$

In contrast to (4.9.4) there is a minus sign for this local truncation error expression since, unlike the FE case, the error term and the $v_{c,n+1}$ both appear on the same side of the equation.

Next, using the Backward Euler approximation

$$v_{c,n+1} \approx v_{c,n} + \frac{\Delta t_n}{RC}[(V_n + \Delta V_n) - v_{c,n+1}] \quad (4.9.15)$$

rearranged in the following manner,

$$v_{c,n+1} - (V_n + \Delta V_n) = \frac{RC}{\Delta t_n}[v_{c,n} - v_{c,n+1}] \quad (4.9.16)$$

we substitute (4.9.16) into (4.9.14) resulting in

$$\begin{aligned}\varepsilon &\le -\frac{1}{2}\left(\frac{\Delta t_n}{RC}\right)[v_{c,n} - v_{c,n+1} + \Delta V_n] \\ &= -\frac{1}{2}\left(\frac{\Delta t_n}{RC}\right)\{[(V_n + \Delta V_n) - v_{c,n+1}] - [V_n - v_{c,n}]\}\end{aligned} \quad (4.9.17)$$

Again, we can express the right-hand side in terms of the capacitance current to obtain

$$\varepsilon \le -\frac{1}{2}\left(\frac{\Delta t_n}{C}\right)\left[i_c(t_{n+1}^-) - i_c(t_n^+)\right] \tag{4.9.18}$$

Qualitatively, equation (4.9.18) is merely the negative of that obtained for the Forward Euler integration error. Quantitatively, the magnitudes of these errors may differ greatly since the actual values of the capacitance currents are computed quite differently. Later in this chapter, we will see that because it is more stable, Backward Euler integration is generally more accurate than Forward Euler.

On a voltage basis, both of the Euler integration approximations have an error component proportional to the input change ΔV_n. Steep ramp input slopes usually drive Euler integration approximations to take very small time steps in order to retain reasonable accuracy.

Trapezoidal Integration

Because we have expanded (in effect) the exact solution exponentials in terms of a smaller multiple of the circuit's time constant, $\frac{\Delta t_n}{2RC}$, we would generally expect Trapezoidal integration to have a smaller local truncation error than either of the Euler approaches. To verify that this is true, we reconsider the exact solution in the form shown in (4.8.11):

$$\left[v_{c,n+1} - (V_n + \Delta V_n) + RC\frac{\Delta V_n}{\Delta t_n}\right]e^{\frac{\Delta t_n}{2RC}} = \left[v_{c,n} - V_n + RC\frac{\Delta V_n}{\Delta t_n}\right]e^{-\frac{\Delta t_n}{2RC}} \tag{4.9.19}$$

To obtain the Trapezoidal approximation we retained the first two terms in the Taylor series expansion of each exponential. To estimate the local truncation error, we retain the next two terms in each of these two Taylor series because cancellations reduce their effects to the same order. After a great deal of algebra the result is

$$\varepsilon \le \frac{1}{3}\left(\frac{\Delta t_n}{2RC}\right)^2 (v_{c,n+1} - v_{c,n} - \Delta V_n) \tag{4.9.20}$$

To verify the above result, substitute the third and fourth terms of the Taylor expansion of the two exponentials and then take the difference between the left hand side and right hand side. Separate out the terms which have a $\frac{\Delta t_n}{2RC}$ multiplier and apply the basic Trapezoidal approximation equation to perform simplification, thus leading to the above expression for ε. Provided that $\Delta t_n < 2RC$, which is a reasonable restriction, this Trapezoidal integration error generally is much smaller than the corresponding Euler errors for a comparable time step. It is proportional to Δt_n^2 along with a dependence on the change in capacitor current during the time step. The truncation error is not as easily computed in terms of capacitor currents as are the Euler errors because of the presence of R^2 in the denominator. We can of course write

$$\varepsilon = \frac{\Delta t_n^2}{12RC^2}[i_c(t_{n+1}^-) - i_c(t_n^+)] \tag{4.9.21}$$

But for a general situation, as opposed to this single series RC circuit, we would have no easy mechanism for deriving a formula independent of R, or other circuit element values. See Section 4.11 for a description of other methods to estimate the local truncation error.

In any of the integration methods, if the truncation error is exceeded, interpolation can be used based on the change in capacitor current to locate a suitable intermediate time point. Another way to approach the problem is to divide the time step in half until a time step with an acceptable local truncation error is found. When circuit simulation programs successively reduce their time steps, they can run into numerical problems. First, the small Δt can cause matrix entries that are very large. Second, the program may hit machine precision limits. To protect against these situations, circuit simulation programs often have a pre-set minimum time step after which they abort with a "Time step too small" error message.

4.10 Stability of One-Step Integration Approximation

In the last section we derived *upper bounds* for the *local* truncation error of the various one-step integration approximations to get a feel for their accuracy. We should recognize that the local truncation error provides us with a worst-case for the error that may be incurred over a single time step of integration. Stability is concerned with whether that accumulated error grows or decays as time evolves through a series of time steps. An easy way to consider stability is in terms of the exact solution, equation (4.7.21):

$$v_{c,n+1} = \underbrace{\left(V_n + \Delta V_n - RC\frac{\Delta V_n}{\Delta t_n}\right)}_{particular\ solution} + \underbrace{\left(v_{c,n} - V_n + RC\frac{\Delta V_n}{\Delta t_n}\right)e^{-\frac{\Delta t_n}{RC}}}_{transient\ solution} \tag{4.10.1}$$

Note that the particular solution is a step-ramp combination since the input forcing function is a step-ramp combination, V, from Figure 4.23.

For example, from the original first-order differential equation for this RC circuit in (4.7.2)

$$\frac{dv_c}{dt} = \frac{1}{RC}V - \frac{1}{RC}v_c \tag{4.10.2}$$

we can evaluate the particular solution, $\bar{v}_c$, as follows.

First, assume a step-ramp form for the particular solution

$$\bar{v}_c = a(t - t_n) + b \tag{4.10.3}$$

which corresponds to the step-ramp input forcing function

$$V = \frac{\Delta V_n}{\Delta t_n}(t - t_n) + V_n \tag{4.10.4}$$

for the open interval $t \in (t_n, t_{n+1})$. Inserting (4.10.3) and (4.10.4) into (4.10.2) and solving for the coefficients a and b results in

$$\bar{v}_c = \frac{\Delta V_n}{\Delta t_n}(t - t_n) + V_n - RC\frac{\Delta V_n}{\Delta t_n} \tag{4.10.5}$$

From (4.10.5) the particular solution at t_{n+1} is

$$\bar{v}_{c,\,n+1} = (V_n + \Delta V_n) - RC\frac{\Delta V_n}{\Delta t_n} \tag{4.10.6}$$

and for t_n it is

$$\bar{v}_{c,\,n} = V_n - RC\frac{\Delta V_n}{\Delta t_n} \tag{4.10.7}$$

Returning to (4.10.1), and rearranging the terms, we can express the exact solution in the following way:

$$\left[v_{c,\,n+1} - (V_n + \Delta V_n) + RC\frac{\Delta V_n}{\Delta t_n}\right] = \left[v_{c,\,n} - V_n + RC\frac{\Delta V_n}{\Delta t_n}\right]e^{-\frac{\Delta t_n}{RC}} \tag{4.10.8}$$

From equations (4.10.6) and (4.10.7) we recognize the particular solution terms on both the right hand side and left hand side of (4.10.8), therefore

$$[v_{c,\,n+1} - \bar{v}_{c,\,n+1}] = [v_{c,\,n} - \bar{v}_{c,\,n}]e^{-\frac{\Delta t_n}{RC}} \tag{4.10.9}$$

The quantities in the LHS and RHS brackets represent the difference between the capacitance voltage value and the steady state solution at time points t_{n+1} and t_n, respectively. The bracketed quantity on the RHS is multiplied by $e^{-(\Delta t_n / RC)}$, which is less than one in magnitude since Δt_n, R, and C are all positive values. So we should expect the approximation to be closer to steady state at the end of the numerical integration time interval than it was at the beginning. In other words, any accumulated error should decay.

We can, therefore, invoke a simple stability criterion which must hold regardless of what integration approximation we choose to use:

$$[v_{c,n+1} - \bar{v}_{c,n+1}] < [v_{c,n} - \bar{v}_{c,n}] \tag{4.10.10}$$

This stability criterion is sufficiently general that it can be invoked even if we were to guess the value of $v_{c,n+1}$. For example, we could jump directly to the steady state solution in one step, ignoring any transient, and certainly be stable. Of course, we may not be very accurate in such a situation.

Classically, integration algorithm stability studies focus on the exponential approximation, and demand that its absolute value be less than one.

Forward Euler

For Forward Euler approximation we use

$$e^{-\frac{\Delta t_n}{RC}} \approx 1 - \frac{\Delta t_n}{RC} \tag{4.10.11}$$

So for stability we must have

$$\left|1 - \frac{\Delta t_n}{RC}\right| < 1 \Rightarrow \Delta t_n < 2RC \tag{4.10.12}$$

Limiting the time step in terms of this restriction is not severe for this series RC circuit, since this criterion imposes a (very reasonable) maximum time step equal to twice the time constant of the circuit. However, we will see later that the equivalent limitation on circuits with multiple time constants can be quite extreme. We should also note the relation between (4.10.12) and (4.10.10) for this simple RC example.

If we apply a unit step of voltage to the series RC circuit, the initial FE current will be $1/R$ if the capacitor is initially uncharged. Let us now apply FE for a maximum allowable time step of $\Delta t = 2RC$. Hence

$$v(\Delta t) = v(0) + \frac{i(0)}{C}\Delta t = 0 + 2 = 2V \tag{4.10.13}$$

Thus the voltage would be as far away from the steady state at the end of the time step as it was before the step was taken; therefore, $\Delta t = 2RC$ would violate both equations (4.10.12) and (4.10.10).

Backward Euler

For Backward Euler approximation we use

$$e^{\frac{\Delta t_n}{RC}} \approx 1 + \frac{\Delta t_n}{RC} \tag{4.10.14}$$

or

$$e^{\frac{-\Delta t_n}{RC}} \approx \frac{1}{1 + \frac{\Delta t_n}{RC}} \tag{4.10.15}$$

So we must have

$$\frac{1}{\left|1 + \frac{\Delta t_n}{RC}\right|} < 1 \tag{4.10.16}$$

which poses no restriction for $\Delta t_n > 0$; hence stability considerations pose no restrictions on the time step for Backward Euler.

Trapezoidal integration

For Trapezoidal integration the exponential approximation is

$$e^{-\frac{\Delta t_n}{RC}} = \frac{e^{-\frac{\Delta t_n}{2RC}}}{e^{\frac{\Delta t_n}{2RC}}} = \frac{1 - \frac{\Delta t_n}{2RC}}{1 + \frac{\Delta t_n}{2RC}} \tag{4.10.17}$$

and

$$\left|\frac{1 - \frac{\Delta t_n}{2RC}}{1 + \frac{\Delta t_n}{2RC}}\right| < 1 \tag{4.10.18}$$

From equation (4.10.18) it is apparent that TR analysis is also stable for any $\Delta t_n > 0$. But for $\Delta t_n \gg 2RC$, the left side of equation (4.10.18) is very close to 1, which means that we approach steady state very slowly. It turns out that for $\Delta t_n \gg 2RC$, the approximation is stable, but oscillatory. It should be noted that these oscillations are purely numerical, since a simple RC circuit cannot oscillate. In the interest of accuracy, time steps that are much larger than the time constant of the circuit being analyzed are not recommended with any integration scheme.

4.11 LTE Estimation via Divided Difference Approximations

We can also derive local truncation error estimates in terms of divided difference approximations for higher order derivative terms. For FE and BE we will find that the error expressions are identical to those in Section 4.9. For TR integration, a truncation error expression is derived that is not dependent on circuit element values.

Forward Euler and Backward Euler

Focusing on the Forward Euler approximation for a single capacitor, from a simple difference equation we can write

$$\hat{v}_{c,\,n+1} \approx v_{c,\,n} + \Delta t_n \dot{v}_{c,\,n} \tag{4.11.1}$$

Assuming that all of the solution points for $t \le t_n$ are exact, we can express the exact capacitor voltage at time t_{n+1} in terms of a Taylor series expansion about $t = t_n$:

$$v_{c,\,n+1} = v_{c,\,n} + \Delta t_n \dot{v}_{c,\,n} + \Delta t_n^2 \frac{\ddot{v}_{c,\,n}}{2} + \Delta t_n^3 \frac{\dddot{v}_{c,\,n}}{6} + \ldots \tag{4.11.2}$$

Recognizing that the first two terms of (4.11.2) correspond to the FE expression in (4.9.5), we can express the error

$$\varepsilon = v_{c,\,n+1} - \hat{v}_{c,\,n+1} \tag{4.11.3}$$

as

$$\varepsilon = \Delta t_n^2 \frac{\ddot{v}_{c,\,n}}{2} + \Delta t_n^3 \frac{\dddot{v}_{c,\,n}}{6} + \ldots \tag{4.11.4}$$

For a convergent series,

$$\varepsilon = \Delta t_n^2 \frac{\ddot{v}_c(\xi)}{2} \tag{4.11.5}$$

where $t_n \le \xi \le t_{n+1}$.

One can easily show that the Backward Euler error is the same as (4.11.5) but of opposite sign

$$\varepsilon = -\Delta t_n^2 \frac{\ddot{v}_c(\xi)}{2} \tag{4.11.6}$$

for some ξ between t_n and t_{n+1}.

Approximating the FE and BE errors requires approximating the second derivative of the capacitor voltage. Similar to the divided difference expressions [Carnahan69] which were used to derive the FE and BE formulas earlier in this chapter,

$$\ddot{v}_c(\xi) \approx \frac{\dot{v}_{c,n+1} - \dot{v}_{c,n}}{\Delta t_n} = \frac{i_{c,n+1} - i_{c,n}}{\Delta t_n C} \tag{4.11.7}$$

Combining (4.11.7) and (4.11.5) we have the following expression for the FE error,

$$\varepsilon \approx \frac{\Delta t_n}{2C}[i_{c,n+1} - i_{c,n}] \tag{4.11.8}$$

and combining (4.11.7) and (4.11.6) we have a BE error of

$$\varepsilon \approx \frac{-\Delta t_n}{2C}[i_{c,n+1} - i_{c,n}] \tag{4.11.9}$$

Notice that these expressions are identical to the FE and BE error expressions derived in the previous section.

Trapezoidal Integration

Next we consider the Trapezoidal integration error. From (4.11.2) we know the exact solution is

$$v_{c,n+1} = v_{c,n} + \Delta t_n \dot{v}_{c,n} + \Delta t_n^2 \frac{\ddot{v}_{c,n}}{2} + \Delta t_n^3 \frac{\dddot{v}_c(\varsigma)}{6} \tag{4.11.10}$$

for $t_n \le \varsigma \le t_{n+1}$. It follows that

$$\dot{v}_{c,n+1} = \dot{v}_{c,n} + \Delta t_n \ddot{v}_{c,n} + \Delta t_n^2 \frac{\dddot{v}_c(\zeta)}{2} \quad (4.11.11)$$

for $t_n \le \zeta \le t_{n+1}$.

Solving for $\ddot{v}_{c,n}$ in (4.11.11)

$$\ddot{v}_{c,n} = \frac{\dot{v}_{c,n+1} - \dot{v}_{c,n}}{\Delta t_n} - \frac{\Delta t_n}{2} \dddot{v}_c(\zeta) \quad (4.11.12)$$

and substituting (4.11.12) into (4.11.10)

$$v_{c,n+1} = v_{c,n} + \Delta t_n \dot{v}_{c,n} + \frac{\Delta t_n^2}{2}\left(\frac{\dot{v}_{c,n+1} - \dot{v}_{c,n}}{\Delta t_n} - \frac{\Delta t_n}{2} \dddot{v}_c(\zeta)\right) + \Delta t_n^3 \frac{\dddot{v}_c(\varsigma)}{6} \quad (4.11.13)$$

Combining the $\dddot{v}$ terms in (4.11.13) and rearranging yields

$$v_{c,n+1} = v_{c,n} + \frac{\Delta t_n}{2}(\dot{v}_{c,n+1} + \dot{v}_{c,n}) - \Delta t_n^3 \frac{\dddot{v}_c(\xi)}{12} \quad (4.11.14)$$

where $t_n \le \xi \le t_{n+1}$.

We can see that the first two terms in (4.11.14) are the TR approximation for the capacitor voltage. Therefore, we know that the error is

$$\varepsilon = -\Delta t_n^3 \frac{\dddot{v}_c(\xi)}{12} \quad (4.11.15)$$

We can approximate (4.11.15) by higher order divided differences. For example, we used the first divided difference formula to derive the FE and BE expressions earlier in this chapter:

$$\dot{v}_c(\xi) \approx v_c[t_{n+1}, t_n] = \frac{v_{c,n+1} - v_{c,n}}{\Delta t_n} \quad (4.11.16)$$

where the term, $v_c[t_{n+1}, t_n]$, refers to the first divided difference using the terminology from [Carnahan69]. Note that $v_c[t_{n+1}, t_n]$ is a first derivative and $v_c[t_{n+1}, t_n, t_{n-1}]$ refers to a second derivative.

The second divided difference expression is given by

$$v_c[t_{n+1}, t_n, t_{n-1}] = \frac{v_c[t_{n+1}, t_n] - v_c[t_n, t_{n-1}]}{\Delta t_n + \Delta t_{n-1}} \tag{4.11.17}$$

where

$$\Delta t_n = t_n - t_{n-1} \tag{4.11.18}$$

and

$$\Delta t_{n-1} = t_{n-1} - t_{n-2} \tag{4.11.19}$$

It follows that the third divided difference expression is

$$v_c[t_{n+1}, t_n, t_{n-1}, t_{n-2}] = \frac{v_c[t_{n+1}, t_n, t_{n-1}] - v_c[t_n, t_{n-1}, t_{n-2}]}{\Delta t_n + \Delta t_{n-1} + \Delta t_{n-2}} \tag{4.11.20}$$

and so on.

In [Carnahan69] it is shown that the kth divided difference is related to the kth derivative by

$$\frac{d^k v_c(\xi)}{dt^k} = k! \cdot v_c[t_{n+1}, t_n, \ldots, t_{n-k+1}] \tag{4.11.21}$$

Using (4.11.21) we can approximate the TR error term in (4.11.15) using the following third order divided difference:

$$\frac{d^3 v_c(\xi)}{dt^3} = v_c[t_{n+1}, t_n, t_{n-1}, t_{n-2}] = \frac{v_c[t_{n+1}, t_n, t_{n-1}] - v_c[t_n, t_{n-1}, t_{n-2}]}{\Delta t_n + \Delta t_{n-1} + \Delta t_{n-2}}$$

$$= \frac{\dfrac{v_c[t_{n+1}, t_n] - v_c[t_n, t_{n-1}]}{\Delta t_n + \Delta t_{n-1}} - \dfrac{v_c[t_n, t_{n-1}] - v_c[t_{n-1}, t_{n-2}]}{\Delta t_{n-1} + \Delta t_{n-2}}}{\Delta t_n + \Delta t_{n-1} + \Delta t_{n-2}}$$

$$= \frac{\left[\dfrac{\dfrac{v_c(t_{n+1}) - v_c(t_n)}{\Delta t_n} - \dfrac{v_c(t_n) - v_c(t_{n-1})}{\Delta t_{n-1}}}{\Delta t_n + \Delta t_{n-1}}\right] - \left[\dfrac{\dfrac{v_c(t_n) - v_c(t_{n-1})}{\Delta t_{n-1}} - \dfrac{v_c(t_{n-1}) - v_c(t_{n-2})}{\Delta t_{n-2}}}{\Delta t_{n-1} + \Delta t_{n-2}}\right]}{\Delta t_n + \Delta t_{n-1} + \Delta t_{n-2}} \tag{4.11.22}$$

The above expression, unlike (4.9.20) and (4.9.21), gives us the truncation error independent of circuit element values. In highly nonlinear circuits, relying on history values is not desirable since variables are prone to change rapidly. To the extent that the above truncation error expression includes circuit values two time steps in the past, it may not work very well with highly nonlinear circuits. See [Rohrer84] for a more general method of estimating trapezoidal truncation errors.

4.12 Inductance

Inductance is the dual of capacitance. Readers might wish to enhance their understanding of this material by deriving appropriate integration and error formulas for a parallel *GL* circuit excited by an independent current source comprised of steps and ramps.

4.13 Summary

In this chapter we have developed stability and accuracy considerations for single-step numerical integration approximations as they apply to circuits with a single energy-storage element. In the following chapter we will introduce state variables so that we can consider similar derivations for the case of circuits containing multiple capacitors and inductors.

4.14 References

[Carnahan69] B. Carnahan, H. A. Luther, and J. O. Wilkes. *Applied Numerical Methods.* John Wiley and Sons, 1969.

[Gear71] C. W. Gear. *Numerical Initial Value Problems in Ordinary Differential Equations.* Prentice-Hall, 1971.

[Brayton72] R. K. Brayton, F. G. Gustavson, and G. D. Hachtel. A New Efficient Algorithm for Solving Differential-Algebraic Systems Using Backward Differentiation Formulas. *Proceedings of the IEEE*, vol. 60, pp. 98-108, 1972.

[Rohrer84] R. A. Rohrer, H. Nosrati and K. Heizer. Quasi-static Control of Explicit Algorithms for Transient Analysis. *IEEE Transactions on Computer Aided Design*, vol. CAD-3(3), pp. 226-234, July 1984.

Chapter 5 Linear Transient Analysis II

We have spent a great deal of time in Chapter 4 discussing the special case of linear transient analysis of circuits containing just one energy storage element. However, we will see that this has not been time wasted. In this chapter, we consider linear circuits which may have many energy storage elements and we will see that the same considerations and comparable manipulations apply. Much of the intuition and many of the results from the previous chapter will now prove to be useful for the general case.

5.1 Multiple Energy Storage Elements

To study the behavior of the various one-step integration approximations when used in conjunction with circuits which may have more than one energy storage element, we use the state variable formulation to simplify the discussion. The standard form of the state equations is

$$\dot{\boldsymbol{x}} = \boldsymbol{A}\boldsymbol{x} + \boldsymbol{B}\boldsymbol{u} \tag{5.1.1}$$

$$\boldsymbol{y} = \boldsymbol{C}\boldsymbol{x} + \boldsymbol{D}\boldsymbol{u} \tag{5.1.2}$$

where $\boldsymbol{x}$ is the $n \times 1$ state vector, $\boldsymbol{u}$ is the $m \times 1$ input vector, and $\boldsymbol{y}$ is the vector containing the outputs of interest. Much of the mathematical literature on the solution of ordinary linear differential equations has been developed in terms of a coupled set of n first-order equations such as (5.1.1).

Although state equations are useful for describing various characteristics of a circuit or set of circuit equations, they are not used in practical programs due to the difficulty in forming them. However, we will see that state equations will help us explain some concepts elegantly in this chapter. Further, the concepts will be applicable to other methods of equation formulation, too. To develop some insight into how a circuit is described by a set of equations such as (5.1.1), consider the linear RLC circuit in Figure 5.1. In order to generate the state equations we attempt to express the circuit exclusively in terms of the two

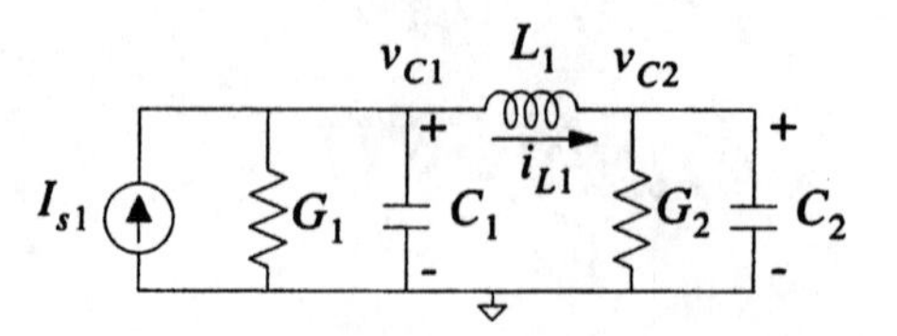

Figure 5.1 RLC circuit example.

capacitor voltages, the inductor current, and the first derivatives of these three variables. To begin, we write the two nodal equations for this circuit:

$$\begin{aligned} G_1 v_{C1} + C_1 \dot{v}_{C1} + i_{L1} &= I_{s1} \\ G_2 v_{C2} + C_2 \dot{v}_{C2} - i_{L1} &= 0 \end{aligned} \qquad \textbf{(5.1.3)}$$

Since the admittance form of the inductor current is an integral expression, we include the inductor current (rather than voltage) as a variable just as we did for the case of ideal independent voltage sources in the MNA equations. Then, as we did for voltage sources, we add an auxiliary equation for the inductor:

$$v_{C1} - v_{C2} - L\dot{i}_{L1} = 0 \qquad \textbf{(5.1.4)}$$

Equations (5.1.3) and (5.1.4) can be combined to form a set of "matrix" equations that describes the circuit in Figure 5.1:

$$\begin{bmatrix} \left[G_1 + C_1 \dfrac{d(\bullet)}{dt}\right] & 0 & 1 \\ 0 & \left[G_2 + C_2 \dfrac{d(\bullet)}{dt}\right] & -1 \\ 1 & -1 & -L_1 \dfrac{d(\bullet)}{dt} \end{bmatrix} \begin{bmatrix} v_{C1} \\ v_{C2} \\ i_{L1} \end{bmatrix} = \begin{bmatrix} I_{s1} \\ 0 \\ 0 \end{bmatrix} \qquad \textbf{(5.1.5)}$$

Or, we can arrange (5.1.3) and (5.1.4) as

$$\begin{bmatrix} C_1 & 0 & 0 \\ 0 & C_2 & 0 \\ 0 & 0 & L_1 \end{bmatrix} \begin{bmatrix} \dot{v}_{C1} \\ \dot{v}_{C2} \\ \dot{i}_{L1} \end{bmatrix} = \begin{bmatrix} -G_1 & 0 & -1 \\ 0 & -G_2 & 1 \\ 1 & -1 & 0 \end{bmatrix} \begin{bmatrix} v_{C1} \\ v_{C2} \\ i_{L1} \end{bmatrix} + \begin{bmatrix} 1 \\ 0 \\ 0 \end{bmatrix} [I_{s1}] \qquad \textbf{(5.1.6)}$$

and then further rearrange (5.1.6) to correspond to the state equation form shown in (5.1.1):

$$\begin{bmatrix} \dot{v}_{C1} \\ \dot{v}_{C2} \\ \dot{i}_{L1} \end{bmatrix} = \begin{bmatrix} C_1 & 0 & 0 \\ 0 & C_2 & 0 \\ 0 & 0 & L_1 \end{bmatrix}^{-1} \begin{bmatrix} -G_1 & 0 & -1 \\ 0 & -G_2 & 1 \\ 1 & -1 & 0 \end{bmatrix} \begin{bmatrix} v_{C1} \\ v_{C2} \\ i_{L1} \end{bmatrix} + \begin{bmatrix} 1 \\ 0 \\ 0 \end{bmatrix} [I_{s1}] \tag{5.1.7}$$

Comparing (5.1.7) with (5.1.1), $\boldsymbol{x}$, the vector of independent state variables, is comprised exclusively of the two capacitor voltages and the inductor current. For this simple circuit it was not difficult to express the equations in terms of these independent variables. In general, however, the formulation of state equations is significantly more involved.

For example, consider the circuit shown in Figure 5.2. We can formulate the state equations in terms of a dc circuit that can be obtained by replacing all capacitances by independent voltage sources and all inductances by independent current sources, as shown in Figure 5.3.

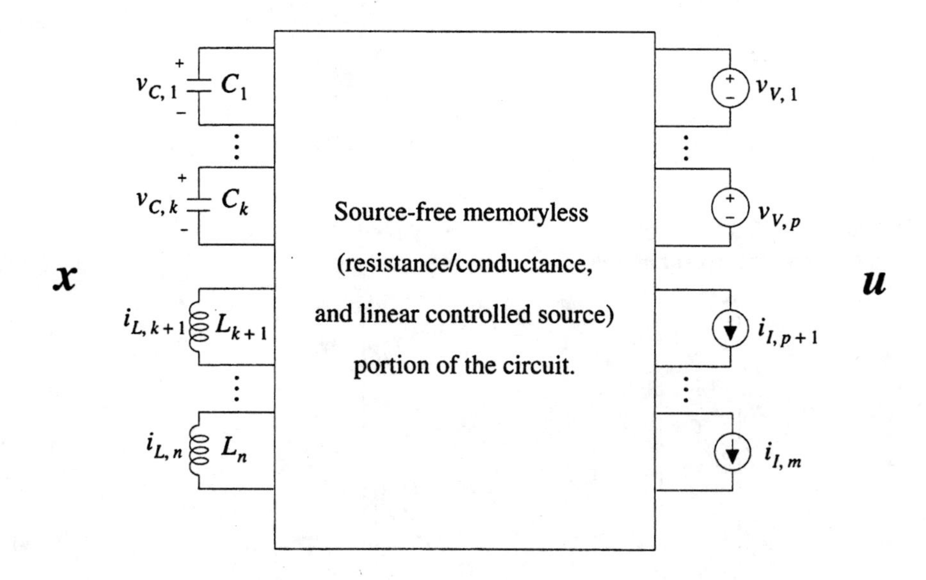

Figure 5.2 Linear circuit with multiple inductors and capacitors.

It may not always be possible to form such a circuit. For example, if the circuit in Figure 5.2 has a loop of voltage sources and capacitors then all of the capacitor voltages in the loop are not independent. We will not discuss how to overcome such situations to obtain state equations, since formulating state equations is not our primary goal here. The details of how to form the state equations for generalized circuits can be found in the literature [Kuh65]. For now we are introducing state variables and their formulation only to add circuit analysis intuition to the mathematical manipulations which come next.

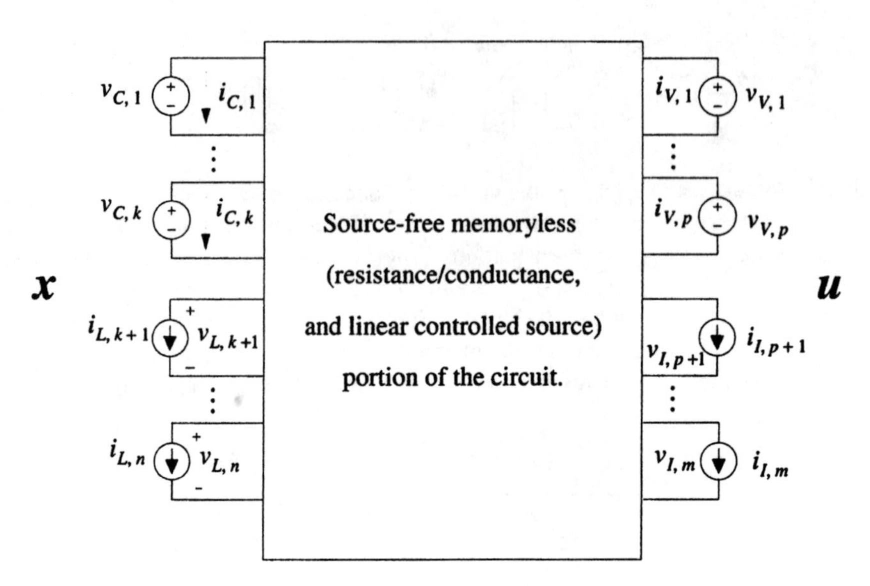

Figure 5.3 The dc circuit used to obtain state equations.

To continue with the our generation of the state equations, the dc circuit in Figure 5.3 is used to generate the hybrid matrix, $\boldsymbol{H}$:

$$\begin{bmatrix} -i_{C,1} \\ \vdots \\ -i_{C,k} \\ v_{L,k+1} \\ \vdots \\ v_{L,n} \\ -i_{V,1} \\ \vdots \\ -i_{V,p} \\ v_{I,p+1} \\ \vdots \\ v_{I,m} \end{bmatrix} = -\begin{bmatrix} \boldsymbol{H}_{CC} & \boldsymbol{H}_{CL} & \boldsymbol{H}_{CV} & \boldsymbol{H}_{CI} \\ \boldsymbol{H}_{LC} & \boldsymbol{H}_{LL} & \boldsymbol{H}_{LV} & \boldsymbol{H}_{LI} \\ \boldsymbol{H}_{VC} & \boldsymbol{H}_{VL} & \boldsymbol{H}_{VV} & \boldsymbol{H}_{VI} \\ \boldsymbol{H}_{IC} & \boldsymbol{H}_{IL} & \boldsymbol{H}_{IV} & \boldsymbol{H}_{II} \end{bmatrix} \begin{bmatrix} v_{C,1} \\ \vdots \\ v_{C,k} \\ -i_{L,k+1} \\ \vdots \\ -i_{L,n} \\ v_{V,1} \\ \vdots \\ v_{V,p} \\ -i_{I,p+1} \\ \vdots \\ -i_{I,m} \end{bmatrix} \quad \textbf{(5.1.8)}$$

$\boldsymbol{H}$ is called a *hybrid port matrix* because it characterizes both port voltages and currents in terms of both currents and voltages. The negative signs denote that the currents in Figure 5.3 are directed oppositely to the traditional reference directions for port currents. Note that each of the 16 entries in $\boldsymbol{H}$ (each of the $\boldsymbol{H}_{ij}$'s), is a submatrix itself. Some of the submatrices of $\boldsymbol{H}$ have dimensions of resistance, some conductance, and some are dimensionless. Each submatrix relates port currents or voltages to the independent-variable currents and voltages, as indicated by the subscripts.

The hybrid matrix is obtained by suppressing certain variables such as node voltages at resistive nodes (nodes without a capacitor or inductor connected to them). The details for formulating the hybrid matrix can be found in [Kuh65]. We do not wish to cover such details here, but we will merely state that these equations can be obtained for all but the most pathological of circuits.

As an example of how one might generate the hybrid matrix consider the simple circuit in Figure 5.4, which contains a voltage source and a capacitor, but no inductors or current sources. The dc equivalent circuit used to generate the hybrid matrix is shown in Figure 5.5.

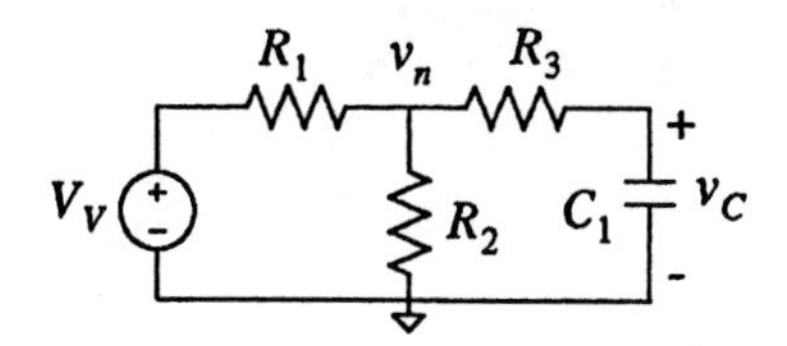

Figure 5.4 RC circuit with an internal resistive node.

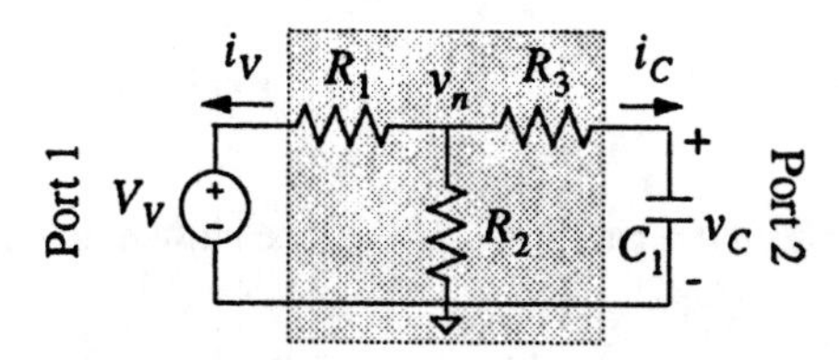

Figure 5.5 The dc equivalent circuit used to eliminate internal resistive node variable.

For the circuit in Figure 5.5 the hybrid equations are

$$\begin{bmatrix} -i_V \\ -i_C \end{bmatrix} = \begin{bmatrix} G_{11} & G_{12} \\ G_{21} & G_{22} \end{bmatrix} \begin{bmatrix} V_V \\ v_C \end{bmatrix} \qquad \textbf{(5.1.9)}$$

Equation (5.1.9) represents the y parameter equations for the two-port circuit in Figure 5.5. The negative signs denote that i_C and i_V are directed oppositely to the traditional reference directions for port currents. We could generate these y parameters by writing the

MNA equations for the dc circuit in Figure 5.5 and then eliminating the node voltage variable, v_n, algebraically. In this example, it can be shown that

$$G_{11} = \frac{\left[1 - \frac{1}{R_1\Delta}\right]}{R_1} \qquad G_{12} = \frac{-1}{R_1 R_3 \Delta}$$

$$G_{21} = \frac{-1}{R_1 R_3 \Delta} \qquad G_{22} = \frac{\left[1 - \frac{1}{R_3\Delta}\right]}{R_3} \tag{5.1.10}$$

$$\Delta = \left(\frac{1}{R_1} + \frac{1}{R_2} + \frac{1}{R_3}\right)$$

Another way of arriving at the same result is by recognizing that the circuit in Figure 5.4 is equivalent to the one shown in Figure 5.6, where we have applied the "Y-delta" or "star-delta" transform to the interconnection of the three resistors. Δ in Figure 5.6 is the same as in (5.1.10). From Figure 5.6, the expressions for G_{11}, G_{12}, G_{21}, and G_{22} can be written by inspection.

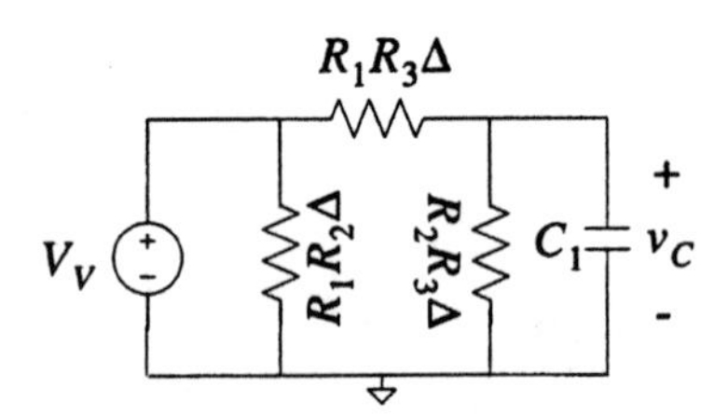

Figure 5.6 Equivalent circuit to the one shown in Figure 5.4.

Once we have the hybrid equations, the state equations follow from the recognition that

$$i_C = C\frac{dv_C}{dt} \quad \text{and} \quad v_L = L\frac{di_L}{dt} \tag{5.1.11}$$

Then, for the general case

$$\dot{x} = Ax + Bu \tag{5.1.12}$$

we have

$$\frac{d}{dt}\begin{bmatrix} v_C \\ i_L \end{bmatrix} = \begin{bmatrix} C^{-1} & 0 \\ 0 & L^{-1} \end{bmatrix}\begin{bmatrix} H_{CC} & -H_{CL} \\ -H_{LC} & H_{LL} \end{bmatrix}\begin{bmatrix} v_C \\ i_L \end{bmatrix} + \begin{bmatrix} C^{-1} & 0 \\ 0 & L^{-1} \end{bmatrix}\begin{bmatrix} H_{CV} & -H_{CI} \\ -H_{LV} & H_{LI} \end{bmatrix}\begin{bmatrix} V_V \\ I_I \end{bmatrix} \tag{5.1.13}$$

For the simple example in Figure 5.5 which is described by equation (5.1.9), we solve for the port current i_C as follows:

$$i_C = -G_{11}v_C - G_{12}V_V \tag{5.1.14}$$

Therefore, we have the following single state equation

$$\dot{v}_C = -C^{-1}G_{11}v_C - C^{-1}G_{12}V_V \tag{5.1.15}$$

which is identical to the form shown in (5.1.1) if $A = -C^{-1}G_{11}$ and $B = -C^{-1}G_{12}$.

For

$$y = Cx + Du \tag{5.1.16}$$

we have in the general case

$$\begin{bmatrix} i_V \\ v_I \end{bmatrix} = \begin{bmatrix} H_{VC} & -H_{VL} \\ -H_{IC} & H_{IL} \end{bmatrix} \begin{bmatrix} v_C \\ i_L \end{bmatrix} + \begin{bmatrix} H_{VV} & -H_{VI} \\ -H_{IV} & H_{II} \end{bmatrix} \begin{bmatrix} V_V \\ I_I \end{bmatrix} \tag{5.1.17}$$

which for the circuit in Figure 5.4 is

$$i_V = -G_{21}v_C - G_{22}V_V \tag{5.1.18}$$

Equation (5.1.17), the input-state-output equation, is simply algebraic. Once we know the input and state values, the outputs can easily be computed. So we will focus our attention on the differential state equation

$$\dot{x} = Ax + Bu \tag{5.1.19}$$

The state of a system at any given time divides the past from the future. If we know the state of a system (values of all the state variables) at any time t_0, we need not know anything about its history prior to that in order to compute its future response. We study the differential state equation just as we did in the first-order case with the simple series RC circuit. First, consider the zero-input case,

$$\dot{x} = Ax \tag{5.1.20}$$

The solution for this coupled set of first-order homogeneous equations is

$$x(t) = e^{At}x_0 \quad \text{for} \quad t \geq 0 \tag{5.1.21}$$

where e^{At} is an $n \times n$ square matrix, and

$$\boldsymbol{x}_0 = \boldsymbol{x}(0) \tag{5.1.22}$$

is the $n \times 1$ vector of specified initial state at $t = 0$. It is most convenient to characterize a square matrix function in terms of the power series (in the matrix) of that function. For example,

$$\boldsymbol{e}^{At} = \mathbf{1} + At + \frac{1}{2}(At)^2 + \frac{1}{6}(At)^3 + \dots \tag{5.1.23}$$

and

$$\frac{d}{dt}(\boldsymbol{e}^{At}) = A + A^2 t + \frac{1}{2}A^3 t^2 + \dots = A\boldsymbol{e}^{At} \tag{5.1.24}$$

It follows that

$$\boldsymbol{e}^{A(t-t_0)} = \boldsymbol{e}^{At}\boldsymbol{e}^{-At_0} \tag{5.1.25}$$

and

$$\boldsymbol{e}^{-At} = [\boldsymbol{e}^{At}]^{-1} \tag{5.1.26}$$

If we wanted an explicit expression for $\boldsymbol{e}^{At}$, we could use the one-sided Laplace transform to obtain it:

$$L\{\boldsymbol{x}(t)\} \equiv \int_o^\infty \boldsymbol{x}(t)\, e^{-st}\, dt \equiv \boldsymbol{X}(s) \tag{5.1.27}$$

Applying the transform to both sides of (5.1.20)

$$\int_o^\infty \dot{\boldsymbol{x}} e^{-st}\, dt = A\int_o^\infty \boldsymbol{x} e^{-st}\, dt \tag{5.1.28}$$

evaluates to

$$\boldsymbol{x}(t)\, e^{-st}\Big|_o^\infty + s\int_o^\infty \boldsymbol{x} e^{-st}\, dt = A\int_o^\infty \boldsymbol{x} e^{-st}\, dt \tag{5.1.29}$$

$$-\boldsymbol{x}_0 + s\boldsymbol{X}(s) = A\boldsymbol{X}(s) \tag{5.1.30}$$

and finally

$$(s\mathbf{1} - A)\boldsymbol{X}(s) = \boldsymbol{x}(0) \tag{5.1.31}$$

where **1** is the unit (or identity) matrix. Solving for $X(s)$ yields

$$X(s) = (s\mathbf{1} - A)^{-1} x_0 \tag{5.1.32}$$

From (5.1.21) and (5.1.32) it is apparent that

$$x(t) = e^{At} x_0 = L^{-1}\{(s\mathbf{1} - A)^{-1} x_0\} \tag{5.1.33}$$

or

$$e^{At} = L^{-1}\{(s\mathbf{1} - A)^{-1}\} \tag{5.1.34}$$

To proceed with (5.1.32), we must solve

$$det(s\mathbf{1} - A) = 0 \tag{5.1.35}$$

to obtain the eigenvalues of A, which are the poles (or natural frequencies) of the circuit.

For example, consider the well known parallel RLC circuit shown in Figure 5.7. Note that the state equations for this circuit are

$$\begin{bmatrix} \dot{v}_C \\ \dot{i}_L \end{bmatrix} = \begin{bmatrix} \frac{1}{C} & 0 \\ 0 & \frac{1}{L} \end{bmatrix} \begin{bmatrix} \frac{-1}{R} & -1 \\ 1 & 0 \end{bmatrix} \begin{bmatrix} v_C \\ i_L \end{bmatrix} + \begin{bmatrix} \frac{1}{C} & 0 \\ 0 & \frac{1}{L} \end{bmatrix} \begin{bmatrix} 1 \\ 0 \end{bmatrix} \begin{bmatrix} I_s \end{bmatrix} \tag{5.1.36}$$

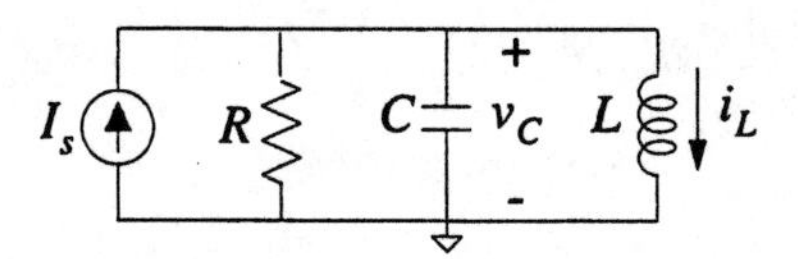

Figure 5.7 Parallel RLC circuit.

The determinant of $s\mathbf{1} - A$ is simply

$$s^2 + \frac{s}{RC} + \frac{1}{LC} \tag{5.1.37}$$

The roots of (5.1.37), hence the poles of the parallel RLC circuit are,

$$\frac{\frac{-1}{RC} \pm \sqrt{\frac{1}{(RC)^2} - \frac{4}{LC}}}{2} \tag{5.1.38}$$

In general, the matrix $(s\mathbf{1} - A)^{-1}$ can be expressed as the matrix of its cofactors divided by its determinant. Partial fraction expansions of the resulting expressions about their root factors yield a set of terms that easily can be inverse Laplace transformed to obtain e^{At}. Except for very small circuits, we would never go to the trouble and expense of computing e^{At} in this manner. Rather, we would approximate that computation in terms of the one-step integration algorithms discussed in Chapter 4. To appreciate such approximations we will study further the nature of e^{At} and the overall solution to the differential state equation.

Suppose that the input u is not zero. Then we try as a solution

$$x(t) = e^{At} z(t) \tag{5.1.39}$$

Note too that

$$x(t_0) = e^{At_0} z(t_0) \tag{5.1.40}$$

implies that

$$z(t_0) = e^{-At_0} x_0 \tag{5.1.41}$$

which we will need later. Substituting (5.1.39) into the differential state equation

$$\dot{x} = Ax + Bu \tag{5.1.42}$$

we obtain

$$Ae^{At} z + e^{At} \dot{z} = Ae^{At} z + Bu \tag{5.1.43}$$

or

$$\dot{z} = e^{-At} Bu \tag{5.1.44}$$

Integrating equation (5.1.44)

$$\int_{t_0}^{t} \dot{z}\, dt = \int_{t_0}^{t} e^{-A\tau} Bu(\tau)\, d\tau \tag{5.1.45}$$

$$z(t) - z(t_0) = \int_{t_0}^{t} e^{-A\tau} Bu(\tau)\, d\tau \qquad (5.1.46)$$

and using (5.1.41)

$$z(t) = e^{-At_0} x_0 + \int_{t_0}^{t} e^{-A\tau} Bu(\tau)\, d\tau \qquad (5.1.47)$$

Then from (5.1.39)

$$x(t) = e^{At}\left[e^{-At_0} x_0 + \int_{t_0}^{t} e^{-A\tau} Bu(\tau)\, d\tau\right] \qquad (5.1.48)$$

Note that we can differentiate (5.1.48) to verify that this solution is correct:

$$\dot{x} = \underbrace{Ae^{At}\left[e^{-At_0} x_0 + \int_{t_0}^{t} e^{-A\tau} Bu(\tau)\, d\tau\right]}_{Ax} + \underbrace{e^{At} e^{-At} Bu(t)}_{Bu} \qquad (5.1.49)$$

Usually we write the general solution in (5.1.48) as

$$x(t) = \underbrace{e^{A(t-t_0)} x_0}_{\text{zero-input response}} + \underbrace{\int_{t_0}^{t} e^{A(t-\tau)} Bu(\tau)\, d\tau}_{\substack{\text{zero-state response} \\ \text{(convolution)}}} \qquad (5.1.50)$$

So far, we have studied (by example) how to obtain state equations. Then, we discussed the meaning of the e^{At} in both the time domain and frequency domain. Then we derived the full solution to state equations in standard form. In the next section, we will restrict the input vector, $u(t)$ to consist of steps and ramps, as we did in Chapter 4, and apply the general solution to circuits with this form of input.

5.2 Step and Ramp Inputs

As with the simple series RC circuit, we confine our attention to the situation in which the input function may only be a sequence of steps and ramps, as shown in Figure 4.23. It should be noted that we are now considering a vector of such inputs, $u(t)$.

We focus our attention on the input segment $t \in (t_n, t_{n+1})$, i.e., $t_n^+ \le t \le t_{n+1}^-$. So, any input step will have occurred prior to time t_n^+ and any subsequent input will occur after time t_{n+1}^{-}. The input is a simple ramp over the interval of interest. We assume that there are no impulses in the inputs, and hence disallow step discontinuities in the state variables. Capacitor voltages and inductor currents cannot change instantaneously, so

$$\boldsymbol{x}(t_n^-) = \boldsymbol{x}(t_n^+) = \boldsymbol{x}(t_n) \equiv \boldsymbol{x}_n \tag{5.2.1}$$

We will consider charge and flux sharing -- in which cases there can be impulses -- later in Section 5.5.

Referring to Figure 5.8, we characterize the input segment on the open time interval $t \in (t_n, t_{n+1})$:

$$\boldsymbol{u}(t) = \boldsymbol{u}_n + (t - t_n)\frac{\Delta \boldsymbol{u}_n}{\Delta t_n} \tag{5.2.2}$$

$$\boldsymbol{u}_n \equiv \boldsymbol{u}(t_n^+) \tag{5.2.3}$$

$$\Delta \boldsymbol{u}_n = \boldsymbol{u}(t_{n+1}^-) - \boldsymbol{u}(t_n^+) \tag{5.2.4}$$

$$\Delta t_n = t_{n+1} - t_n \tag{5.2.5}$$

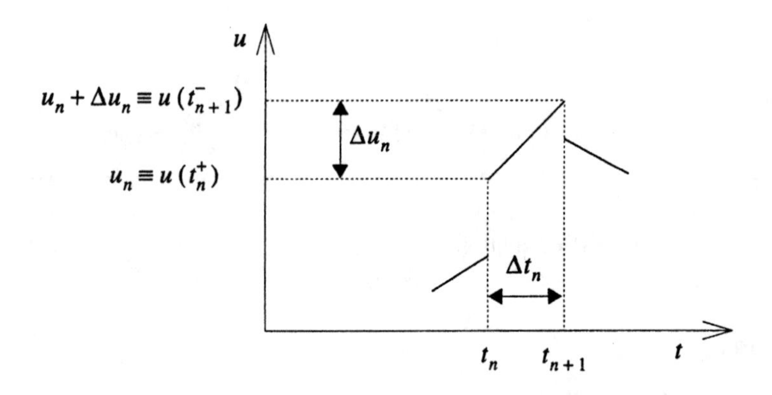

Figure 5.8 The form of one input signal from $\boldsymbol{u}(t)$.

Under these circumstances (equations (5.2.2) through (5.2.5)) we can obtain an exact solution for the state at time $t_n^+ \le t \le t_{n+1}^-$ in terms of the state at t_n using (5.1.50):

$$x(t) = e^{A(t-t_n)}\left[x(t_n) + A^{-1}(Bu_n + A^{-1}B\frac{\Delta u_n}{\Delta t_n})\right] \\ -A^{-1}\left[B(u_n + \frac{(t-t_n)}{\Delta t_n}\Delta u_n) + A^{-1}B\frac{\Delta u_n}{\Delta t_n}\right] \tag{5.2.6}$$

The above equation was derived by substituting (5.2.2) into (5.1.50), integrating by parts and indulging in some tedious algebra. Setting $t = t_{n+1}$,

$$x(t_{n+1}) = e^{A\Delta t_n}\left[x(t_n) + A^{-1}(Bu_n + A^{-1}B\frac{\Delta u_n}{\Delta t_n})\right] \\ -A^{-1}\left[B(u_n + \Delta u_n) + A^{-1}B\frac{\Delta u_n}{\Delta t_n}\right] \tag{5.2.7}$$

Note once again that A^{-1} may not exist if there are inductance loops or in other degenerate cases. Using the notation

$$x_{n+1} \equiv x(t_{n+1}) \\ x_n \equiv x(t_n) \tag{5.2.8}$$

it is most convenient to rewrite equation (5.2.7) as

$$x_{n+1} + A^{-1}\left[B(u_n + \Delta u_n) + A^{-1}B\frac{\Delta u_n}{\Delta t_n}\right] \\ = e^{A\Delta t_n}\left[x_n + A^{-1}(Bu_n + A^{-1}B\frac{\Delta u_n}{\Delta t_n})\right] \tag{5.2.9}$$

Equation (5.2.9) shows explicitly the exponential transformation from a deviation from the initial steady state

$$x_{nss} = -A^{-1}\left[Bu_n + A^{-1}B\frac{\Delta u_n}{\Delta t_n}\right] \tag{5.2.10}$$

to the final one

$$x_{n+1,ss} = -A^{-1}\left[B(u_n + \Delta u_n) + A^{-1}B\frac{\Delta u_n}{\Delta t_n}\right] \tag{5.2.11}$$

just as we saw for the single RC circuit case in Chapter 4. Equations (5.2.10) and (5.2.11) are the steady state terms at time t_n and t_{n+1} respectively, in a manner similar to that shown for the single RC case in equations (4.10.6) and (4.10.7).

5.3 One-Step Integration Approximations

Similar to our analysis for the simple series RC circuit, we can obtain various approximations to the exact solution in terms of the Taylor series expansion of the matrix exponential

$$e^{A\Delta t_n} = \mathbf{1} + \Delta t_n \mathbf{A} + \frac{1}{2}(\Delta t_n \mathbf{A})^2 + \frac{1}{6}(\Delta t_n \mathbf{A})^3 + \ldots \tag{5.3.1}$$

Forward Euler

For the Forward Euler integration approximation

$$e^{A\Delta t_n} \approx \mathbf{1} + \Delta t_n \mathbf{A} \tag{5.3.2}$$

we have

$$\mathbf{x}_{n+1} \approx (\mathbf{1} + \Delta t_n \mathbf{A})\left[\mathbf{x}_n + \mathbf{A}^{-1}(\mathbf{B}\mathbf{u}_n + \mathbf{A}^{-1}\mathbf{B}\frac{\Delta \mathbf{u}_n}{\Delta t_n})\right] - \mathbf{A}^{-1}\left[\mathbf{B}(\mathbf{u}_n + \Delta \mathbf{u}_n) + \mathbf{A}^{-1}\mathbf{B}\frac{\Delta \mathbf{u}_n}{\Delta t_n}\right] \tag{5.3.3}$$

which reduces to

$$\mathbf{x}_{n+1} \approx \mathbf{x}_n + \Delta t_n (\mathbf{A}\mathbf{x}_n + \mathbf{B}\mathbf{u}_n) \tag{5.3.4}$$

Equation (5.3.4) makes intuitive sense since $\mathbf{A}\mathbf{x}_n + \mathbf{B}\mathbf{u}_n$ is simply $\dot{\mathbf{x}}$ evaluated at t_n^+. Of course, this result is the same as the one we obtained earlier on an individual element basis:

$$\begin{bmatrix} v_C(t_{n+1}) \\ i_L(t_{n+1}) \end{bmatrix} = \begin{bmatrix} v_C(t_n) \\ i_L(t_n) \end{bmatrix} + \Delta t_n \begin{bmatrix} C^{-1} & 0 \\ 0 & L^{-1} \end{bmatrix} \begin{bmatrix} i_C(t_n^+) \\ v_L(t_n^+) \end{bmatrix} \tag{5.3.5}$$

Next, we can estimate the local truncation error vector of the FE approximation:

$$\begin{aligned} \varepsilon_n &\approx \frac{1}{2}(\Delta t_n \mathbf{A})^2 \left[\mathbf{x}_n + \mathbf{A}^{-1}(\mathbf{B}\mathbf{u}_n + \mathbf{A}^{-1}\mathbf{B}\frac{\Delta \mathbf{u}_n}{\Delta t_n})\right] \\ &= \frac{1}{2}\Delta t_n [\mathbf{A}(\mathbf{x}_{n+1} - \mathbf{x}_n) + \mathbf{B}\Delta \mathbf{u}_n] \end{aligned} \tag{5.3.6}$$

which can be rewritten as

$$\begin{aligned}\varepsilon_n &\approx \frac{1}{2}\Delta t_n [\{Ax_{n+1} + B(u_n + \Delta u_n)\} - \{Ax_n + Bu_n\}] \\ &\approx \frac{1}{2}\Delta t_n [\dot{x}_{n+1} - \dot{x}_n]\end{aligned} \tag{5.3.7}$$

Expression (5.3.7) can be interpreted as

$$\varepsilon_n \approx \frac{1}{2}\Delta t_n \begin{bmatrix} C^{-1} & 0 \\ 0 & L^{-1} \end{bmatrix} \left\{ \begin{bmatrix} i_C(t_{n+1}^-) \\ v_L(t_{n+1}^-) \end{bmatrix} - \begin{bmatrix} i_C(t_n^+) \\ v_L(t_n^+) \end{bmatrix} \right\} \tag{5.3.8}$$

Equation (5.3.6) shows that the error is proportional to the changes in the state variables and the changes in the inputs. Equation (5.3.7) shows that the error is proportional to the changes in the state variable derivatives. We note in particular that large Δu_n values may dictate small Δt_n values to maintain reasonable error sizes. On the other hand, we can accommodate a step function change with no such error problem merely by ensuring that we compute $i_c(t_n^+)$ and $v_L(t_n^+)$ immediately after the onset of the step. With a Forward Euler approximation such a computation typically is not very costly. So it may be better with Forward Euler to approximate a very steep ramp input in terms of a step function rather than the opposite.

Backward Euler

For a Backward Euler approximation we write the exact solution as

$$x_n = e^{-A\Delta t_n}\left[x_{n+1} + A^{-1}\{B(u_n + \Delta u_n) + A^{-1}B\frac{\Delta u_n}{\Delta t_n}\}\right] - A^{-1}\left[Bu_n + A^{-1}B\frac{\Delta u_n}{\Delta t_n}\right] \tag{5.3.9}$$

We then assume $e^{-A\Delta t_n} \approx 1 - \Delta t_n A$ to get

$$x_n \approx (1 - \Delta t_n A)\left[x_{n+1} + A^{-1}\{B(u_n + \Delta u_n) + A^{-1}B\frac{\Delta u_n}{\Delta t_n}\}\right] - A^{-1}\left[Bu_n + A^{-1}B\frac{\Delta u_n}{\Delta t_n}\right] \tag{5.3.10}$$

which reduces to

$$x_n \approx x_{n+1} - \Delta t_n [Ax_{n+1} + B(u_n + \Delta u_n)] \tag{5.3.11}$$

Then, rearranging the terms in (5.3.11) yields

$$\begin{aligned} \boldsymbol{x}_{n+1} &= \boldsymbol{x}_n + \Delta t_n [\boldsymbol{A}\boldsymbol{x}_{n+1} + \boldsymbol{B}(\boldsymbol{u}_n + \Delta \boldsymbol{u}_n)] \\ &= \boldsymbol{x}_n + \Delta t_n \dot{\boldsymbol{x}}_{n+1} \end{aligned} \tag{5.3.12}$$

Again, we can interpret this result on an individual energy storage element basis as

$$\begin{bmatrix} v_C(t_{n+1}) \\ i_L(t_{n+1}) \end{bmatrix} = \begin{bmatrix} v_C(t_n) \\ i_L(t_n) \end{bmatrix} + \Delta t_n \begin{bmatrix} C^{-1} & 0 \\ 0 & L^{-1} \end{bmatrix} \begin{bmatrix} i_C(t_{n+1}^-) \\ v_L(t_{n+1}^-) \end{bmatrix} \tag{5.3.13}$$

Next, we see that the Backward Euler local truncation error is (qualitatively) equal to and opposite the sign of the Forward Euler approximation:

$$\begin{aligned} \varepsilon_n &\approx -\frac{1}{2}(\Delta t_n \boldsymbol{A})^2 \left[\boldsymbol{x}_{n+1} + \boldsymbol{A}^{-1} \{ \boldsymbol{B}(\boldsymbol{u}_n + \Delta \boldsymbol{u}_n) + \boldsymbol{A}^{-1}\boldsymbol{B}\frac{\Delta \boldsymbol{u}_n}{\Delta t_n} \} \right] \\ &= -\frac{1}{2}\Delta t_n [\boldsymbol{A}(\boldsymbol{x}_{n+1} - \boldsymbol{x}_n) + \boldsymbol{B}\Delta \boldsymbol{u}_n] \\ &= -\frac{1}{2}\Delta t_n [\{ \boldsymbol{A}\boldsymbol{x}_{n+1} + \boldsymbol{B}(\boldsymbol{u}_n + \Delta \boldsymbol{u}_n) \} - [\boldsymbol{A}\boldsymbol{x}_n + \boldsymbol{B}\boldsymbol{u}_n]] \\ &= -\frac{1}{2}\Delta t_n [\dot{\boldsymbol{x}}_{n+1} - \dot{\boldsymbol{x}}_n] \end{aligned} \tag{5.3.14}$$

As before, the local truncation error estimate is shown to be proportional both to the changes in the state variables and to the changes in the input values. From the final expression we can compute the local truncation error to be

$$\varepsilon_n \approx -\frac{1}{2}\Delta t_n \begin{bmatrix} C^{-1} & 0 \\ 0 & L^{-1} \end{bmatrix} \left\{ \begin{bmatrix} i_C(t_{n+1}^-) \\ v_L(t_{n+1}^-) \end{bmatrix} - \begin{bmatrix} i_C(t_n^+) \\ v_L(t_n^+) \end{bmatrix} \right\} \tag{5.3.15}$$

Trapezoidal

For a Trapezoidal integration approximation we start with the exact solution in the following form

$$
\begin{aligned}
&e^{-\frac{1}{2}A\Delta t_n}\left[x_{n+1} + A^{-1}\{B(u_n + \Delta u_n) + A^{-1}B\frac{\Delta u_n}{\Delta t_n}\}\right] \\
&\quad = e^{\frac{1}{2}A\Delta t_n}\left[x_n + A^{-1}\{Bu_n + A^{-1}B\frac{\Delta u_n}{\Delta t_n}\}\right]
\end{aligned} \tag{5.3.16}
$$

Again, using the first-order truncation of the Taylor series for both matrix exponentials, we obtain

$$
\begin{aligned}
&(1 - \tfrac{1}{2}\Delta t_n A)\left[x_{n+1} + A^{-1}\{B(u_n + \Delta u_n) + A^{-1}B\frac{\Delta u_n}{\Delta t_n}\}\right] \\
&\quad \approx (1 + \tfrac{1}{2}\Delta t_n A)\left[x_n + A^{-1}\{Bu_n + A^{-1}B\frac{\Delta u_n}{\Delta t_n}\}\right]
\end{aligned} \tag{5.3.17}
$$

or,

$$
x_{n+1} - \frac{1}{2}\Delta t_n[Ax_{n+1} + B(u_n + \Delta u_n)] \approx x_n + \frac{1}{2}\Delta t_n(Ax_n + Bu_n) \tag{5.3.18}
$$

Finally, rearranging (5.3.18),

$$
\begin{aligned}
x_{n+1} &\approx x_n + \frac{1}{2}\Delta t_n[\{Ax_n + Bu_n\} + \{Ax_{n+1} + B(u_n + \Delta u_n)\}] \\
&\approx x_n + \frac{1}{2}\Delta t_n[\dot{x}_n + \dot{x}_{n+1}]
\end{aligned} \tag{5.3.19}
$$

In terms of the individual energy storage elements we have

$$
\begin{bmatrix} v_C(t_{n+1}) \\ i_L(t_{n+1}) \end{bmatrix} = \begin{bmatrix} v_C(t_n) \\ i_L(t_n) \end{bmatrix} + \frac{1}{2}\Delta t_n \begin{bmatrix} C^{-1} & 0 \\ 0 & L^{-1} \end{bmatrix} \left\{ \begin{bmatrix} i_C(t_n^+) \\ v_L(t_n^+) \end{bmatrix} + \begin{bmatrix} i_C(t_{n+1}^-) \\ v_L(t_{n+1}^-) \end{bmatrix} \right\} \tag{5.3.20}
$$

To estimate the local truncation error, we take the next two terms in the Taylor series expansions of both exponentials.

$$\varepsilon_n \approx \left[\frac{1}{2}\left(-\frac{1}{2}\Delta t_n A\right)^2 + \frac{1}{6}\left(-\frac{1}{2}\Delta t_n A\right)^3\right]\left[x_{n+1} + A^{-1}\{B(u_n + \Delta u_n) + A^{-1}B\frac{\Delta u_n}{\Delta t_n}\}\right]$$
$$-\left[\frac{1}{2}\left(-\frac{1}{2}\Delta t_n A\right)^2 + \frac{1}{6}\left(\frac{1}{2}\Delta t_n A\right)^3\right]\left[x_n + A^{-1}(Bu_n + A^{-1}B\frac{\Delta u_n}{\Delta t_n})\right]$$
$$= \frac{1}{12}\Delta t_n^2 A\,[A(x_{n+1} - x_n) + B\Delta u_n]$$
$$= \frac{1}{12}\Delta t_n^2 A\,[\{Ax_{n+1} + B(u_n + \Delta u_n)\} - \{Ax_n + Bu_n\}] \tag{5.3.21}$$
$$= \frac{1}{6}\Delta t_n A\left[\frac{1}{2}\Delta t_n\{(Ax_{n+1} + B(u_n + \Delta u_n)) - (Ax_n + Bu_n)\}\right]$$
$$= \frac{1}{6}\Delta t_n A\left[\frac{1}{2}\Delta t_n\{\dot{x}_{n+1} - \dot{x}_n\}\right]$$

We see once again that the local truncation error is proportional to the changes in state variable values and input values across the time interval of interest, as well as to the square of Δt_n. Overall, the error is proportional to Δt_n^3 since the aforementioned changes are proportional to Δt_n. From equation (5.3.21) we can write

$$\varepsilon_n \approx \frac{1}{6}\Delta t_n A\left(\frac{1}{2}\Delta t_n\begin{bmatrix} C^{-1} & 0 \\ 0 & L^{-1}\end{bmatrix}\left\{\begin{bmatrix} i_C(t_{n+1}^-) \\ v_L(t_{n+1}^-)\end{bmatrix} - \begin{bmatrix} i_C(t_n^+) \\ v_L(t_n^+)\end{bmatrix}\right\}\right) \tag{5.3.22}$$

We consider the computation of this quantity as follows. First define an auxiliary vector

$$x_T \equiv \begin{bmatrix} v_{CT} \\ i_{LT}\end{bmatrix} \equiv \frac{1}{2}\Delta t_n\begin{bmatrix} C^{-1} & 0 \\ 0 & L^{-1}\end{bmatrix}\left\{\begin{bmatrix} i_C(t_{n+1}^-) \\ v_L(t_{n+1}^-)\end{bmatrix} - \begin{bmatrix} i_C(t_n^+) \\ v_L(t_n^+)\end{bmatrix}\right\} \tag{5.3.23}$$

And then use that vector x_T to define a set of excitation sources in place of the original energy storage elements as shown in Figure 5.9. Then

$$\varepsilon_n \approx \frac{1}{6}\Delta t_n A\begin{bmatrix} v_{CT} \\ i_{LT}\end{bmatrix}$$
$$= \frac{1}{6}\Delta t_n A x_T$$
$$= \frac{1}{6}\Delta t_n \dot{x}_T \tag{5.3.24}$$
$$= \frac{1}{6}\Delta t_n\begin{bmatrix} C^{-1} & 0 \\ 0 & L^{-1}\end{bmatrix}\begin{bmatrix} i_{CT} \\ v_{LT}\end{bmatrix}$$

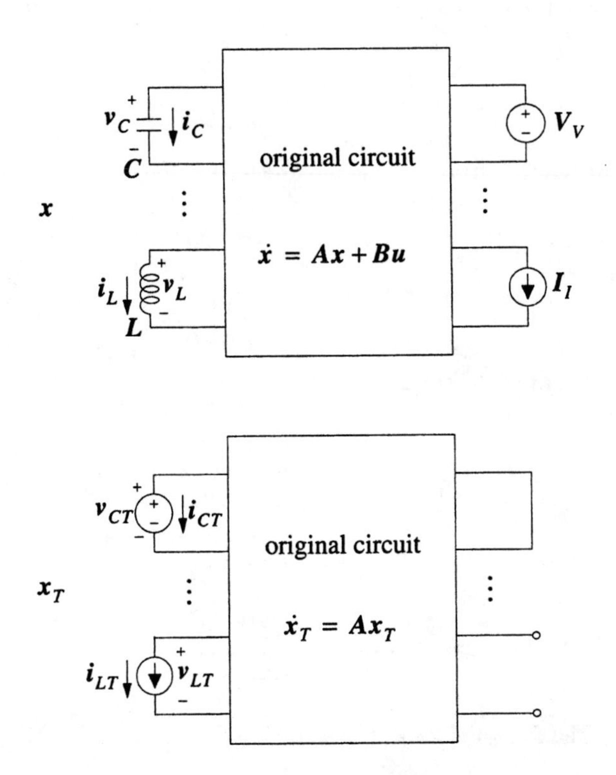

Figure 5.9 Calculating the auxiliary vector to measure the local truncation error.

This auxiliary computation usually is not performed since there are other means to estimate the local truncation error for Trapezoidal integration [Nagel75]. For example, given that the TR approximation is

$$x_{n+1} = x_n + \frac{\Delta t_n}{2}(x'_n + x'_{n+1}) \tag{5.3.25}$$

We can express the derivative at t_{n+1} in terms of a Taylor series expansion

$$x'_{n+1} = x'_n + \Delta t_n x''_n + \frac{\Delta t_n^2}{2} x'''_n + \ldots \tag{5.3.26}$$

Combining (5.3.25) and (5.3.26), the TR approximation is

$$x_{n+1} = x_n + \frac{\Delta t_n}{2}\left[x'_n + \left(x'_n + \Delta t_n x''_n + \frac{\Delta t_n^2}{2} x'''_n + \ldots\right)\right] \tag{5.3.27}$$

However, the full Taylor series expansion is

$$x_{n+1} = x_n + \Delta t_n x_n' + \frac{\Delta t_n^2}{2} x_n'' + \frac{\Delta t_n^3}{6} x_n''' + \frac{\Delta t_n^4}{24} x_n'''' + \ldots \tag{5.3.28}$$

and hence the local truncation error can be approximated by

$$\varepsilon_n = \frac{1}{6}\Delta t_n^3 x_n''' - \frac{1}{4}\Delta t_n^3 x_n''' + \frac{\Delta t_n^4}{24} x_n'''' + \ldots = -\frac{\Delta t_n^3}{12} x_n'''(\xi) \tag{5.3.29}$$

where $t_n \le \xi \le t_n + \Delta t_n$. To estimate this error, we approximate $x_n'''(\xi)$ by a divided difference formula. The first divided difference DD_1 is

$$DD_1(t_{n+1}) = \frac{x_{n+1} - x_n}{\Delta t_{n+1}} \tag{5.3.30}$$

which approximates $x_n'(\xi)$. The second divided difference

$$DD_2(t_{n+1}) = \frac{DD_1(t_{n+1}) - DD_1(t_n)}{\Delta t_n + \Delta t_{n+1}} \tag{5.3.31}$$

approximates $x_n''(\xi)$. The third divided difference

$$DD_3(t_{n+1}) = \frac{DD_2(t_{n+1}) - DD_2(t_n)}{\Delta t_n + \Delta t_{n+1}} \tag{5.3.32}$$

approximates $x_n'''(\xi)$. It is apparent that the kth divided difference requires the retention of k previous time point solutions. So for TR integration, we would need the solutions at the time points t_{n-2}, t_{n-1}, t_n, and t_{n+1} to estimate the local truncation error for the time step from t_n to t_{n+1}.

5.4 Stability

We can discuss one-step integration approximations as before in terms of the exact solution (5.2.9) as we did for the single RC circuit case in Chapter 4:

$$
\begin{aligned}
&x_{n+1} + A^{-1}\left[B\,(u_n + \Delta u_n) + A^{-1}B\frac{\Delta u_n}{\Delta t_n}\right] \\
&= e^{A\Delta t_n}\left[x_n + A^{-1}\,(Bu_n + A^{-1}B\frac{\Delta u_n}{\Delta t_n})\right]
\end{aligned}
\tag{5.4.1}
$$

Equation (5.4.1) represents a transformation from the initial steady state error

$$x_n + A^{-1}\left[Bu_n + A^{-1}B\frac{\Delta u_n}{\Delta t_n}\right] \tag{5.4.2}$$

to the final steady state error

$$x_{n+1} + A^{-1}\left[B\,(u_n + \Delta u_n) + A^{-1}B\frac{\Delta u_n}{\Delta t_n}\right] \tag{5.4.3}$$

by a factor of $e^{A\Delta t_n}$.

If the final steady state error vector is smaller than the initial, then we can pronounce the integration algorithm stable:

$$\left\| x_{n+1} + A^{-1}\left[B\,(u_n + \Delta u_n) + A^{-1}B\frac{\Delta u_n}{\Delta t_n}\right]\right\| < \left\| x_n + A^{-1}\left[Bu_n + A^{-1}B\frac{\Delta u_n}{\Delta t_n}\right]\right\| \tag{5.4.4}$$

Given a vector

$$y \equiv [y_1, y_2, \ldots, y_n]^T \tag{5.4.5}$$

we can measure the size (or length) of the vector in any of the following ways.

$$\|y\|_1 \equiv \sum_{k=1}^{n} |y_k| \tag{5.4.6}$$

$$\|y\|_2 \equiv \sqrt{\sum_{k=1}^{n} |y_k|^2} \tag{5.4.7}$$

$$\|y\|_\infty = \max_{k=1}^{n} \{\,|y_k|\} \tag{5.4.8}$$

All three of these length measures (norms) work; it is merely a matter of convenience as to which to use. The euclidean, $\|\boldsymbol{y}\|_2$, is the most natural, but the $\|\boldsymbol{y}\|_\infty$ norm often is easier to compute and apply in practice. It is possible, of course, to reject as possibly unstable a result based on the $\|\boldsymbol{y}\|_\infty$ norm that may be shown to be stable on the other bases.

Classically, stability studies focus on the approximation of $e^{A\Delta t_n}$. Provided that the original circuit is asymptotically stable, we would expect

$$\lim_{k \to \infty} \left[e^{A\Delta t_n}\right]^k \to 0 \tag{5.4.9}$$

which would be the result of a uniform time step. For stability of an integration approximation we demand the same behavior.

For Forward Euler approximation we have then

$$e^{A\Delta t_n} \approx 1 + \Delta t_n A \tag{5.4.10}$$

So we would ask that

$$\lim_{k \to \infty} (1 + \Delta t_n A)^k \to 0 \tag{5.4.11}$$

too.

We recognize that A can be diagonalized [Strang86] as

$$A = T\Lambda T^{-1} \tag{5.4.12}$$

where Λ is the Jordon canonical form for A and T is a normalized modal matrix. Then

$$\begin{aligned} 1 + \Delta t_n A &= TT^{-1} + \Delta t_n T\Lambda T^{-1} \\ &= T(1 + \Delta t_n \Lambda)\, T^{-1} \end{aligned} \tag{5.4.13}$$

and it follows that

$$(1 + \Delta t_n A)^k = T(1 + \Delta t_n \Lambda)^k T^{-1} \tag{5.4.14}$$

Under the reasonable assumption that the eigenvalues of A, its λs, are distinct, from (5.4.11) we must have

$$\lim_{k \to \infty} (1 + \Delta t_n \lambda)^k \to 0 \text{ for all } \lambda \tag{5.4.15}$$

for all of the eigenvalues of A in order to guarantee stability. If the eigenvalues are not distinct the same result holds; it is merely more difficult to derive. From (5.4.15) we conclude that

$$|1 + \Delta t_n \lambda| < 1 \quad \text{for all} \quad \lambda \tag{5.4.16}$$

Suppose that the eigenvalues are complex numbers. Consider one eigenvalue

$$\lambda = \alpha + j\beta \tag{5.4.17}$$

Then (5.4.16) becomes

$$|1 + \Delta t_n \alpha + j\Delta t_n \beta| < 1 \tag{5.4.18}$$

$$(1 + \Delta t_n \alpha)^2 + (\Delta t_n \beta)^2 < 1 \tag{5.4.19}$$

$$1 + 2\Delta t_n \alpha + (\Delta t_n \alpha)^2 + (\Delta t_n \beta)^2 < 1 \tag{5.4.20}$$

$$2\Delta t_n \alpha + \Delta t_n^2 (\alpha^2 + \beta^2) < 0 \tag{5.4.21}$$

$$\Delta t_n < -\frac{2\alpha}{\alpha^2 + \beta^2} \tag{5.4.22}$$

$$\Delta t_n < -2Re\left(\frac{1}{\lambda}\right) \tag{5.4.23}$$

for all of the eigenvalues of A.

We note that this stability criterion is very restrictive for large real eigenvalues. Complex eigenvalues that have large imaginary parts and small real parts (high-Q poles) will necessitate small time steps, and likewise with eigenvalues that have large real parts and small imaginary parts. From (5.4.23) it is apparent that the largest pole (or smallest time constant or highest natural frequency), is the most restrictive and limits the time step possible in a stable Forward Euler approximation. The *region of stability* of an integration method is defined as that subset of the complex plane such that if $\Delta t_n \lambda$ is inside the region of stability for all λ, then the method of integration is guaranteed to be stable. Thus the *region of stability* of Forward Euler integration is limited.

For Backward Euler approximation we have

$$e^{A\Delta t_n} \approx (1 - \Delta t_n A) \tag{5.4.24}$$

so we must demand

$$\left| \frac{1}{1 - \Delta t_n \lambda} \right| < 1 \tag{5.4.25}$$

which leads to

$$\Delta t_n (\alpha^2 + \beta^2) > 2\alpha \tag{5.4.26}$$

$$\Delta t_n > 2Re\,(\frac{1}{\lambda}) \tag{5.4.27}$$

for all of the eigenvalues of A. Since $\alpha < 0$ for an originally stable system, this condition poses no restriction for $\Delta t_n > 0$. The region of stability thus includes the entire left half of the complex plane and hence Backward Euler integration is said to be *A-stable.*

For the Trapezoidal approximation we have

$$e^{A \Delta t_n} \approx (1 - \frac{1}{2}\Delta t_n A)^{-1} (1 + \frac{1}{2}\Delta t_n A) \tag{5.4.28}$$

so we must require

$$\left| \frac{(1 + \frac{1}{2}\Delta t_n \lambda)}{(1 - \frac{1}{2}\Delta t_n \lambda)} \right| < 1 \tag{5.4.29}$$

which leads to

$$-2\Delta t_n \alpha > 0 \tag{5.4.30}$$

or

$$-2\Delta t_n Re\,(\lambda) > 0 \tag{5.4.31}$$

for all of the eigenvalues of A. This condition poses no restriction for an originally stable system. The region of stability thus includes the entire left half of the complex plane and hence Trapezoidal integration is A-stable. However, Trapezoidal integration is oscillatory, as discussed in Section 4.10.

For originally stable systems only the Forward Euler integration approximation poses a time step restriction due to stability. And that restriction can be severe especially in cases of "stiff systems." A stiff system is one with natural frequencies widely spread in values.

5.5 Limitations of One-Step Integration Models

The one-step integration techniques described in this chapter are implemented efficiently in terms of the companion models in Figures 4.8 through 4.11 and in Figures 4.13 through 4.15. As for the case of stamps in Chapter 2, however, we must determine when these models can and cannot be used.

For instance, in addition to all of its other limitations, Forward Euler integration cannot handle loops of capacitors or cutsets of inductors. This is apparent from the FE companion models which would result in loops of voltage sources and cutsets of current sources from circuits containing capacitor loops and inductor cutsets, respectively. We should add that these loop and cutset situations are not a problem for the implicit integration algorithms, such as BE and TR, which are characterized by companion models that contain resistors or conductors.

The only other restriction on these integration models is due to our assumption in equation (5.2.1) that there are no impulses in the circuit and that the state variables are continuous. In situations where we have loops comprised solely of independent voltage sources and capacitors, or cutsets comprised solely of independent current sources and inductors, the assumption in equation (5.2.1) may be violated.

For example, consider the circuit shown in Figure 5.10 in which an independent voltage source and a capacitor form a loop. If the input voltage V_S is a step function then the capacitor voltage v_C is discontinuous.

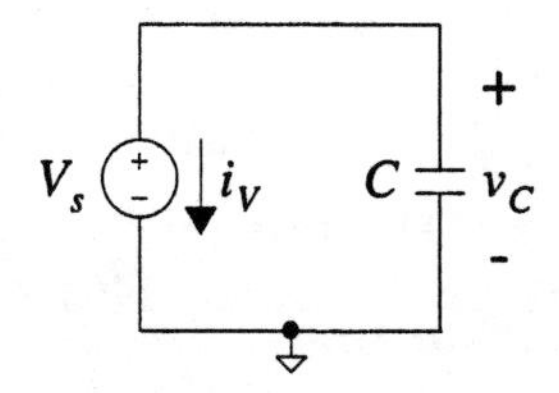

Figure 5.10 Capacitor and voltage source loop.

The voltage v_C in Figure 5.9 violates equation (5.2.1) because v_C is not a state variable. The state equation assumptions made in Section 5.1 assume that the state variables are independent. When there are loops of voltage sources and capacitors, the capacitor voltages are not independent and one of the capacitors must be omitted as a state variable. A similar argument applies to cutsets comprised solely of independent current sources and inductors.

If we take the voltage source current as a variable, for the example in Figure 4.28, then we have

$$i_V = -C\frac{dV_s}{dt} \tag{5.5.1}$$

So step function input voltage sources which form loops solely with capacitors give rise to impulse currents, which cause step changes in the capacitor voltages. Such situations do not occur often, but we must be careful to watch for them and handle the potential impulse currents separately. Charge sharing is a classic example of such a situation, in which two capacitances of unequal initial voltages are suddenly switched to be in parallel. An impulse of current must flow to instantly equilibrate the voltages. This topic will be discussed further in our treatment of nonlinear transient analysis and switches in Chapter 10.

Finally, we mention that multi-step methods take into account more than one past time point in computing the present integration approximation. In nonlinear circuits, these methods must be used carefully, since more history is not always useful in predicting the future. In the case of a highly nonlinear element there can be a large change in operating point with a small change in the voltage across it, rendering the multi-step approximation possibly less useful than a one-step approximation.

This chapter developed single-step integration methods and studied their stability in circuits with multiple energy storage elements. The state variable formulation provided a convenient mathematical framework for this discussion. However, the methods themselves are applicable in any equation formulation method, as was discussed earlier in the chapter.

5.6 References

[Kuh65] E. S. Kuh and R. A. Rohrer. The State Variable Approach to Network Analysis. *Proceedings of the IEEE*, vol. 53, pp. 672-686, July 1965.

[Nagel75] L. W. Nagel. *SPICE2, A Computer Program to Simulate Semiconductor Circuits*. Technical Report ERL-M520, UC-Berkeley, May 1975.

[Strang86] G. Strang. *An Introduction to Applied Mathematics*. Wellesley-Cambridge Press, Wellesley, MA, 1986.

Chapter 6

Frequency Domain Analysis and Moment-Matching Methods

So far we have covered in detail the techniques for evaluating the time domain responses of lumped, linear, time-invariant circuits. In this chapter we consider various approaches for analyzing the same class of circuits in the frequency domain in terms of magnitude and phase, and poles and zeros. In addition, dominant pole analysis, based upon moment matching, is covered in detail. Dominant pole approximations have been widely used for analog circuit design, digital circuit delay modeling, and RLC interconnect analysis. Dominant pole approximations capture the salient features of linear circuit behavior without resorting to expensive analysis by "exact methods."

6.1 Small Signal ac Analysis

As we mentioned in Chapter 1, a straightforward approach to small signal ac analysis is to first obtain the dc bias point for the nonlinear circuit, and then replace all nonlinear elements by their linearized equivalents at the appropriate bias points. This is most conveniently explained in terms of the circuit example in Figure 6.1.

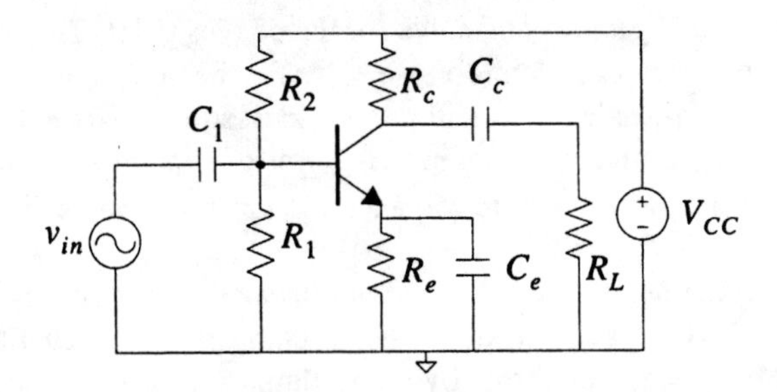

Figure 6.1 Simple BJT amplifier.

This familiar common emitter amplifier circuit is biased using resistors R_1 and R_2 so that the BJT operates in the forward active mode. To simplify the example, we use the ideal Ebers-Moll model to represent the dc characteristics of this transistor as it operates in the forward active mode. For this mode we need not consider the base-collector diode, as shown in Figure 6.2, since it is reverse biased, or turned off. Although we will discuss device modeling in more detail in later chapters, we should point out that we have included a zero-valued voltage source in series with the diode which acts as an ammeter to measure the controlling current I_E. The complete BJT model also includes nonlinear device capacitances C_{BE}, C_{BC}, and C_{JS} (more on this topic later).

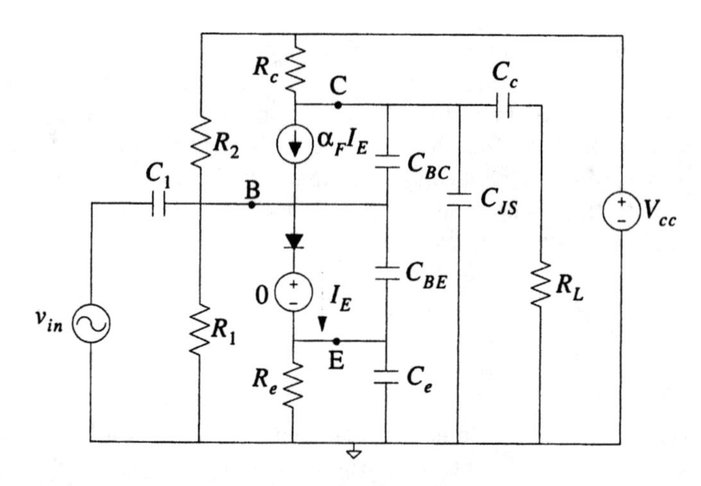

Figure 6.2 Amplifier circuit using the ideal Ebers-Moll model for the BJT.

To begin, we open all capacitors and solve for the dc bias point. The base-emitter diode is the only nonlinear dc element, therefore, we solve for the dc node voltages in terms of a one-dimensional Newton-Raphson algorithm as discussed in Chapter 1. Graphically, the diode operating point is computed by iterating in terms of straight-line tangent approximations to its $i-v$ characteristics. At convergence the tangent straight line passes through the bias point as shown in Figure 6.3.

Since the straight-line tangent for the final nonlinear iteration passes through the bias point, it is the small signal model for the diode. To complete the ac analysis circuit model we replace the energy storage elements by their complex-valued immittance equivalents shown in Figure 6.4. (Immittance is a combined term for impedance and admittance.) Therefore, the circuit model for ac analysis uses the final Newton-Raphson companion model (see Figure 6.3) and is shown in Figure 6.5.

Note that our example does not contain any inductors or nonlinear capacitors. If it did, inductors would be represented by their complex impedance values and nonlinear capacitors by their complex admittance at the appropriate dc bias point.

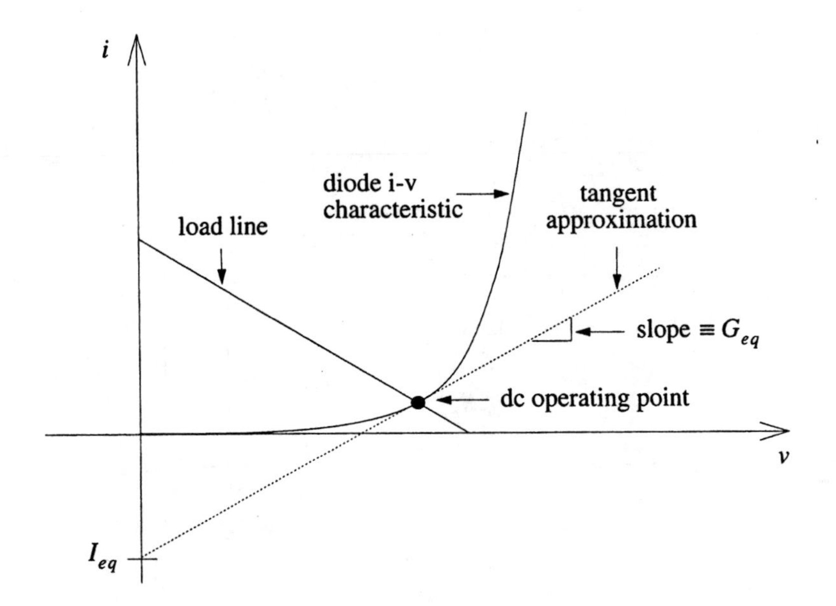

Figure 6.3 Newton-Raphson straight-line tangent at convergence.

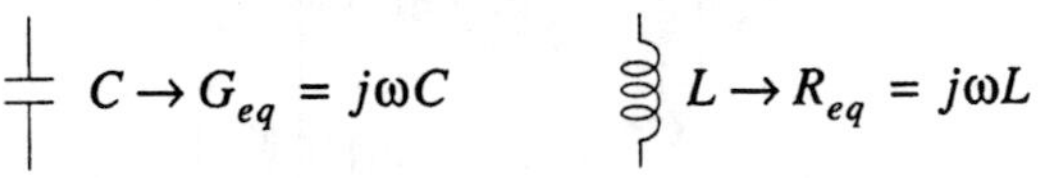

$$C \rightarrow G_{eq} = j\omega C \qquad L \rightarrow R_{eq} = j\omega L$$

Figure 6.4 The complex immittance equivalents of small signal C and L models.

Once the energy storage elements are replaced by their frequency domain models, it is merely a matter of solving the resulting complex linear circuit for various frequency values of interest. Any formulation of the circuit equations can be employed. If Modified Nodal Analysis (MNA) were used, and there were inductances in the circuit, they would be treated as complex impedances to provide a dc compatible set of equations as $\omega \rightarrow 0$.

Inserting an impedance value in the MNA matrix requires an auxiliary equation just as for the case of independent voltage sources. The inductor current is added as a variable, I_{kl} as shown in Figure 6.6, and the auxiliary equation for this inductor from node k to node l is:

$$v_k - v_l - j\omega L i_{kl} = 0 \tag{6.1.1}$$

The corresponding stamp for the inductor in Figure 6.6 is:

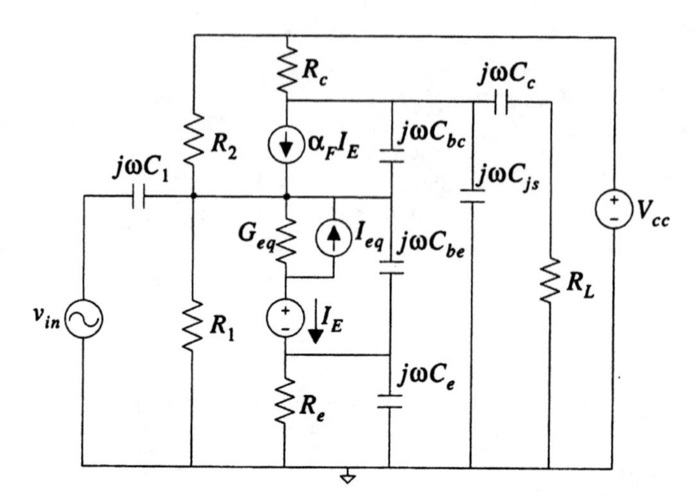

Figure 6.5 Small signal ac circuit model for the common emitter amplifier.

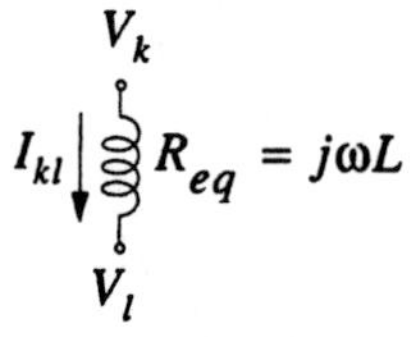

V_k

I_{kl} $R_{eq} = j\omega L$

V_l

Figure 6.6 An inductor connected between nodes k and l.

$$
\begin{array}{c}
k \\ l \\ \\ \\
\end{array}
\left[\begin{array}{ccc|c}
 & & & +1 \\
 & & & -1 \\
 & & & \\
\hline
+1 & & -1 & -j\omega L
\end{array}\right]
\left[\begin{array}{c} \\ \\ \\ \hline I_{kl} \end{array}\right]
=
\left[\begin{array}{c} \\ \\ \\ \hline 0 \end{array}\right]
\qquad \begin{array}{cc} k & l \end{array}
\tag{6.1.2}
$$

So, if we wanted a Bode plot with 10 points per decade over 10 decades of frequency, we would need to solve 100 complex-valued linear circuits involving 100 LU Factorizations, and 100 Forward and Back Substitutions. Such an approach, while straightforward, is computationally expensive. Moreover, it may not provide as much design-oriented insight as a pole/zero analysis.

6.2 Pole/Zero Analysis

Given the one-sided Laplace Transform

$$F(s) \equiv L\{f(t)\} \equiv \int_{o}^{\infty} e^{-st} f(t)\,dt \tag{6.2.1}$$

we can apply it directly to lumped, linear, time-invariant capacitance and inductance elements to obtain the complex frequency domain models shown in Figure 6.7 and Figure 6.8, respectively.

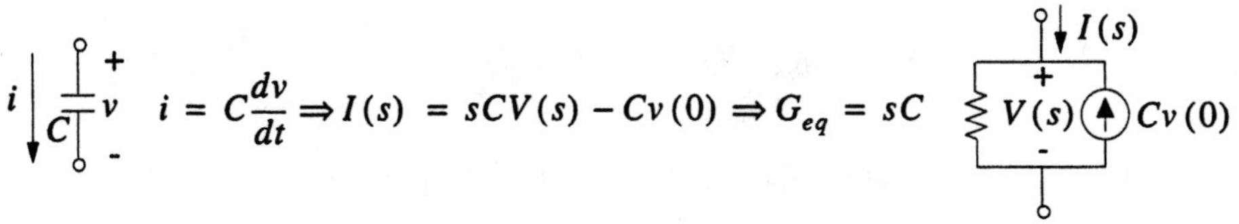

$$i = C\frac{dv}{dt} \Rightarrow I(s) = sCV(s) - Cv(0) \Rightarrow G_{eq} = sC$$

Figure 6.7 The frequency domain model for a capacitor, including the initial condition.

$$v = L\frac{di}{dt} \Rightarrow V(s) = sLI(s) - Li(0) \Rightarrow R_{eq} = sL$$

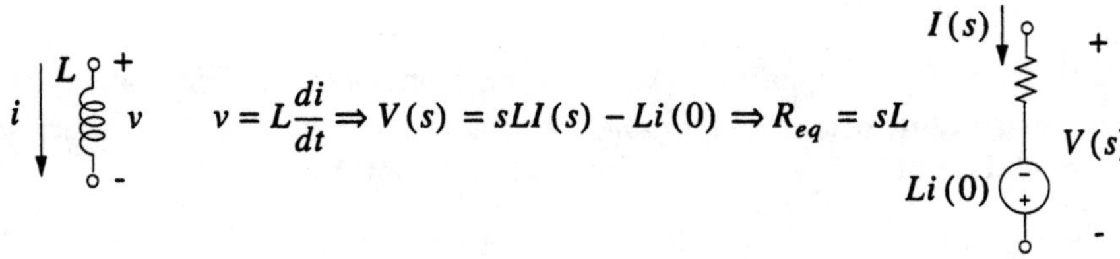

Figure 6.8 The frequency domain model for an inductor, including the initial condition.

For the purposes of frequency domain analysis, we can omit the initial conditions while calculating the circuit poles. The Laplace Transform then provides a linear model for the frequency dependent components in terms of the complex frequency s. Replacing the capacitors and inductors with frequency domain models, the resulting linear circuit may be solved algebraically to obtain the frequency domain response.

In general, we seek the circuit function $H(s)$, shown in Figure 6.9, which is often expressed as a ratio of polynomials in s:

$$H(s) = \frac{Q(s)}{P(s)} \tag{6.2.2}$$

The roots of the denominator polynomial

$$P(s) = (s-p_1)(s-p_2)\dots(s-p_n) = 0 \tag{6.2.3}$$

are the poles (or natural frequencies) of the circuit. Both the time and frequency domain responses are obtainable from the poles.

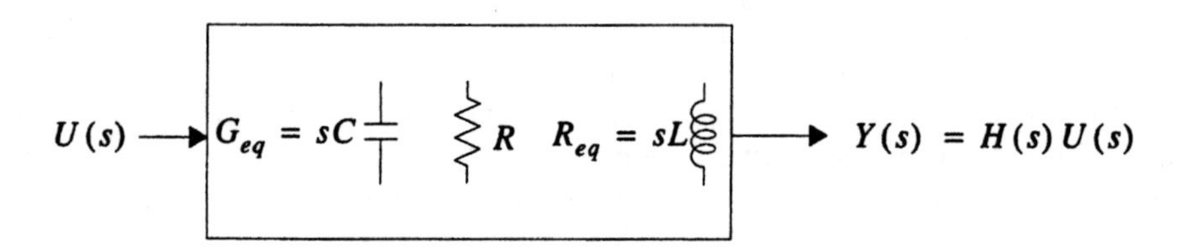

Figure 6.9 The circuit function $H(s)$.

An easy way to deal with the circuit function $H(s)$ is in terms of its partial fraction expansion. For the case of n distinct poles

$$H(s) = \sum_{l=1}^{n} \frac{k_l}{s-p_l} \tag{6.2.4}$$

where k_l is the residue that corresponds to pole p_l. From these, and knowing the input function, we can easily obtain the zero-state transient response from the inverse Laplace Transform. And the frequency response is even easier to come by:

$$H(j\omega) = \sum_{l=1}^{n} \frac{k_l}{j\omega-p_l} \tag{6.2.5}$$

So, if we could determine $H(s)$ and its poles efficiently, then to obtain a frequency plot would be a simple and efficient post-processing task. We'll see that it is difficult to obtain $H(s)$ and its poles efficiently in general, so instead we attempt to find effective approximations to these quantities.

6.3 *Laplace Transform of the State Equations*

The state equations for a circuit are the most convenient form for discussing the circuit function $H(s)$ and its poles. From Chapter 5 we expressed the standard form of the state equations as

$$\dot{\boldsymbol{x}} = \boldsymbol{A}\boldsymbol{x} + \boldsymbol{B}\boldsymbol{u} \tag{6.3.1}$$

where $\boldsymbol{x}$ is the $n \times 1$ state vector, and $\boldsymbol{u}$ the $m \times 1$ input vector. Upon applying the Laplace Transform to these matrix equations we have

$$s\boldsymbol{X}(s) - \boldsymbol{x}(0) = \boldsymbol{A}\boldsymbol{X}(s) + \boldsymbol{B}\boldsymbol{U}(s) \tag{6.3.2}$$

or

$$(s\boldsymbol{1} - \boldsymbol{A})\boldsymbol{X}(s) = \boldsymbol{x}(0) + \boldsymbol{B}\boldsymbol{U}(s) \tag{6.3.3}$$

and

$$\boldsymbol{X}(s) = (s\boldsymbol{1} - \boldsymbol{A})^{-1}[\boldsymbol{x}(0) + \boldsymbol{B}\boldsymbol{U}(s)] \tag{6.3.4}$$

Neglecting the initial conditions, that is, with $\boldsymbol{x}(0) = \boldsymbol{0}$, we have

$$\boldsymbol{X}(s) = (s\boldsymbol{1} - \boldsymbol{A})^{-1}\boldsymbol{B}\boldsymbol{U}(s) \tag{6.3.5}$$

From equation (6.3.5) and Figure 6.9 we recognize that if we treat the set of state variables as the outputs, the circuit function is the matrix

$$\boldsymbol{H}(s) = (s\boldsymbol{1} - \boldsymbol{A})^{-1}\boldsymbol{B} \tag{6.3.6}$$

We recognize that the poles of $\boldsymbol{H}(s)$ are the roots of the characteristic polynomial

$$P(s) = det(s\boldsymbol{1} - \boldsymbol{A}) = 0 \tag{6.3.7}$$

or the eigenvalues of the matrix $\boldsymbol{A}$.

There are means available for finding the eigenvalues of a matrix, and some of them have been translated into operations that can be performed directly on a circuit without actually forming the state equations. But even if we don't have to formulate the state equations, solving for all of the eigenvalues of $\boldsymbol{A}$ by direct or indirect methods is inefficient for large problems.

Circuit simulators which provide pole/zero analysis typically use Muller's root finding algorithm [Muller58]. Using the complex-frequency dependent immittances (admittances or impedances) in Figure 6.9, the equations formulated via Modified Nodal Analysis are:

$$\boldsymbol{Y}(s)\,\boldsymbol{V}(s) = \boldsymbol{J}(s) \tag{6.3.8}$$

where $\boldsymbol{Y}(s)$ is the complex-valued nodal admittance matrix, $\boldsymbol{V}(s)$ is the matrix of node voltages and $\boldsymbol{J}(s)$ the vector of input stimuli.

We know from Cramer's Rule that the response at node i can be obtained from

$$V_i = \frac{det\,\boldsymbol{T}}{det\,\boldsymbol{Y}} \tag{6.3.9}$$

where

$$\boldsymbol{T} = \begin{bmatrix} Y_{1,1} \cdots & Y_{1,i-1} & J_1 & Y_{1,i+1} \cdots & Y_{1,n} \\ Y_{2,1} \cdots & Y_{2,i-1} & J_2 & Y_{2,i+1} \cdots & Y_{2,n} \\ \vdots & \vdots & \vdots & \vdots & \vdots \\ Y_{n,1} \cdots & Y_{n,i-1} & J_n & Y_{n,i+1} \cdots & Y_{n,n} \end{bmatrix} \tag{6.3.10}$$

It is apparent from (6.3.9) that the roots of $|\boldsymbol{Y}(s)|$ must be the poles for the circuit response functions. So for a value $s = p_i$, where p_i is a circuit pole, $|\boldsymbol{Y}(s)| = 0$ and therefore $\boldsymbol{Y}(s = p_i)$ is singular.

Muller's algorithm iteratively searches for points in the s plane where $|\boldsymbol{Y}(s)|$ is singular. Starting with three points in the s plane to evaluate $|\boldsymbol{Y}(s)|$, an interpolating polynomial is formed to search for roots of $|\boldsymbol{Y}(s)|$. The search is continued iteratively until $|\boldsymbol{Y}(s)|$ is sufficiently small. To determine if a matrix is singular at a point an LU factorization is attempted. When the determinant of $\boldsymbol{L}$ becomes sufficiently small, it is assumed that the point represents a pole.

Obviously, determining all of the circuit poles can be inefficient. Especially since some of them make an insignificant contribution to the circuit performance, what we may seek instead is an efficient means to obtain those few "dominant poles" that adequately characterize circuit behavior. One of the most effective procedures for approximating a set of dominant poles (and zeros) is *moment matching*.

6.4 Moments of the Impulse Response and Linear Delay Estimation

To motivate our subsequent studies we digress briefly to consider the qualitative behavior of a circuit in the time domain. First, consider the ideal delay element shown in Figure 6.10. The Laplace Transform of the input function $f(t)$ is

$$F(s) = L\{f(t)\} = \int_0^{\infty} f(t)\, e^{-st}\, dt \tag{6.4.1}$$

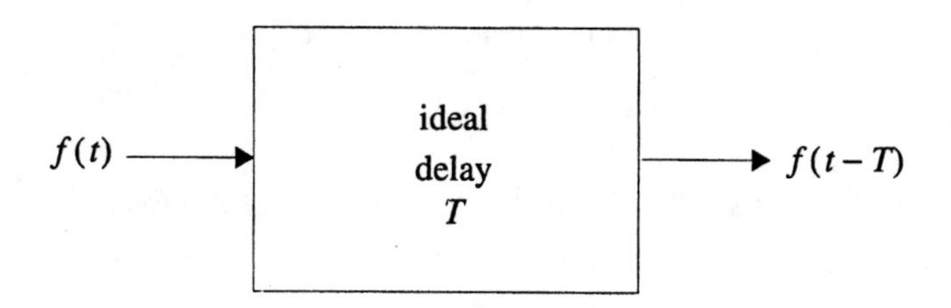

Figure 6.10 Ideal delay function.

The Laplace Transform of the output is

$$\begin{aligned} L\{f(t-T)\} &= \int_0^{\infty} f(t-T)\, e^{-st}\, dt \\ &= \int_{-T}^{\infty} f(\tau)\, e^{-s(\tau+T)}\, d\tau \\ &= e^{-sT} \int_0^{\infty} f(\tau)\, e^{-s\tau}\, d\tau \qquad [f(\tau) = 0 \text{ for } \tau < 0] \\ &= e^{-sT} F(s) \end{aligned} \tag{6.4.2}$$

It is apparent from (6.4.2) that an ideal delay of time T is equivalent in the frequency domain to multiplication by e^{-sT}. An ideal delay element is characterized in the frequency domain by the system transfer function

$$H(s) = e^{-sT} \tag{6.4.3}$$

Given the transfer function for an ideal delay system such as the one in Figure 6.10, the delay can be calculated by differentiating the system function and evaluating it at $s = 0$:

$$H'(s) = \frac{d}{ds}[H(s)] = -Te^{-sT} \tag{6.4.4}$$

Therefore,

$$H'(s)|_{s=0} = -T \tag{6.4.5}$$

From this observation, Elmore [Elmore48] postulated that the delay for a circuit characterized by a transfer function $H(s)$ could be approximated by:

$$T_d \approx -H'(0) \tag{6.4.6}$$

Elmore also showed that this value, T_d, is the first *moment* of the impulse response. This can be shown by first starting with the definition of the Laplace Transform of $h(t)$

$$H(s) = \int_0^{\infty} h(t) e^{-st} dt \tag{6.4.7}$$

and expanding e^{-st} about $s = 0$ to yield

$$\begin{aligned} H(s) &= \int_0^{\infty} h(t)\left[1 - st + \frac{1}{2}s^2t^2 - \frac{1}{6}s^3t^3 + \dots\right] dt \\ &= \sum_{k=0}^{\infty} \frac{(-1)^k}{k!} s^k \int_0^{\infty} t^k h(t)\, dt \end{aligned} \tag{6.4.8}$$

Taking the derivative of (6.4.8) with respect to s and evaluating at $s = 0$, results in

$$H'(0) = -\int_0^{\infty} th(t)\, dt \tag{6.4.9}$$

We define the *first moment* of a function $h(t)$ as

$$m_1 \equiv -\int_0^{\infty} th(t)\, dt \tag{6.4.10}$$

Therefore

$$H'(0) = m_1 \tag{6.4.11}$$

(Our definition of the moments is slightly different from the classical definition; see the last paragraph of this section.) The first moment can be thought of as the (negative of) the mean of $h(t)$. That is, if we treat $h(t)$ as a probability density function (since the total area under the unit impulse response for a system with unity dc gain is 1.0), the first moment is the (negative of the) "average time" and the (negative of the) Elmore delay.

In general, we define the qth *moment* of $h(t)$ as

$$m_q \equiv \frac{(-1)^q}{q!} \int_0^\infty t^q h(t)\, dt \tag{6.4.12}$$

Thus from (6.4.8),

$$H(s) = \sum_{k=0}^{\infty} s^k m_k \tag{6.4.13}$$

Expanding $H(s)$ about $s = 0$ directly in the frequency domain,

$$\begin{aligned} H(s) &= H(0) + sH'(0) + \frac{1}{2}s^2 H''(0) + \frac{1}{6}s^3 H'''(0) + \ldots \\ &= \sum_{k=0}^{\infty} \frac{s^k}{k!} H^{(k)}(0) \end{aligned} \tag{6.4.14}$$

where $H^{(k)}(0)$ is the kth derivative of $H(s)$ evaluated at $s = 0$. Comparing coefficients of s^k between (6.4.13) and (6.4.14), we find

$$m_k = \frac{1}{k!} H^{(k)}(0) \tag{6.4.15}$$

Note that our definition of moments from (6.4.12) is slightly different from the classical definition

$$m_k = \int_0^\infty t^q h(t)\, dt \tag{6.4.16}$$

but our definition will simplify notation throughout the rest of the chapter and is consistently used according to our definition in this book.

Elmore used only the first moment of the impulse response to generate a simple delay expression for wideband amplifiers. Over thirty years later, this first moment value was used for estimating logic gate delays in terms of simple RC tree circuit models [Penfield81].

6.5 The Elmore Delay and RC Trees

Consider the CMOS inverter in Figure 6.11 driving a similar inverter through a long length of interconnect. The interconnect is modeled as several lumped RC segments, thereby approximating the actual distributed resistance and capacitance of the conducting path. For efficiency, the driving gate is sometimes modeled as a linear resistor in order to simplify the delay analysis, as shown in Figure 6.12. Notice that the driver is modeled by a voltage source step function with a 600Ω resistance, and the metal interconnect and the inverter load at the end of the path are modeled by the Rs and Cs as shown. One could perform a transient analysis on this circuit to determine the delay, but as we will demonstrate shortly, the first moment of the impulse response, or the Elmore delay, can be calculated very efficiently for this circuit model.

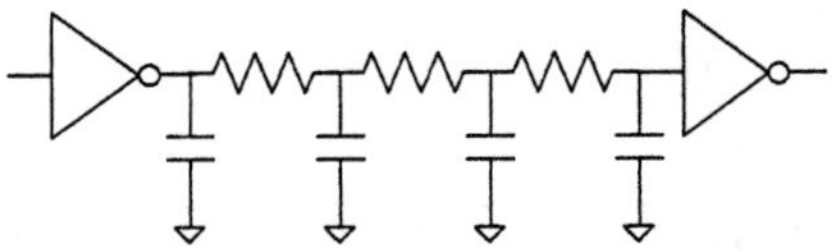

Figure 6.11 A CMOS inverter driving a similar inverter through a long stretch of interconnect.

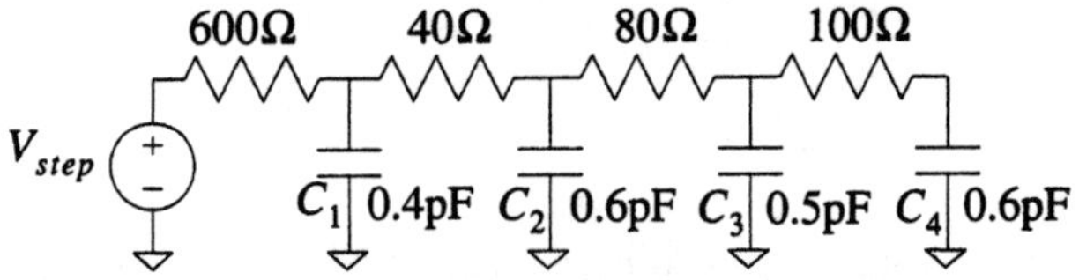

Figure 6.12 The RC circuit model to calculate the delay for the circuit in Figure 6.11.

The impulse response and the step response for the RC circuit in Figure 6.12 are shown in Figure 6.13. Note that the impulse response has been scaled by 10^9 in order to plot it on the same scale as the step response. The mean of this impulse response, or the Elmore delay for this circuit, is 1.476 ns. Notice that the essence of Elmore's approximation is that the 50 percent point of the step response (which is the median point of the unit impulse response) can be approximated by the centroid, or the mean of the impulse response. Of course we would expect this approximation to be accurate when $h(t)$ is symmetric, and we would envision it becoming inaccurate as $h(t)$ becomes asymmetric.

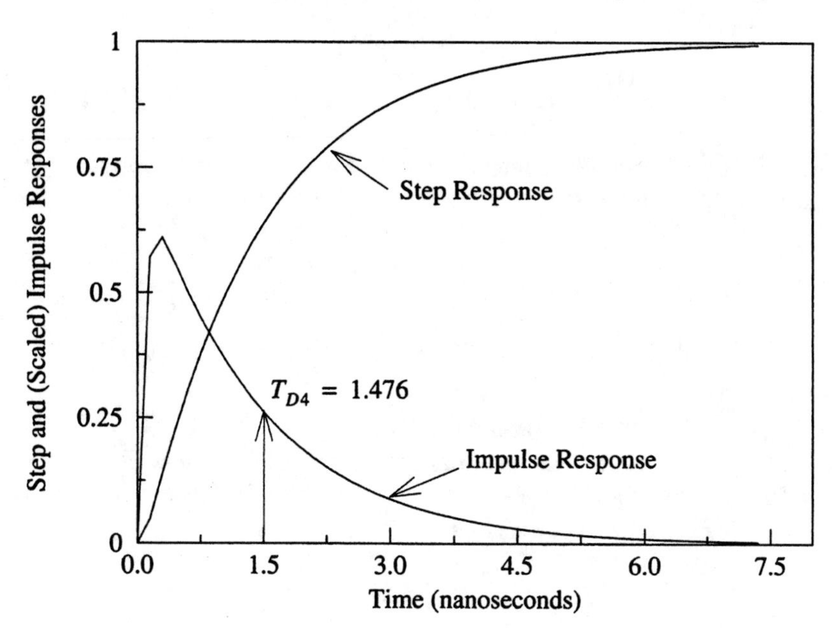

Figure 6.13 The step response and the impulse response (scaled by 10^9) for the voltage at capacitor C_4 in Figure 6.12. The Elmore delay value represents the mean of the impulse response.

The first moment of the impulse response (the Elmore delay), T_d, is used more often as a *dominant time constant approximation* than as a 50 percent delay estimate for RC trees. The relation between moments and circuit time-constants will be studied in detail in subsequent sections.

6.6 Moments of the Impulse Response

It is usually more efficient to analyze linear circuits such as the one in Figure 6.12 in the frequency domain as opposed to the time domain. We can express the transfer function of this circuit (for $V_{in} = \delta(t)$) as

$$H(s) = \frac{V_{out}(s)}{V_{in}(s)} = V_{out}(s) \tag{6.6.1}$$

Let

$$H(s) = \frac{1 + a_1 s + a_2 s^2 + \dots + a_n s^n}{1 + b_1 s + b_2 s^2 + \dots + b_m s^m} \quad \textbf{(6.6.2)}$$

where $m > n$. For a 4th order RC circuit, $m = 4$ and $n \leq 3$. We can factor the numerator and the denominator of (6.6.2) to display the poles and zeros explicitly:

$$H(s) = \frac{a_n (s - z_1)(s - z_2) \dots (s - z_n)}{b_m (s - p_1)(s - p_2) \dots (s - p_m)} = K \frac{(1 - \frac{s}{z_1})(1 - \frac{s}{z_2}) \dots (1 - \frac{s}{z_n})}{(1 - \frac{s}{p_1})(1 - \frac{s}{p_2}) \dots (1 - \frac{s}{p_m})} \quad \textbf{(6.6.3)}$$

where K is the dc gain, which is 1.0 for this RC tree.

Coming up with the rational form in (6.6.2) or all of the poles and zeros in (6.6.3) is difficult for a large circuit. We will instead approximate the transfer function in the complex frequency domain as a series in powers of s:

$$H(s) = m_0 + m_1 s + m_2 s^2 + m_3 s^3 + \dots \quad \textbf{(6.6.4)}$$

From the previous section, we know that the coefficients of the power series terms, the m_j's, are the moments of the impulse response. To consider the relation between these moments and the poles of $H(s)$, we expand (6.6.2) about $s = 0$, or equivalently, divide the denominator into the numerator in (6.6.2), yielding $H(s)$ as an infinite series in powers of s:

$$\begin{aligned} H(s) &= 1 + (a_1 - b_1) s + (a_2 - b_2 - b_1 a_1 + b_1^2) s^2 \\ &\quad + (a_3 - b_3 - a_1 b_2 + 2 b_1 b_2 - a_2 b_1 + a_1 b_1^2 - b_1^3) s^3 + \dots \end{aligned} \quad \textbf{(6.6.5)}$$

Comparing (6.6.4) and (6.6.5) we can recognize the m_j coefficients as a function of the numerator coefficients (a_j's) and the denominator coefficients (b_j's) of the transfer function in (6.6.2). Alternately, for our 4th order circuit, from (6.6.2) and (6.6.4), we have:

$$(1 + b_1 s + b_2 s^2 + b_3 s^3 + b_4 s^4)(m_0 + m_1 s + m_2 s^2 + \dots) = 1 + a_1 s + a_2 s^2 + a_3 s^3 \quad \textbf{(6.6.6)}$$

Collecting the first four powers of s ($s^0 \rightarrow s^3$) in (6.6.6), we can express the numerator polynomial coefficients in terms of the m_j's and the b_j's:

$$
\begin{aligned}
1 &= m_0 \\
a_1 &= m_0 b_1 + m_1 \\
a_2 &= m_0 b_2 + m_1 b_1 + m_2 \\
a_3 &= m_0 b_3 + m_1 b_2 + m_2 b_1 + m_3
\end{aligned}
\tag{6.6.7}
$$

Note that for this RC tree example, the dc gain is 1, hence, $a_0 = 1$. We will show later that in general, a_0 is equal to m_0.

The next four powers of s $(s^4 \to s^7)$ in (6.6.6) express the coefficients of the pole polynomial in terms of the m_j's:

$$
\begin{aligned}
0 &= m_0 b_4 + m_1 b_3 + m_2 b_2 + m_3 b_1 + m_4 \\
0 &= m_1 b_4 + m_2 b_3 + m_3 b_2 + m_4 b_1 + m_5 \\
0 &= m_2 b_4 + m_3 b_3 + m_4 b_2 + m_5 b_1 + m_6 \\
0 &= m_3 b_4 + m_4 b_3 + m_5 b_2 + m_6 b_1 + m_7
\end{aligned}
\tag{6.6.8}
$$

We have shown that if we had the first eight m_j's for this 4th order system (more on how to get these m_j's in the next section) we could uniquely specify the poles and the zeros for this circuit. That is, we can rearrange (6.6.8) as a matrix problem to determine the b_j's:

$$
\begin{bmatrix} m_0 & m_1 & m_2 & m_3 \\ m_1 & m_2 & m_3 & m_4 \\ m_2 & m_3 & m_4 & m_5 \\ m_3 & m_4 & m_5 & m_6 \end{bmatrix}
\begin{bmatrix} b_4 \\ b_3 \\ b_2 \\ b_1 \end{bmatrix}
= -
\begin{bmatrix} m_4 \\ m_5 \\ m_6 \\ m_7 \end{bmatrix}
\tag{6.6.9}
$$

Once the b_j's are obtained we can calculate the four poles for this circuit by finding the roots of a 4th order polynomial. In RC circuits, all the poles lie on the negative real axis. For this example, the four poles are: -0.72296572, -15.038701, -55.767850, and -119.30382, all in units of -10^9 Hertz. These values are identical to those obtained by solving for the eigenvalues of the state matrix A in (6.3.7). Once we know the moments and the b_j's, (6.6.7) can be used to find the a_j's. Then by finding the roots of a 3rd order polynomial, we can find the zeros.

Notice that in general, for a qth order circuit, the first $2q$ moments can uniquely specify the circuit poles and zeros. But all of this is predicated on the calculation of the moment values, which is addressed in the next section.

6.7 Efficiently Computing Moments for RC Trees

In the previous section we showed how the first eight moments for our sample circuit could be used to calculate the poles and zeros for a 4th order circuit. This approach is useful only if the moments are easy to compute. In this section we demonstrate the ease with which these coefficients can be obtained for our lumped linear RC tree example. Later in this chapter we will demonstrate the ease of moment calculations for generalized RLC circuits.

Returning to our example in Figure 6.12, the impulse response in the complex frequency domain can be analyzed in terms of the circuit in Figure 6.14, where capacitors have been replaced by their complex admittances. Let us assume that each of the capacitor voltages (which in this circuit are also the node voltages) is expressed in terms of an infinite series in powers of s as shown in the figure. The superscripts for the m_j's in Figure 6.14 denote that all of the m_j's are different from one node to the next.

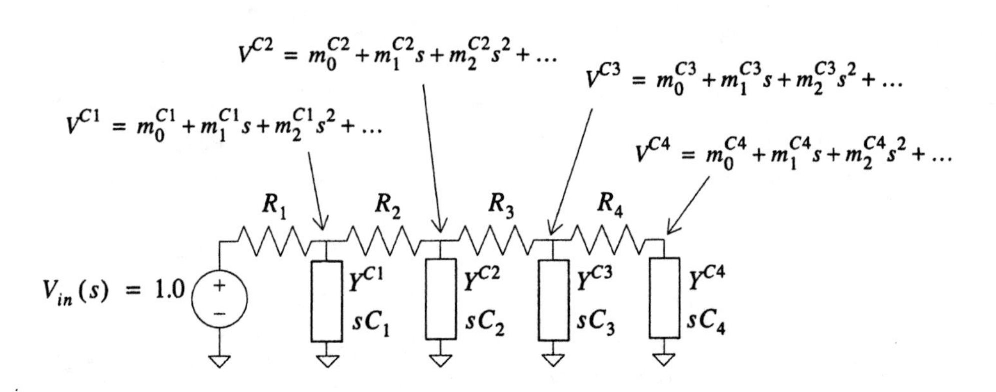

Figure 6.14 The s-domain representation of the RC circuit of Figure 6.12 in terms of complex admittances.

Expressing the capacitor voltages in this way and knowing the capacitor admittances, we can write similar expressions for the capacitor currents. Moreover, knowing the capacitor currents, we can replace the complex admittances by current sources using the Substitution Theorem, as shown in Figure 6.15. The m_j^k terms are the only unknowns in Figure 6.15.

Referring to Figure 6.15, we can solve for the m_0's for all of the capacitor voltages by setting $s = 0$. Since there are no constant terms (s^0 terms) in the capacitor currents (they are open for $s = 0$), we set the current sources in Figure 6.15 to zero and solve for the m_0's using the dc equivalent circuit in Figure 6.16. For this RC tree, the m_0's are all equal to 1.0. Note that this procedure for replacing capacitors by zero valued current sources to calculate the m_0's holds for all circuit topologies. When there are inductors in the circuit, they are replaced by zero valued voltage sources when calculating the m_0 terms for their current responses. More on RLC circuits in Section 6.10.

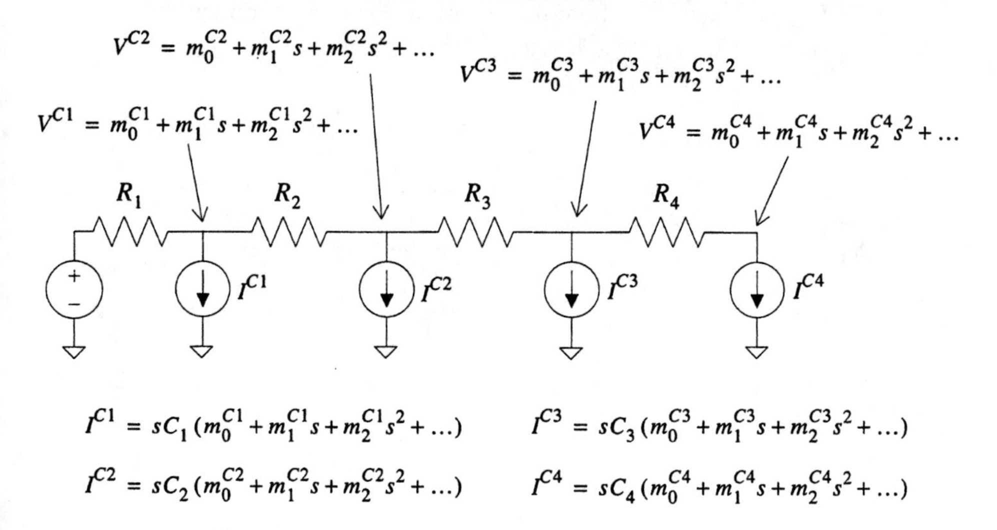

$V^{C2} = m_0^{C2} + m_1^{C2}s + m_2^{C2}s^2 + \dots$
$V^{C3} = m_0^{C3} + m_1^{C3}s + m_2^{C3}s^2 + \dots$
$V^{C1} = m_0^{C1} + m_1^{C1}s + m_2^{C1}s^2 + \dots$
$V^{C4} = m_0^{C4} + m_1^{C4}s + m_2^{C4}s^2 + \dots$
R_1
R_2
R_3
R_4
I^{C1}
I^{C2}
I^{C3}
I^{C4}
$I^{C1} = sC_1(m_0^{C1} + m_1^{C1}s + m_2^{C1}s^2 + \dots)$
$I^{C3} = sC_3(m_0^{C3} + m_1^{C3}s + m_2^{C3}s^2 + \dots)$
$I^{C2} = sC_2(m_0^{C2} + m_1^{C2}s + m_2^{C2}s^2 + \dots)$
$I^{C4} = sC_4(m_0^{C4} + m_1^{C4}s + m_2^{C4}s^2 + \dots)$

Figure 6.15 A circuit equivalent to Figure 6.12 assuming the node voltages solutions of the form shown.

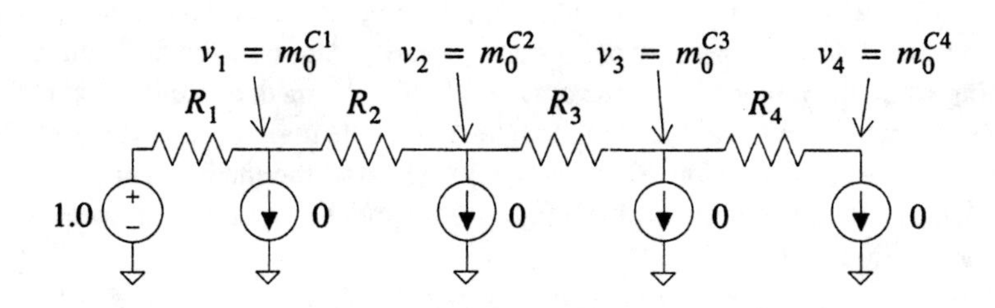

Figure 6.16 The dc equivalent circuit used to calculate the m_0's for all of the capacitor voltages.

Referring back to Figure 6.15, we now solve for the s^1 coefficients, i.e., for the m_1^k's. The s^1 terms in the current sources have m_0 coefficients, which are now known. Therefore, we can evaluate the m_1's of the voltage responses by setting each of the respective current sources equal to $C_k m_0^k$, and solving for the node voltages, which are the m_1^k's. The voltage input is a constant, so it does not affect the calculation of any of the terms other than the m_0's. Subsequent moments are calculated from Figure 6.15 following the same recursion. All of these moments are calculated from a dc equivalent circuit, as shown in Figure 6.17. To generate a complete transfer function for this 4th order circuit we would calculate the first eight moments from the circuit in Figure 6.17 recursively. Once we know the moments, we know how to find the poles and zeros of $H(s)$.

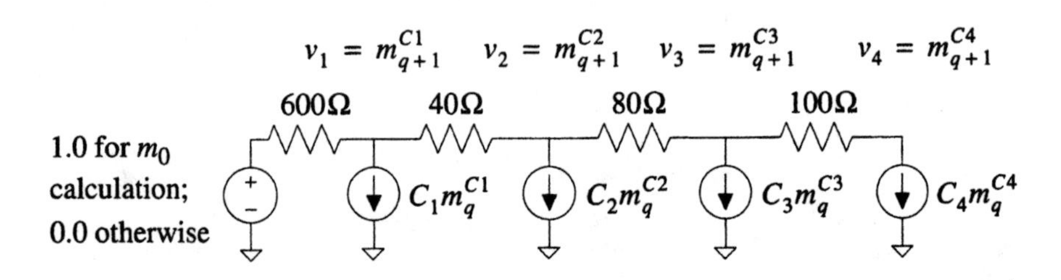

Figure 6.17 The dc equivalent circuit used to calculate the moments for the RC circuit in Figure 6.12.

6.8 Dominant Pole Approximations

We have seen that moments can be computed by recursively solving a simple dc circuit. It is this ease with which moments are calculated that makes them so useful. Elmore was attempting to model the 50 percent point delay of unity gain transfer function circuits by determining when the impulse response area was half consumed. Since the first moment of the impulse response is so readily calculated, he treated the non-negative impulse response as a probability density function (PDF) and approximated the median point by the mean value. However, the first moment of the impulse response is more often used as a dominant pole approximation.

Consider once again Elmore's unit step response delay approximation, T_d:

$$T_d = \int_0^\infty th(t)\,dt \qquad \text{when} \qquad m_0 = \int_0^\infty h(t)\,dt = 1.0 \tag{6.8.1}$$

From the transfer function coefficients in (6.6.5), it follows that the Elmore delay is:

$$T_d = -m_1 = b_1 - a_1 \tag{6.8.2}$$

when $m_0 = 1.0$.

From equation (6.6.3) we observe that terms b_1 and a_1 are the sum of the reciprocal poles (circuit time constants) and the sum of the reciprocal zeros respectively:

$$b_1 = -\sum_{j=1}^{m} \frac{1}{p_j} \qquad a_1 = -\sum_{j=1}^{n} \frac{1}{z_j} \tag{6.8.3}$$

Therefore, returning to (6.8.2), if there are no low frequency zeros (all the z_j's are large), the numerator coefficients, including a_1, are small and

$$T_d \approx b_1 \tag{6.8.4}$$

Further, if one of the time constants (or poles) is dominant:

$$\frac{1}{p_1} \gg \frac{1}{p_j} \qquad \text{for} \qquad j = 2, 3, \ldots, m \tag{6.8.5}$$

then

$$T_d \approx -\frac{1}{p_1} \tag{6.8.6}$$

A single RC ladder circuit such as the one in Figure 6.12 has no finite zeros for the response at C_4 (the end of the ladder). Therefore, we would expect that the Elmore delay value is a reasonable dominant time constant approximation at this node:

$$T_d = 1.476 \text{ ns} \tag{6.8.7}$$

The actual first time constant is

$$\tau_1 = -(p_1)^{-1} = 1.383 \text{ ns} \tag{6.8.8}$$

Assuming a single pole response of the form

$$v(t) = 1 + ke^{pt} \tag{6.8.9}$$

we fit the first order model as a dominant pole approximation by setting $p = -(T_d)^{-1}$ and $k = -1.0$ in (6.8.9). The waveform approximation is compared with the exact 4-pole response in Figure 6.18.

From Figure 6.18 we can see that the dominant pole approximation is reasonable, although slightly optimistic near $t = 0$ and then pessimistic for large values of t. Other nodes in the RC circuit of Figure 6.12 do have finite zeros. Thus we would expect less

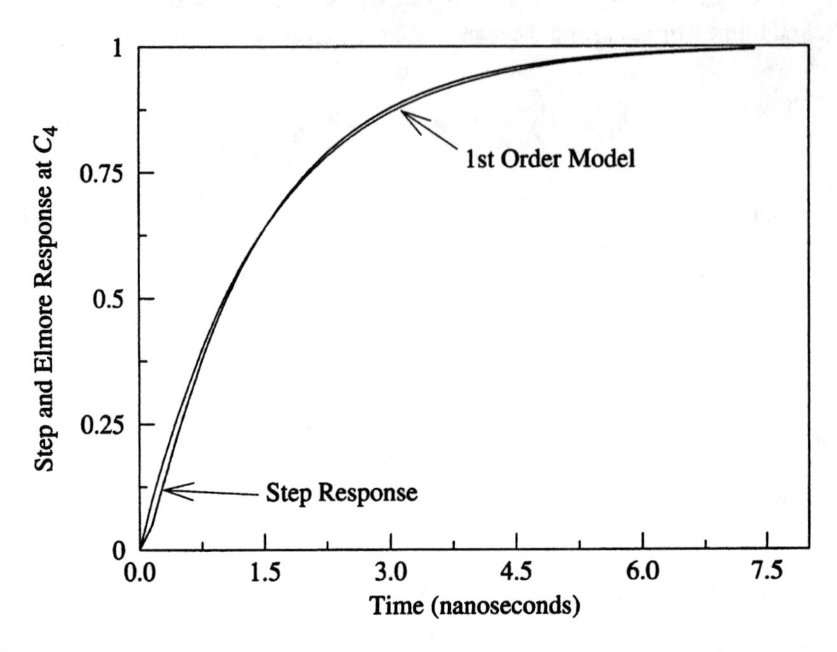

Figure 6.18 Comparison of the step response and the first order response at C_4 in the circuit of Figure 6.12 using the Elmore value as a dominant time-constant approximation.

accuracy in their dominant pole approximations, since T_d varies for different node voltages.

At C_1, the Elmore delay value is 1.260 ns. Using this value in equation (6.8.9) results in the first order approximation shown in Figure 6.19. Notice that the approximation is now pessimistic near $t = 0$ and becomes optimistic as t becomes larger. This change in sign of the error is due to the low frequency zeros for the voltage response at capacitor C_1.

It is difficult to know when a single pole dominates the low frequency behavior of a circuit. Even for these simple RC tree circuits, when the model deviates from a simple RC ladder there can be a large number of low frequency zeros at any of the nodes. For this reason Penfield and Rubenstein established bounds for the step response delay of this important class of RC circuits [Penfield81]. When there are dominant zeros or several dominant poles, or if we are dealing with more general RLC circuits, then higher order moments of $h(t)$ are important, too.

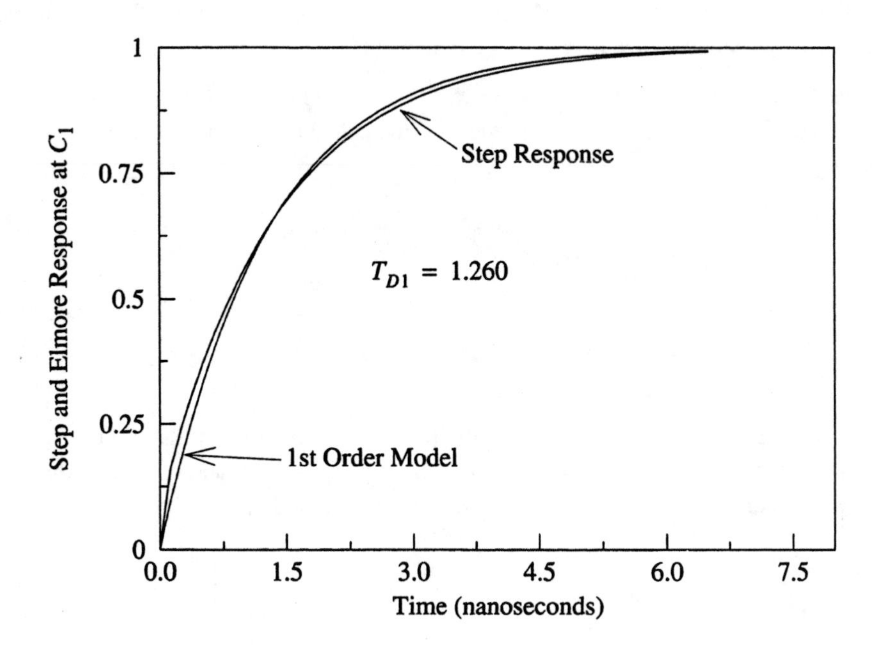

Figure 6.19 Comparison of the step response and the first order response at C_1 in the circuit of Figure 6.12 using the Elmore value as a dominant time-constant approximation.

6.9 Dominant Poles via Moment Matching

The first order approximations shown in the previous section are actually forms of moment matching. That is, even though this 4th order circuit is described exactly by the first eight moments, we can derive a unique first order approximation by matching the first two moments. To clarify, consider once again the RC tree example in Figure 6.12. We can, by inspection of this simple circuit, calculate the first two moments for the voltage response at C_4:

$$m_0^{C4} = 1.000$$

$$m_1^{C4} = -1.476 \tag{6.9.1}$$

Note that we have scaled the Rs to be in kΩ and the Cs to be pF so that the units of time would be nanoseconds when calculating the moments.

A first order approximation, $h(t) = \hat{k}_1 e^{\hat{p}_1 t}$, is obtained by matching the first two moments of the approximate model to those of the actual circuit. In the time domain, the moments of $h(t)$ are:

$$m_0 = \int_0^\infty h(t)\,dt = \int_0^\infty \hat{k}_1 e^{\hat{p}_1 t}dt = -\frac{\hat{k}_1}{\hat{p}_1}$$
$$m_1 = -\int_0^\infty th(t)\,dt = -\int_0^\infty t\hat{k}_1 e^{\hat{p}_1 t}dt = -\frac{\hat{k}_1}{\hat{p}_1^2} \quad (6.9.2)$$

This relation between the moments and the poles and residues can also be obtained by expanding the partial fraction expression in (6.2.4) about $s = 0$, and collecting the powers of s. Matching moments in pole-residue form requires solving:

$$\frac{\hat{k}_1}{\hat{p}_1} = -m_0 = -1.000$$
$$\frac{\hat{k}_1}{\hat{p}_1^2} = -m_1 = 1.476 \times 10^{-9} \quad (6.9.3)$$

Evaluating (6.9.3) yields

$$\frac{1}{\hat{p}_1} = \frac{-m_1}{-m_0} = -1.476 \times 10^{-9}$$
$$\hat{k}_1 = \frac{-1.0}{\frac{1}{\hat{p}_1}} = \frac{1}{1.476} \times 10^9 \quad (6.9.4)$$

which when integrated produces a unit step response identical to the dominant pole approximation waveform in Figure 6.18.

The above describes a first order AWE (Asymptotic Waveform Evaluation) approximation [Pillage90]. In general, AWE is a qth order extension of this dominant pole approach in that $2q$ moments are used to generate q^{th} order approximations, where q is less than m (the order of the actual circuit).

The first four moments for our RC tree example are:

$$\begin{aligned} m_0 &= 1.000 \\ m_1 &= -1.476 \\ m_2 &= 2.048 \\ m_3 &= -2.834 \end{aligned} \tag{6.9.5}$$

A second order approximation of the form

$$h(t) = k_1 e^{p_1 t} + k_2 e^{p_2 t} \tag{6.9.6}$$

is characterized by matching the first four moments of (6.9.6) to those in (6.9.5):

$$\begin{aligned} \frac{\hat{k}_1}{\hat{p}_1} + \frac{\hat{k}_2}{\hat{p}_2} &= -1.000 \\ \frac{\hat{k}_1}{\hat{p}_1^2} + \frac{\hat{k}_2}{\hat{p}_2^2} &= 1.476 \\ \frac{\hat{k}_1}{\hat{p}_1^3} + \frac{\hat{k}_2}{\hat{p}_2^3} &= -2.048 \\ \frac{\hat{k}_1}{\hat{p}_1^4} + \frac{\hat{k}_2}{\hat{p}_2^4} &= 2.834 \end{aligned} \tag{6.9.7}$$

Since evaluating (6.9.7) directly requires a nonlinear analysis, we prefer to apply the moment matching formulas from Section 6.6 which permit us to first calculate the coefficients of the polynomial for the poles from (6.6.9) via moment matching. That is, while all of the moment-matching arguments in Section 6.6 were for an mth order approximation for an mth order circuit, they are also the moment matching equations for a qth order approximation. The only difference here is that we are assuming that we have a qth order system, while it is m^{th} order in reality, with $q < m$. Applying (6.6.9),

$$\begin{bmatrix} 1.000 & -1.476 \\ -1.476 & 2.048 \end{bmatrix} \begin{bmatrix} b_2 \\ b_1 \end{bmatrix} = -\begin{bmatrix} 2.048 \\ -2.834 \end{bmatrix} \tag{6.9.8}$$

yields the characteristic polynomial for the approximate poles, the $\hat{p}$'s,

$$b_2 \hat{p}^2 + b_1 \hat{p} + 1 = 0 \tag{6.9.9}$$

from which we obtain the two approximate poles:

$$\hat{p}_1 = -0.7227$$
$$\hat{p}_2 = -15.96 \qquad \textbf{(6.9.10)}$$

Notice that the first approximate pole is now extremely close to the first actual pole (-0.72296572), while the second approximate pole is attempting to model the effects of all the remaining poles. This type of pole convergence is usual with moment matching. We should add that it is important to maintain as much precision as possible when solving this matrix problem, and we use four digits here only to demonstrate the approach. When we use double precision for the moment calculations, the second approximate pole is -13.7791, quite different from our result in (6.9.10) where we used four decimal digits of accuracy. The importance of numerical precision will be covered in detail in Section 6.12.

Once the poles are known, the zeros can be calculated starting with a formula similar to (6.6.7). In this case, however, we prefer to go directly to the time domain response by solving for the residues using the approximate pole values and any two equations from (6.9.7).

This second order transient waveform estimate at C_4 is shown plotted with the exact response in Figure 6.20. We could use six moments to obtain a third order approximation, but that procedure is hardly necessary given the accuracy of the second order estimate. We should point out, however, that one usually has to calculate the next order of approximation in order to test the accuracy. In this case, there is little difference between second and third order, hence the approximations would conclude at this level.

Of course it may not seem that significant that a 4th order system can be accurately approximated by one of second order; however, low orders of approximation ($q = 1$ to $q = 4$) are often the range of necessity for big circuits too. Fortunately, even a circuit with over a hundred thousand state variables (poles) usually has only a handful of dominant poles in terms of which its behavior can be adequately characterized.

Just as it is possible to compute the first moment of the impulse response (the Elmore delay) with linear complexity for RC trees, higher order moments can be calculated for any RLC tree topology with the same complexity [Pillage90, Ratzlaff94]. But moment matching, as implemented in Asymptotic Waveform Evaluation, applies to other circuit topologies, too. Controlled sources can be part of the linear circuit models for applications such as pole/zero analysis. We will show in the next section that calculating the moments for general circuits is a simple extension of the preceding discussion.

To summarize the key points of dominant pole approximations for our RC tree example:

1. $2q$ moments are easily computed by $2q$ recursive dc solutions of a simple circuit.

2. Once the moments are obtained, the coefficients of the approximate q^{th} order denominator polynomial of $H(s)$ can be computed using a linear set of equations such as in (6.6.9).

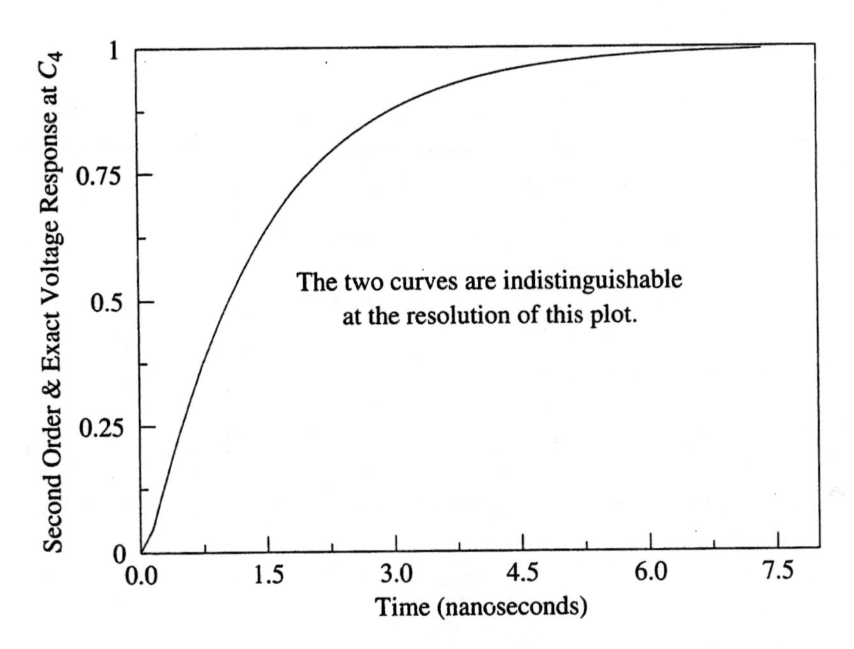

Figure 6.20 Comparison of a second order approximation with the exact response for the voltage at C_4 in Figure 6.12.

3. The poles can be determined by finding the roots of that polynomial.
4. The residues can be obtained from an expression of the form shown in (6.9.7).
5. A time-domain response can be computed as a sum of q exponentials.

6.10 Computing Moments for Generalized Circuits

To consider general RLC circuits, we return to the differential state equations for a circuit driven by a single impulse function, $\delta(t)$:

$$\dot{x} = Ax + B\delta(t) \tag{6.10.1}$$

Note that we are considering the transfer function from a single input to, in this case, all the states. If there are multiple inputs, the response to each can be computed separately and then combined via superposition.
Applying the Laplace Transform to these equations yields

$$sX(s) - x(0) = AX(s) + B \tag{6.10.2}$$

Assuming

$$x(0) = 0 \tag{6.10.3}$$

then

$$\begin{aligned} X(s) &= (s\mathbf{1} - A)^{-1} B \\ &= [(\mathbf{1} - sA^{-1})(-A)]^{-1} B \\ &= -A^{-1}(\mathbf{1} - sA^{-1})^{-1} B \end{aligned} \tag{6.10.4}$$

Expanding $(\mathbf{1} - sA^{-1})^{-1}$ about $s = 0$:

$$X(s) = -A^{-1}(\mathbf{1} + sA^{-1} + s^2A^{-2} + s^3A^{-3} + \dots) B \tag{6.10.5}$$

The vector of coefficients of the powers of s are directly related to the moments of the impulse response at all the state variables:

$$\begin{aligned} m_0 &= -A^{-1}B \\ m_1 &= -A^{-2}B = A^{-1}m_0 \\ m_2 &= -A^{-3}B = A^{-1}m_1 \\ &\vdots \\ m_q &= -A^{-(q+1)}B = A^{-1}m_{q-1} \end{aligned} \tag{6.10.6}$$

We can obtain the moments (for the series coefficients) for any output variable of interest by selecting the appropriate combination of state variable moments. LU factoring A once, we can apply its LU factors recursively to obtain successively higher moments. We should note, however, that we're assuming that the A matrix is nonsingular. Such is not the case when there are cutsets of capacitances or loops of inductances. In such cases, the circuit under consideration does not have a unique dc solution. We can get around this problem by imposing charge and flux conservation constraints as described in Section 2.7. In addition, there is a circuit transformation we will show later in this chapter that also overcomes potential problems with singular A matrices.

We must also point out that the recursive Forward and Back Substitutions are equivalent to raising A^{-1} to higher and higher powers. If A has a large spread in eigenvalues, such computations can be numerically ill-conditioned. This problem can also be addressed on the same basis as the potentially singular A matrix.

The above recursive relation is useful, but we would rather not formulate the state matrix A if we can avoid it. We know that A has the following form:

$$A = \begin{bmatrix} C & 0 \\ 0 & L \end{bmatrix}^{-1} \begin{bmatrix} H_{CC} & H_{CL} \\ H_{LC} & H_{LL} \end{bmatrix} \tag{6.10.7}$$

such that

$$A^{-1} = \begin{bmatrix} H_{CC} & H_{CL} \\ H_{LC} & H_{LL} \end{bmatrix}^{-1} \begin{bmatrix} C & 0 \\ 0 & L \end{bmatrix} \tag{6.10.8}$$

where H is the dc hybrid matrix which we've described in terms of four submatrix components.

Referring to (6.10.6), only the dc hybrid matrix H need actually be inverted (LU factored) to calculate the moments. Inverting H for a circuit such as the one in Figure 6.21 is equivalent to performing the dc analysis in Figure 6.22. All capacitances are replaced by independent current sources and all inductances by independent voltage sources.

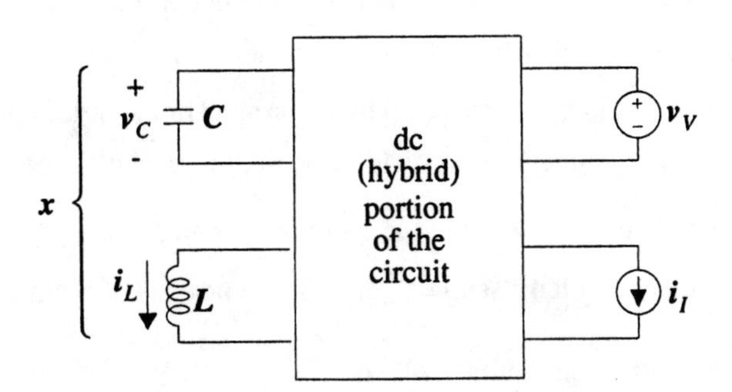

Figure 6.21 Circuit with energy storage elements and independent sources separated from the "dc portion."

To obtain the vector of m_0 moments, we solve

$$m_0 = -A^{-1}B = -[H]^{-1} \begin{bmatrix} C & 0 \\ 0 & L \end{bmatrix} [B] \tag{6.10.9}$$

Equation (6.10.9) is equivalent to setting the independent sources, v_V and i_I, equal to 1 in Figure 6.22 and solving for the open circuit capacitor voltages and short circuit inductor

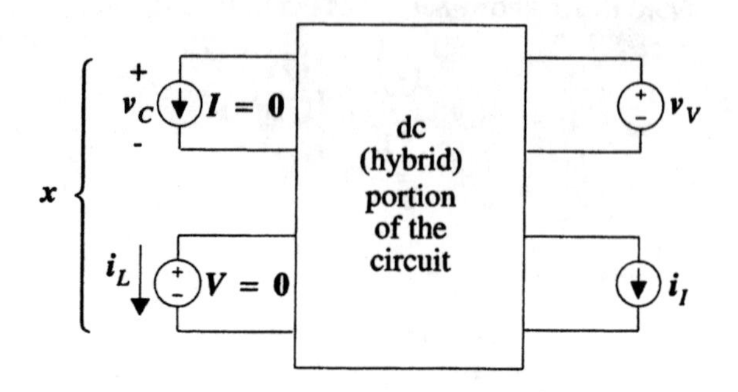

Figure 6.22 The dc analysis equivalent of inverting $\boldsymbol{H}$.

currents. The m_0 moments are the initial conditions, or the open-circuit capacitance voltages and the short-circuit inductance currents in the absence of initial conditions.

From the same dc circuit all subsequent sets of moments can be obtained as follows:

1. Set independent sources equal to zero.
2. Set each capacitor current source equal to the product of the capacitance value C and the corresponding element of the previous moment vector $\boldsymbol{m}_j^C$.
3. Set each inductor voltage source equal to the product of the inductance value L and the corresponding element of the previous moment vector $\boldsymbol{m}_j^L$.
4. Solve for the voltages across the capacitor current sources $(\boldsymbol{m}_{j+1}^C)$ and the currents through the inductor voltage sources $(\boldsymbol{m}_{j+1}^L)$, the next set of moments.

These steps are summarized in Figure 6.23.

Note the similarity between the above recursion and that described for an RC tree in Section 6.7. We need not formulate the state equations to solve the dc circuit defined in Figure 6.22 and Figure 6.23. We can use any dc circuit analysis scheme to solve for the voltages of the capacitor current sources and the currents of the inductor voltage sources. We note that the analysis in Figure 6.23 poses no problem when the original circuit has capacitance loops or inductance cutsets. But there will be a problem when the original circuit has capacitance-current source cutsets or inductance-voltage source loops. These are the situations for which the dc circuit may not have a unique solution -- the $\boldsymbol{A}$ matrix is singular. This problem will be addressed a little later in the chapter.

For the simple RC tree example in Figure 6.12, replacing capacitors with current sources results in the dc equivalent circuit shown in Figure 6.15. The zeroth moment, m_0, is obtained by setting $V_{in} = 1$ (which is equivalent to setting the proper entry in $\boldsymbol{u}$ equal

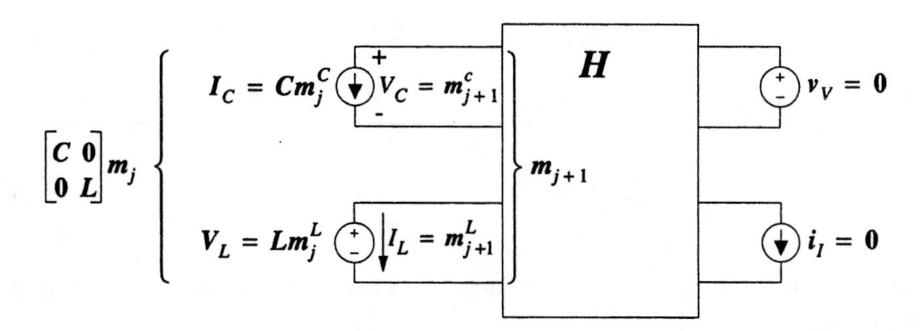

Figure 6.23 The dc circuit used to solve recursively for the sets of moments.

to 1) and solving for the dc voltages when all of the capacitor current sources are equal to zero (which is equivalent to setting $\dot{x} = \mathbf{0}$). Then, using the same dc circuit and solution (the LU factors are unchanged) we set $V_{in} = \mathbf{0}$ (which is equivalent to setting all entries of $\mathbf{u}$ equal to zero) and set the current sources equal to the product of the capacitance and the previous moment for that capacitance voltage (which is equivalent to setting $\dot{x} = \mathbf{m}_0$) and solving for the current source voltages as shown previously for the RC tree example in Figure 6.12 and the dc equivalent circuit in Figure 6.17. The procedure is repeated recursively until we have enough moments.

6.11 Generalized Moment Matching

Now that we have generalized the calculation of moments to all lumped, linear, time-invariant circuits, we would like to do the same for the moment matching procedure and equations. Under the assumption that there are no repeated roots, we can expect the lth state variable to have an impulse response characterized by

$$x_l(t) = \sum_{j=1}^{q} k_j e^{p_j t} \tag{6.11.1}$$

which in the frequency domain corresponds to

$$X_l(s) = \sum_{j=1}^{q} \frac{k_j}{s - p_j} = \frac{a_0 + a_1 s + a_2 s^2 + \ldots + a_{q-1} s^{q-1}}{1 + b_1 s + b_2 s^2 + \ldots + b_m s^q} \tag{6.11.2}$$

Recall that

$$X_l(s) = m_{0,l} + m_{1,l}s + m_{2,l}s^2 + \dots + m_{2q-1,l}s^{2q-1} \tag{6.11.3}$$

We can expand each of the partial fraction terms in (6.11.2) as series in powers of s:

$$\sum_{j=1}^{q} \frac{k_j}{s-p_j} = \sum_{j=1}^{q} \left[-\frac{k_j}{p_j}\left(1 + \frac{s}{p_j} + \frac{s^2}{p_j^2} + \frac{s^3}{p_j^3} + \dots\right)\right] \tag{6.11.4}$$

Then, with the $2q$ unknowns -- q poles and the q corresponding residues -- we arrange $2q$ moment matching equations:

$$\begin{aligned}
-\left(\frac{k_1}{p_1} + \frac{k_2}{p_2} + \dots + \frac{k_q}{p_q}\right) &= m_{0,l} \\
-\left(\frac{k_1}{p_1^2} + \frac{k_2}{p_2^2} + \dots + \frac{k_q}{p_q^2}\right) &= m_{1,l} \\
\vdots \qquad\qquad & \quad \vdots \\
-\left(\frac{k_1}{p_1^{2q}} + \frac{k_2}{p_2^{2q}} + \dots + \frac{k_q}{p_q^{2q}}\right) &= m_{2q-1,l}
\end{aligned} \tag{6.11.5}$$

The subscript l refers to the lth state variable, or the lth component of each moment vector.

If we can solve these $2q$ equations in $2q$ unknowns, we will have a low order approximation which matches the first $2q$ moments of the original circuit function. Note that, contrary to what prevails in the actual system, different state variables for the low order approximating systems may have different approximating poles. Recall that the dominant pole approximations (the Elmore delays) varied for different nodes in the RC tree example in the previous section. Moreover, in the RC tree example, we noticed that different nodes have a different number of low frequency zeros, therefore, we would expect that the accuracy of our approximations would be different for different state variables. If we use a convergence criterion to determine the order of our approximations, the orders for two different nodes may end up being different.

To proceed to the solution of equations (6.11.5) we rewrite them as follows:

$$\begin{bmatrix} 1 & 1 & \cdots & 1 \\ \frac{1}{p_1} & \frac{1}{p_2} & \cdots & \frac{1}{p_q} \\ \vdots & \vdots & & \vdots \\ \frac{1}{p_1^{q-1}} & \frac{1}{p_2^{q-1}} & \cdots & \frac{1}{p_q^{q-1}} \end{bmatrix} \begin{bmatrix} \frac{1}{p_1} & 0 & 0 & \cdots & 0 & 0 \\ 0 & \frac{1}{p_2} & 0 & \cdots & 0 & 0 \\ \vdots & \vdots & \vdots & & \vdots & \vdots \\ 0 & 0 & 0 & \cdots & \frac{1}{p_{q-1}} & 0 \\ 0 & 0 & 0 & \cdots & 0 & \frac{1}{p_q} \end{bmatrix} \begin{bmatrix} k_1 \\ k_2 \\ \vdots \\ k_q \end{bmatrix} = - \begin{bmatrix} m_{0,l} \\ m_{1,l} \\ \vdots \\ m_{q-1,l} \end{bmatrix} \tag{6.11.6}$$

and

$$\begin{bmatrix} 1 & 1 & \cdots & 1 \\ \frac{1}{p_1} & \frac{1}{p_2} & \cdots & \frac{1}{p_q} \\ \vdots & \vdots & & \vdots \\ \frac{1}{p_1^{q-1}} & \frac{1}{p_2^{q-1}} & \cdots & \frac{1}{p_q^{q-1}} \end{bmatrix} \begin{bmatrix} \frac{1}{p_1} & 0 & \cdots & 0 & 0 \\ 0 & \frac{1}{p_2} & 0 \cdots & 0 & 0 \\ \vdots & \vdots & \vdots & \vdots & \vdots \\ 0 & 0 & 0 \cdots & \frac{1}{p_{q-1}} & 0 \\ 0 & 0 & 0 \cdots & 0 & \frac{1}{p_q} \end{bmatrix}^{q+1} \begin{bmatrix} k_1 \\ k_2 \\ \vdots \\ k_q \end{bmatrix} = - \begin{bmatrix} m_{q,l} \\ m_{q+1,l} \\ \vdots \\ m_{2q-1,l} \end{bmatrix} \tag{6.11.7}$$

We recognize a Vandermonde matrix, $\boldsymbol{V}$, and a diagonal matrix, Λ, on the left hand side of both of these equations. Upon the introduction of some convenient vector and matrix notation, we rewrite (6.11.6) and (6.11.7) as

$$\boldsymbol{V}\Lambda\boldsymbol{k} = -\boldsymbol{m}_l \tag{6.11.8}$$

$$\boldsymbol{V}\Lambda^{q+1}\boldsymbol{k} = -\boldsymbol{m}_h \tag{6.11.9}$$

where $\boldsymbol{m}_l$ represents the first q moments (0 through $q-1$) and $\boldsymbol{m}_h$ represents the higher order moments (q through $2q-1$).

Solving (6.11.8) we have

$$\boldsymbol{k} = -\Lambda^{-1}\boldsymbol{V}^{-1}\boldsymbol{m}_l \tag{6.11.10}$$

and therefore using (6.11.10) we can rewrite (6.11.9) as

$$V\Lambda^q V^{-1} m_l = m_h \tag{6.11.11}$$

To continue toward the solution we note that a Vandermonde matrix

$$V = \begin{bmatrix} 1 & 1 & \dots & 1 \\ \lambda_1 & \lambda_2 & \dots & \lambda_q \\ \vdots & \vdots & & \vdots \\ \lambda_1^{q-1} & \lambda_2^{q-1} & \dots & \lambda_q^{q-1} \end{bmatrix} \tag{6.11.12}$$

is a modal matrix for a related system in *companion form*:

$$\hat{A} = \begin{bmatrix} 0 & 1 & 0 & \dots & 0 \\ 0 & 0 & 1 & \dots & 0 \\ \vdots & \vdots & \vdots & & \vdots \\ 0 & 0 & 0 & \dots & 1 \\ -b_q & -b_{q-1} & -b_{q-2} & \dots & -b_1 \end{bmatrix} \tag{6.11.13}$$

The matrix $\hat{A}$ is said to have the *Forbenius form*, with only super-diagonal and last row elements. This companion form arises from a qth order linear system description of the form

$$x^{(q)} + b_1 x^{(q-1)} + \dots + b_{q-2}\ddot{x} + b_{q-1}\dot{x} + b_q x = bu \tag{6.11.14}$$

if we define

$$\begin{aligned} x_1 &\equiv x \\ x_2 &\equiv \dot{x} = \dot{x}_1 \\ x_3 &\equiv \ddot{x} = \dot{x}_2 \\ &\vdots \\ x_q &= x^{(q-1)} = \dot{x}_{q-1} \\ \dot{x}_q &= x^{(q)} = bu - b_q x_1 - b_{q-1} x_2 - \dots - b_1 x_q \end{aligned} \tag{6.11.15}$$

Then

$$\dot{x} = \hat{A}x + \begin{bmatrix} 0 \\ 0 \\ \vdots \\ b \end{bmatrix} u \tag{6.11.16}$$

with the state vector

$$x = [x_1, x_2, \ldots, x_q]^T = [x, \dot{x}, \ldots, x^{(q-1)}]^T \tag{6.11.17}$$

Such a linear system can be used to obtain the state-space representation of a frequency domain transfer function in the time domain. The system has as many states as the order of the denominator polynomial. Note that successive multiplication by s in the frequency domain corresponds to successive derivatives in the time domain, resulting in equations like those in (6.11.15).

Given the companion form $\hat{A}$, it has an eigenvector

$$[1, \lambda_k, \lambda_k^2, \ldots, \lambda_k^{q-1}]^T \tag{6.11.18}$$

with the eigenvalue λ_k provided that

$$\lambda_k^q + b_1\lambda_k^{q-1} + \ldots + b_{q-2}\lambda_k^2 + b_{q-1}\lambda_k + b_q = 0 \tag{6.11.19}$$

So, we know that

$$V^{-1}\hat{A}V = \Lambda \tag{6.11.20}$$

where Λ is the diagonal matrix of eigenvalues.

Rearranging (6.11.20) we have

$$\hat{A} = V\Lambda V^{-1} \tag{6.11.21}$$

and

$$\hat{A}^q = V\Lambda^q V^{-1} \tag{6.11.22}$$

Therefore we can substitute (6.11.22) into (6.11.11) to obtain

$$\hat{A}^q m_l = m_h \tag{6.11.23}$$

By repeatedly raising $\hat{A}$ to higher powers and indulging in a bit more algebra, we obtain the equivalent set of equations

$$\begin{bmatrix} m_0 & m_1 & m_2 & \dots & m_{q-1} \\ m_1 & m_2 & m_3 & \dots & m_q \\ m_2 & m_3 & m_4 & \dots & m_{q+1} \\ \vdots & \vdots & \vdots & & \vdots \\ m_{q-1} & m_q & m_{q+1} & \dots & m_{2q-2} \end{bmatrix} \begin{bmatrix} b_q \\ b_{q-1} \\ \vdots \\ b_1 \end{bmatrix} = - \begin{bmatrix} m_q \\ m_{q+1} \\ \vdots \\ m_{2q-1} \end{bmatrix} \tag{6.11.24}$$

Notice that this set of matrix equations is identical in form to those we derived earlier in Section 6.6 for a 4th order system (see equation (6.6.9)).

We first solve this set of linear equations for the characteristic polynomial coefficients $\{b_q, b_{q-1}, b_{q-2}, \dots, b_1\}$. From these coefficients we form the related characteristic polynomial,

$$b_q p^q + b_{q-1} p^{q-1} + b_{q-2} p^{q-2} + \dots + b_1 p + 1 = 0 \tag{6.11.25}$$

and obtain its q roots. We can reverse the characteristic equation to solve for the reciprocal roots, hence the time constants:

$$b_q + b_{q-1}\tau + b_{q-2}\tau^2 + \dots + b_1\tau^{q-1} + \tau^q = 0 \tag{6.11.26}$$

From the roots of (6.11.25), which are the p_k's, we can return to

$$\boldsymbol{k} = -\Lambda^{-1} \boldsymbol{V}^{-1} \boldsymbol{m}_l \tag{6.11.27}$$

to obtain the corresponding residues.

To summarize the steps involved in obtaining a low order approximation to a high order circuit response via moment matching:

1. Find as many moments $(2q)$ as necessary or desirable from the recursive application of $\boldsymbol{A}^{-1}$ (the solution of a related dc circuit).
2. Obtain the set of polynomial coefficients $\{b_q, b_{q-1}, b_{q-2}, \dots, b_1\}$ from the solution of q equations in q unknowns in terms of the $2q$ moments $\{m_0, m_1, \dots, m_{2q-1}\}$ using (6.11.24).
3. Obtain the roots of the resulting characteristic equation:

$$b_q p^q + b_{q-1} p^{q-1} + b_{q-2} p^{q-2} + \dots + b_1 p + 1 = 0 \tag{6.11.28}$$

4. Find corresponding residues by solving the related Vandermonde moment equations (q equations in q unknown):

$$V\Lambda k = -m_l \tag{6.11.29}$$

5. Compute the required time-domain response as a sum of q exponentials.

The above description is a more formal way of describing the procedure presented at the end of section 6.9.

6.12 Practical (Numerical) Considerations

The moment matching described above is a form of *Padé approximation* [Baker75]. Suppose we have a ratio of polynomials

$$\hat{H}(s) = \frac{a_{q-1}s^{q-1} + a_{q-2}s^{q-2} + a_{q-3}s^{q-3} + \ldots + a_1 s + a_0}{b_q s^q + b_{q-1}s^{q-1} + b_{q-2}s^{q-2} + \ldots + b_1 s + 1} \tag{6.12.1}$$

and we wish to match it to

$$H(s) = m_0 + m_1 s + m_2 s^2 + m_3 s^3 + \ldots + m_{2q-1}s^{2q-1} \tag{6.12.2}$$

Then, just as we did for the 4th order circuit in Section 6.6, we have

$$\begin{aligned}
a_0 &= m_0 \\
a_1 &= m_0 b_1 + m_1 \\
a_2 &= m_0 b_2 + m_1 b_1 + m_2 \\
\vdots & \qquad \vdots \\
0 &= m_0 b_q + m_1 b_{q-1} + m_2 b_{q-2} + \ldots + m_q \\
0 &= m_1 b_q + m_2 b_{q-1} + m_3 b_{q-2} + \ldots + m_{q+1} \\
\vdots & \qquad \vdots \\
0 &= m_{q-1} b_q + m_q b_{q-1} + m_{q+1} b_{q-2} + \ldots + m_{2q-1}
\end{aligned} \tag{6.12.3}$$

The last q of these equations are the same as the ones we obtained previously for the coefficients of the characteristic polynomial, the roots of which are the approximate dominant poles. The first q of these equations are for the coefficients of the numerator polynomial, the roots of which are the approximate zeros. We should note that other Padé approximations of this order can be obtained by raising the number of finite poles sought and correspondingly lowering the number of finite zeros. But the one we have here with one fewer zero than pole corresponds to what we did earlier in terms of poles and residues.

The Padé approximation seems straightforward, so the reader might wonder why we did not just present the results on that basis in the first place. The reason is simply that the Padé approximation, as powerful as it can be, is fraught with danger. The pole-residue approach that we discussed earlier is better for pointing out some of its problems and how to overcome them. When the Padé approximation works, which it does in most cases, it works well, but when it fails, it can do so spectacularly.

Consider, for example, a stable system with two real (negative valued) poles,

$$H(s) = \frac{k_1}{s+\sigma_1} + \frac{k_2}{s+\sigma_2} \tag{6.12.4}$$

and the following one pole Padé approximation to it:

$$\hat{H}(s) = \frac{\hat{k}}{s+\hat{\sigma}} \tag{6.12.5}$$

From the above we have

$$\begin{aligned} \frac{\hat{k}}{\hat{\sigma}} &= \frac{k_1}{\sigma_1} + \frac{k_2}{\sigma_2} \\ \frac{\hat{k}}{\hat{\sigma}^2} &= \frac{k_1}{\sigma_1^2} + \frac{k_2}{\sigma_2^2} \end{aligned} \tag{6.12.6}$$

so

$$\hat{\sigma} = \frac{\dfrac{k_1}{\sigma_1} + \dfrac{k_2}{\sigma_2}}{\dfrac{k_1}{\sigma_1^2} + \dfrac{k_2}{\sigma_2^2}} \tag{6.12.7}$$

Even though both σ_1 and σ_2 are negative, left-half plane poles, it is easy to conceive of values of k_1 and k_2 which would render $\hat{\sigma}$ positive, a right-half plane pole leading to an unstable and incorrect result.

In all cases for which the poles returned by the Padé approximation are suspect, it is good practice to then attempt a higher order approximation. Experience has shown that Padé approximation yields two kinds of poles. The first kind are good or genuine poles, which continue to appear with consistent values as higher orders of approximation are undertaken. The second kind are "bogus poles" which bounce around in value and should be rejected.

For passive networks, such as large, linear interconnect circuits, we can of course categorically reject right half plane poles, which are impossible in such cases. We can even force the solution of the moment matching equations to yield only left half plane poles by means of constrained optimization. But if we want to use Padé approximation for active circuits, which actually may possess right half plane poles, we cannot afford such a cavalier strategy. The presence of a right half plane pole for an active circuit is the most vital piece of information we could have regarding the potential performance of that circuit, so we would not want to suppress or ignore it.

If we could take the Padé approximation to a sufficiently high order we could avoid false poles and always obtain excellent results. But the very nature of Padé approximation may preclude that. The problem manifests itself in equation (6.11.5), which is reproduced here for convenience:

$$\begin{aligned} -\left(\frac{k_1}{p_1} + \frac{k_2}{p_2} + \ldots + \frac{k_q}{p_q}\right) &= m_{0,l} \\ -\left(\frac{k_1}{p_1^2} + \frac{k_2}{p_2^2} + \ldots + \frac{k_q}{p_q^2}\right) &= m_{1,l} \\ \vdots \qquad\qquad &\qquad \vdots \\ -\left(\frac{k_1}{p_1^{2q}} + \frac{k_2}{p_2^{2q}} + \ldots + \frac{k_q}{p_q^{2q}}\right) &= m_{2q-1,l} \end{aligned} \tag{6.12.8}$$

If all of the poles are approximately the same magnitude, then these equations may be easily solved on a computer with reasonable floating-point precision. But if they are not, the higher order moments may represent numerical noise rather than useful information. We can take some steps to overcome such problems, but in practice a Padé approximation is limited to finding about ten approximate dominant poles on a computer with reasonable precision.

Frequency scaling

Since we start with a first order approximation, *frequency scaling* can be used to get to higher orders before encountering numerical problems. At the outset, we have

$$\frac{k_1}{p_1} = m_0$$
$$\frac{k_1}{p_1^2} = m_1 \tag{6.12.9}$$

so

$$p_1 = \frac{m_0}{m_1} \tag{6.12.10}$$

We can scale the frequency of this dominant pole to magnitude 1 for subsequent computations by multiplying all circuit capacitance and inductance values by p_1:

$$sC \to \left(\frac{s}{p_1}\right) p_1 C \qquad sL \to \left(\frac{s}{p_1}\right) p_1 L \tag{6.12.11}$$

This method is classical frequency scaling as used in filter design. Frequency scaling keeps the first terms in (6.12.8) reasonable. In situations for which we may only be able to obtain one or two good poles without frequency scaling, we are usually able to obtain three or four with it.

Expansions about $s = \infty$

When approximating the time domain waveforms by matching moments, the largest error will occur near $t = 0$. In [Pillage90], one point from the expansion about $s = \infty$ was used for a more accurate starting waveform at $t = 0$. Combining an arbitrary number of points from expansions about $s = 0$ and $s = \infty$ was described in [Huang90]. To include nonlinear elements in Padé type approximations, some researchers have tried using mostly points from an expansion about $s = \infty$, and only one point from $s = 0$ to force stability [Lin92]. Perhaps most successful, however, has been the combination of expansions from various points in the s plane when a large number of poles are required. Frequency shifting allows us to expand about various points along the real axis in order to obtain pole convergence. It also alleviates some of the numerical conditioning problems.

Frequency shifting

Referring to (6.12.8) it is obvious that the spread in magnitudes of the poles can be the source of numerical problems. Expanding about $s = 0$ results in relatively small pole values raised to high powers which can wash out the effects of other poles on the moment values. There are situations where this effect is beneficial. If we are only looking for truly dominant poles, then those that are much larger in magnitude and have relatively small residues are not significant anyway. This situation tends to prevail for a large class of linear interconnect circuit problems. But for active circuits, non-dominant poles near the unity gain frequency can have significant effects on their frequency response characteristics like gain and phase margins. These non-dominant poles can sometimes be computed through *frequency shifting*.

From classical filter design, it is well known that all circuit poles can be shifted uniformly by an amount α in the s plane upon the addition of a parallel conductance proportion to each capacitance and a series resistance proportional to each inductance as shown in Figure 6.24. The constant of proportionality is α.

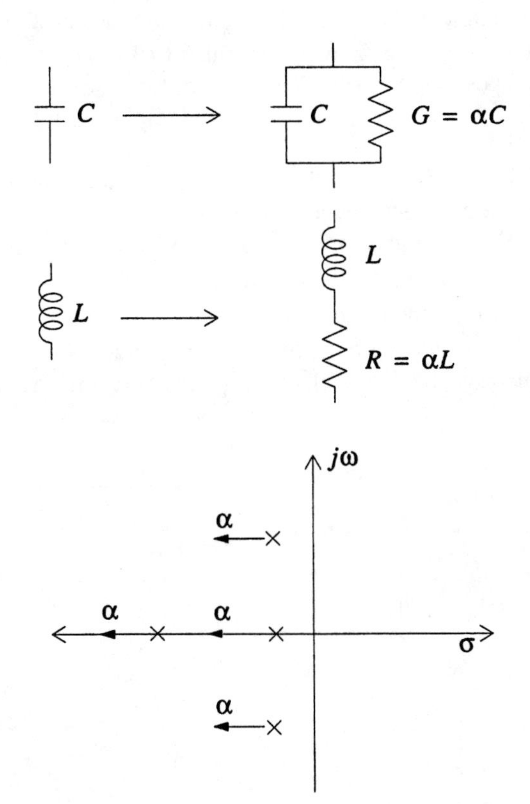

Figure 6.24 Frequency shifting to obtain higher order moments.

Suppose that a circuit has poles $p_1 = -1$ and $p_2 = -10^6$, then we may shift them by -10^6 to obtain $\hat{p}_1 = -1000001$ and $\hat{p}_2 = -2000000$. Their original ratio was 10^6, but after shifting it is approximately 2, and now we are more likely to find both. Alternately, we could first determine the small pole and then apply the shift. Using a sequence of such shifts, we can even attempt to compute clusters of poles. Frequency shifting works well, often allowing us to obtain as many as ten genuine poles -- from twenty good moments. It is non-trivial, however, to determine the best frequency shift. For many analog problems the high frequency range of interest is obvious. For linear interconnect, analysis by frequency shifting is beneficial, but not as necessary. This type of expansion shift has been successfully applied for both real [Huang90] and complex frequency points [Chiprout93].

Frequency shifting has another advantage; it breaks capacitance cutsets and inductance loops, which would otherwise have rendered the A matrix singular. So it is a simple and straightforward means for mapping the related dc circuit into one that is easier to analyze. Capacitance cutsets and inductance loops give rise to poles at the origin in the original circuit. If a circuit has a pole at the origin, it is impossible to expand its circuit function in a series about $s = 0$. Once we have frequency shifted all such poles away from the origin, the series expansion poses no problem. We must not forget, of course, to shift the approximate dominant poles that we obtain for the altered circuit back to the right to obtain the actual answers.

Frequency shifting poses yet another advantage. Since all capacitances are to be replaced by independent current sources and inductances by independent voltage sources, we can convert all of the energy storage elements to Norton equivalents to calculate their moments as shown in Figure 6.25. There is no fear of the equivalent G's becoming either 0 or ∞. Therefore, we can employ straightforward Nodal Analysis to find the moments. In practice we have found this approach to work on frequency-shifted circuits for which Modified Nodal Analysis failed to work on their unshifted versions.

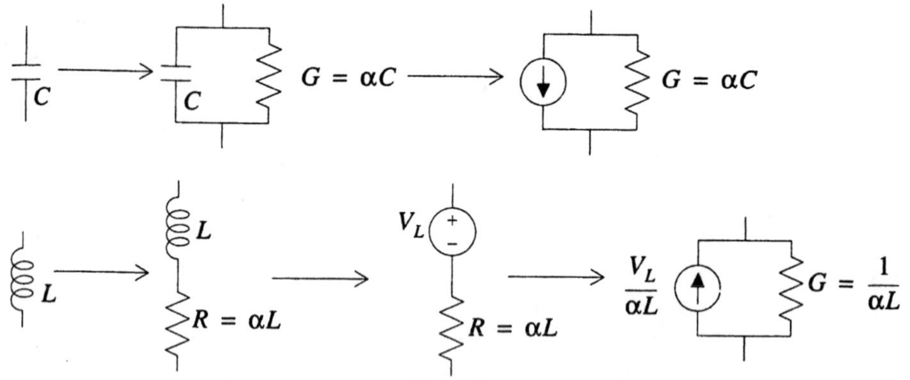

Figure 6.25 Energy storage element models with frequency shifting.

Moment shifting

If one is interested in only a few poles or more accuracy in the pole values with frequency shifting, *moment shifting* can be applied to remove the large magnitude pole effects and make the poles converge to the actual low frequency poles (or whatever frequency to which one has shifted). Since the poles are invariant to the input forcing function, we can multiply any series of moments by powers of s^{-1} and still expect the moment matching equations to hold. In effect, what we would be matching is the transient portion of some forced response (s^{-1} is a step, s^{-2} is a ramp, etc.), but the poles are invariant to this change.

With a wide spread in the pole values, since the moments are related to the sum of the powers of p^{-1}, as we use a higher approximation order the lower poles begin to dominate the moments. At some point, with finite machine precision, only the first pole affects the higher order moments.

We can demonstrate this pole convergence using the first six moments for the response at C_1 of the RC tree in Figure 6.12:

$$\begin{aligned} m_0 &= 1.000 \\ m_1 &= -1.260 \\ m_2 &= 1.737 \\ m_3 &= -2.402 \\ m_4 &= 3.322 \\ m_5 &= -4.595 \end{aligned} \tag{6.12.12}$$

For an impulse response, the dominant time constant (p^{-1}) is m_1/m_0. For a step response, the dominant time constant is m_2/m_1, etc. As we use a higher order set of two moments to find the dominant time constant approximation, we find that it converges to the exact first time constant for this circuit, as summarized below:

$$\begin{aligned} \left|\frac{m_1}{m_0}\right| &= 1.260 \\ \left|\frac{m_2}{m_1}\right| &= 1.379 \\ \left|\frac{m_3}{m_2}\right| &= 1.383 \\ \left|\frac{m_4}{m_3}\right| &= 1.383 \end{aligned} \tag{6.12.13}$$

After only two moment shifts, the first time constant has converged to the exact value within four significant digits.

6.13 Sensitivity Analysis

It is sometimes useful to have the dominant pole and zero sensitivities in addition to the approximate values. In Section 3.6 we showed how one can calculate the sensitivities of the solution vector, $\boldsymbol{x}$, for a set of equations $\boldsymbol{Mx} = \boldsymbol{b}$, in terms of the parameter values in the $\boldsymbol{M}$ matrix and the $\boldsymbol{b}$ vector. This procedure was shown to be useful for determining the sensitivities of dc voltages and currents in terms of circuit parameter values. Since moments are obtained by LU factoring a dc equivalent circuit, we can evaluate the sensitivities of the moments following an approach similar to that outlined in Section 3.6. From these moment sensitivities, the pole and zero sensitivities can be calculated as described in [Lee92]. From those, we can make sensitivity statements about the transient waveforms of the circuit.

Referring back to the moment expressions in state variable form in (6.10.6), we can write a recursive expression for the jth set of moments as

$$\boldsymbol{m}_j = -\boldsymbol{A}^{-(j+1)}\boldsymbol{B} \tag{6.13.1}$$

Using the identity

$$\frac{\partial \boldsymbol{A}^{-1}}{\partial e} = -\boldsymbol{A}^{-1}\frac{\partial \boldsymbol{A}}{\partial e}\boldsymbol{A}^{-1} \tag{6.13.2}$$

we can write the sensitivity of the jth moment with respect to some scalar parameter e using the chain rule:

$$\frac{\partial \boldsymbol{m}_j}{\partial e} = (\boldsymbol{A}^{-1}\frac{\partial \boldsymbol{A}}{\partial e}\boldsymbol{A}^{-j-1} + \ldots + \boldsymbol{A}^{-j-1}\frac{\partial \boldsymbol{A}}{\partial e}\boldsymbol{A}^{-1})\boldsymbol{B} - \boldsymbol{A}^{-j-1}\frac{\partial \boldsymbol{B}}{\partial e} \tag{6.13.3}$$

Focusing on the jth moment at one particular node,

$$\begin{aligned}\frac{\partial m_j}{\partial e} &= \boldsymbol{c}^T\left[(\boldsymbol{A}^{-1}\frac{\partial \boldsymbol{A}}{\partial e}\boldsymbol{A}^{-j-1} + \ldots + \boldsymbol{A}^{-j-1}\frac{\partial \boldsymbol{A}}{\partial e}\boldsymbol{A}^{-1})\boldsymbol{B} - \boldsymbol{A}^{-j-1}\frac{\partial \boldsymbol{B}}{\partial e}\right] \\ &= \left[\sum_{i=0}^{j}\boldsymbol{c}^T\boldsymbol{A}^{-i-1}\frac{\partial \boldsymbol{A}}{\partial e}\boldsymbol{A}^{-(j-i+1)}\boldsymbol{B}\right] - \boldsymbol{c}^T\boldsymbol{A}^{-j-1}\frac{\partial \boldsymbol{B}}{\partial e}\end{aligned} \tag{6.13.4}$$

where $\boldsymbol{c}$ is a column vector with all zeros and a 1.0 in the row corresponding to the node of interest.

To explain all of the terms in (6.13.4), we begin with the original solution for the vector of zeroth moments, $\boldsymbol{m}_0 = \boldsymbol{A}^{-1}\boldsymbol{B}$. The zeroth moment sensitivities for one particular node would be (from (6.13.4)):

$$\frac{\partial m_0}{\partial e} = \boldsymbol{c}^T\left[(\boldsymbol{A}^{-1}\frac{\partial \boldsymbol{A}}{\partial e}\boldsymbol{A}^{-1})\boldsymbol{B} - \boldsymbol{A}^{-1}\frac{\partial \boldsymbol{B}}{\partial e}\right] \tag{6.13.5}$$

The $\frac{\partial \boldsymbol{A}}{\partial e}$ and $\frac{\partial \boldsymbol{B}}{\partial e}$ terms in (6.13.5) are simply the derivatives of the element stamps (more on this topic in Chapter 9). These matrices are zero everywhere except for those matrix positions that are a function of e. The $\boldsymbol{A}^{-1}\frac{\partial \boldsymbol{B}}{\partial e}$ term is easily evaluated using the $\frac{\partial \boldsymbol{B}}{\partial e}$ stamps and the LU factors used to obtain the original solution $\boldsymbol{m}_0 = \boldsymbol{A}^{-1}\boldsymbol{B}$. The first term in (6.13.5) post-multiplies the $\frac{\partial \boldsymbol{A}}{\partial e}$ stamps by the vector $\boldsymbol{m}_0$, and pre-multiplies them by row vector $\boldsymbol{c}^T\boldsymbol{A}^{-1}$.

The $\boldsymbol{c}^T\boldsymbol{A}^{-1}$ term represents the adjoint circuit solution which we introduced in Chapter 3 and which will be covered in detail in Chapter 9. The adjoint system is

$$\boldsymbol{A}^T\boldsymbol{y} = \boldsymbol{c} \tag{6.13.6}$$

and the solution of this adjoint system is

$$\boldsymbol{y} = (\boldsymbol{A}^T)^{-1}\boldsymbol{c} = (\boldsymbol{A}^{-1})^T\boldsymbol{c} \tag{6.13.7}$$

Therefore, if we solve the nominal circuit $\boldsymbol{m}_0 = \boldsymbol{A}^{-1}\boldsymbol{B}$ and the adjoint circuit described by (6.13.7), we can readily evaluate the m_0 sensitivities:

$$\frac{\partial m_0}{\partial e} = \boldsymbol{y}^T\frac{\partial \boldsymbol{A}}{\partial e}\boldsymbol{m}_0 - \boldsymbol{y}^T\frac{\partial \boldsymbol{B}}{\partial e} \tag{6.13.8}$$

We should point out, as we did in Chapter 3, that the adjoint solution is obtained using the original LU factors, therefore it requires only one additional Forward and Back Substitution.

The higher order sensitivities can also be calculated recursively using the original and adjoint solutions. In summary,

$$\frac{\partial m_j}{\partial e} = \left(\sum_{i=0}^{j} y_i^T \frac{\partial A}{\partial e} m_{j-i}\right) - y_j^T \frac{\partial B}{\partial e} \tag{6.13.9}$$

where y_i is the vector for the ith adjoint solution:

$$y_i = (A^{-i-1})^T c \tag{6.13.10}$$

For a qth order approximation, we calculate $2q$ moments and the $2q$ sensitivities for the response node(s) of interest using (6.13.9) and (6.13.10). To obtain the characteristic polynomial for the approximate poles we solve equation (6.11.24)

$$\begin{bmatrix} m_0 & m_1 & m_2 & \dots & m_{q-1} \\ m_1 & m_2 & m_3 & \dots & m_q \\ m_2 & m_3 & m_4 & \dots & m_{q+1} \\ \vdots & \vdots & \vdots & & \vdots \\ m_{q-1} & m_q & m_{q+1} & \dots & m_{2q-2} \end{bmatrix} \begin{bmatrix} b_q \\ b_{q-1} \\ \vdots \\ b_1 \end{bmatrix} = - \begin{bmatrix} m_q \\ m_{q+1} \\ \vdots \\ m_{2q-1} \end{bmatrix} \tag{6.13.11}$$

which we will refer to as

$$m_m b = m_h \tag{6.13.12}$$

We evaluate (6.13.12) by inverting m_m

$$b = m_m^{-1} m_h \tag{6.13.13}$$

To consider the sensitivity of the characteristic polynomial with respect to some parameter e, we evaluate (6.13.13) via the chain rule:

$$\begin{aligned} \frac{\partial b}{\partial e} &= m_m^{-1} \frac{\partial m_h}{\partial e} + \frac{\partial}{\partial e}(m_m^{-1}) m_h \\ &= m_m^{-1} \frac{\partial m_h}{\partial e} + \left(-m_m^{-1} \frac{\partial m_m}{\partial e} m_m^{-1}\right) m_h \\ &= m_m^{-1} \left(\frac{\partial m_h}{\partial e} - \frac{\partial m_m}{\partial e} b\right) \end{aligned} \tag{6.13.14}$$

Equation (6.13.14) relates the sensitivity of the characteristic polynomial coefficients to the sensitivities of the $2q$ moments. For a qth order approximation the characteristic polynomial for the AWE poles is

$$P(s) = \sum_{k=0}^{q} b_k s^k \quad (6.13.15)$$

Referring to [Frank78], we can evaluate (6.13.15) at a root, p_j, and apply the chain rule to obtain the root sensitivity:

$$\left.\frac{\partial P}{\partial e}\right|_{s=p_j} = 0 = \sum_{k=0}^{q} \left(\frac{\partial b_k}{\partial e} p_j^k + k b_k \frac{\partial p_j}{\partial e} p_j^{k-1}\right) \quad (6.13.16)$$

Or, rearranging (6.13.16) in terms of the pole sensitivity of interest,

$$\frac{\partial p_j}{\partial e} = -\frac{\sum_{k=0}^{q} \frac{\partial b_k}{\partial e} p_j^k}{\sum_{l=0}^{q} l b_l p_j^{l-1}} \quad (6.13.17)$$

Using the moment sensitivities and equations (6.13.14) and (6.13.17) we can easily generate the pole sensitivities. For analog circuit design applications we would want to know the zero sensitivities too. They are obtained using an expression similar to (6.13.17) along with the sensitivities of the numerator coefficients in (6.12.1), $\frac{\partial a_j}{\partial e}$. These numerator sensitivities are available by directly differentiating the expressions in (6.12.3). These sensitivities are a function of the moment sensitivities and the denominator coefficient sensitivities, both of which are known.

If we seek the time domain sensitivities, we can calculate the residue sensitivities directly. The residues are obtained by solving

$$\boldsymbol{k} = -\Lambda^{-1} \boldsymbol{V}^{-1} \boldsymbol{m}_l \quad (6.13.18)$$

Therefore, following the steps used to generate (6.13.14) from (6.13.13),

$$\frac{\partial \boldsymbol{k}}{\partial e} = -\Lambda^{-1} \boldsymbol{V}^{-1} \left(\frac{\partial \boldsymbol{m}_l}{\partial e} - \frac{\partial [-\boldsymbol{V}\Lambda]}{\partial e} \boldsymbol{k}\right) \quad (6.13.19)$$

where $\frac{\partial [-\boldsymbol{V}\Lambda]}{\partial e}$ is calculated using the pole sensitivities.

6.14 Conclusions

We began our discussion with small-signal ac analysis and then studied moment matching. Moment matching methods are relatively new mechanisms for approximating the frequency domain response for large linear circuits. These methods find a low order rational polynomial approximation of the actual impulse response in the frequency domain. The approximations yield valuable time domain information, too. Various techniques are used to avoid numerical problems. Further, sensitivity calculations are possible with only a small computational overhead. These methods are very efficient and they have been made to work on a wide variety of linear circuits.

6.15 References

[Elmore48] W. C. Elmore. The Transient Analysis of Damped Linear Networks with Particular Regard to Wideband Amplifiers. *Journal of Applied Physics*, vol. 19(1), 1948.

[Muller56] D. E. Muller. A Method for Solving Algebraic Equations Using an Automated Computer. In *Mathematical Tables and Other Aids to Computation (MTAC)*, vol. 10, pp. 208-215, 1956.

[Baker75] G. A. Baker, Jr. *Essentials of Padé Approximants*. Academic Press, 1975.

[Frank78] P. M. Frank. *Introduction to System Sensitivity Theory.* Academic Press, 1978.

[Penfield81] P. Penfield and J. Rubenstein. Signal Delay in RC Tree Networks. In *Proceedings of the 19th Design Automation Conference*, 1981.

[Pillage90] L. T. Pillage and R. A. Rohrer. Asymptotic Waveform Evaluation for Timing Analysis. *IEEE Transactions on Computer Aided Design of ICs and Systems,* April 1990.

[Huang90] X. Huang, V. Raghavan, and R. A. Rohrer. AWEsim: A Program for Efficient Analysis of Linear(ized) Circuits. *Proceedings of the IEEE International Conference on Computer-Aided Design (ICCAD)*, November 1990.

[Lee92] J. Y. Lee, X. Huang, and R. A. Rohrer. Pole and Zero Sensitivity Calculation in Asymptotic Waveform Evaluation. *IEEE Transactions on Computer Aided Design of ICs and Systems,* May 1992.

[Lin92] S. Lin and E. S. Kuh. Transient Simulation of Lossy Interconnect. *Proceedings of the 29th Design Automation Conference*, 1992.

[Chiprout93] E. Chiprout and M. Nakhla. Transient Waveform Estimation of High-Speed MCM Networks Using Complex Frequency Hopping. *Proceedings of IEEE MCM Conference*, March 1993.

[Ratzlaff94] C. Ratzlaff and L. T. Pillage. RICE: Rapid Interconnect Circuit Evaluator using AWE. *IEEE Transactions on Computer Aided Design of ICs and Systems,* June 1994.

Chapter 7

Sparse Matrices and Some of Their Implications

Circuit elements are usually connected to four or fewer nodes. Similarly, a small fraction of all pairwise combinations of nodes has an element connecting them. Thus matrices describing circuits are generally *sparse*. *Sparse matrices* are matrices that contain a high proportion of zero entries. We can make use of this sparsity to accelerate circuit analysis. With *sparse matrix methods*, zero-valued matrix entries are not stored (resulting in a savings of memory) and, to the extent possible, not manipulated (resulting in a savings of CPU time). In fact, while dense matrix LU factorization or Gaussian elimination has an order of complexity n^3 in the size of the matrix, for sparse matrices typical of circuits, the run time grows only as about $n^{1.5}$. Very large sparse matrix solutions may show an asymptotic complexity as low as $n^{1.1}$.

7.1 Introduction

For the circuit example that we worked earlier in Chapter 1 through Chapter 3, shown here again in Figure 7.1, we had the Nodal Analysis equations

$$\begin{bmatrix} 2 & -1 & 0 & 0 \\ -1 & 3 & -1 & 0 \\ 0 & -1 & 3 & -1 \\ 0 & 0 & -1 & 2 \end{bmatrix} \begin{bmatrix} v_1 \\ v_2 \\ v_3 \\ v_4 \end{bmatrix} = \begin{bmatrix} 1 \\ 0 \\ 0 \\ 1 \end{bmatrix} \qquad \textbf{(7.1.1)}$$

This is a sparse matrix because it has plenty of zero-valued elements. Every node is not connected to every other node in this circuit, and for Nodal Analysis nonzero-valued elements result only from direct connections. For much larger circuits with graphs that are fairly planar, there are relatively few direct connections between pairs of nodes, and the nodal admittance matrices are extremely sparse.

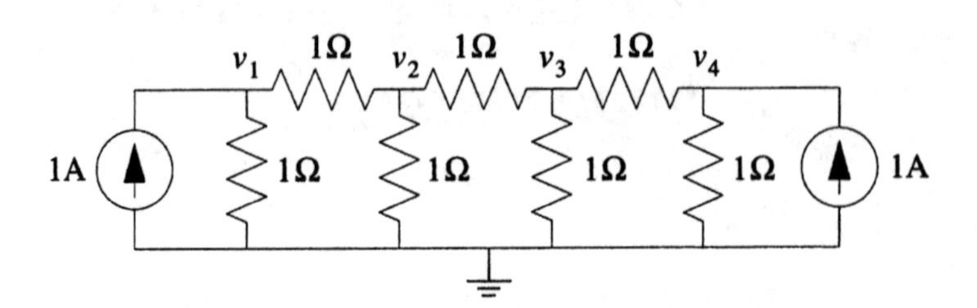

Figure 7.1 Circuit example from earlier chapters.

Generally speaking, the larger the circuit, the more sparse is its nodal admittance matrix. We can figure that there are usually about two to four branches incident on non-datum nodes, with a reasonable average being three. But even for a high of four nonzero-valued elements for each nodal admittance matrix row, at $n = 40$ non-datum nodes the nodal admittance matrix would be 90 percent sparse. At $n = 100$, it would be 96 percent sparse, and at $n = 400$ non-datum nodes it would be 99 percent sparse. Hence circuits of any reasonable size have very sparse nodal admittance matrices and the sparsity increases with circuit size.

In this chapter we will discuss means of taking advantage of matrix sparsity to reduce storage requirements and to reduce the number of floating point operations entailed in LU factorization and Forward and Back Substitution. We will start our study in terms of the simple example introduced above, and then extend it to cover other more general situations.

7.2 Sparse Nodal Admittance Matrices

Because the nodal admittance matrices that represent reasonably sized circuits typically are very sparse, we seek mechanisms whereby we need neither to store nor to process their zero-valued elements. To reduce storage, we simply avoid storing zero-valued elements. By default, every element of a sparse matrix is a zero unless explicitly stored as a nonzero. To reduce the number of operations in dealing with sparse matrices, we recognize the following, which provides the basic motivation for sparse matrix algorithms:

zero-element * zero-element = zero-element
zero-element * nonzero-element = zero-element
zero-element + zero-element = zero-element
zero-element + nonzero-element = that nonzero-element

As we process the elements of a sparse matrix, it is in our interest to preserve its sparsity. We have already unwittingly accomplished this goal in the example that we worked in Chapter 1 through Chapter 3. To obtain the upper triangular $\boldsymbol{U}$ matrix, we did not process the lower triangular zeros of the original $\boldsymbol{Y}$ matrix, and we did not alter the upper triangular zeros of the original $\boldsymbol{Y}$ matrix either, as shown in Figure 7.2.

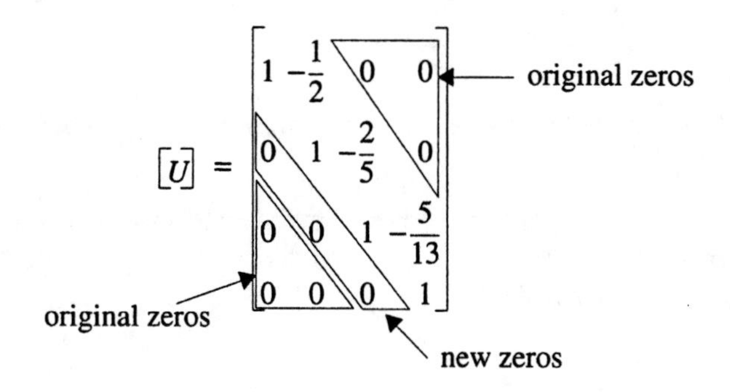

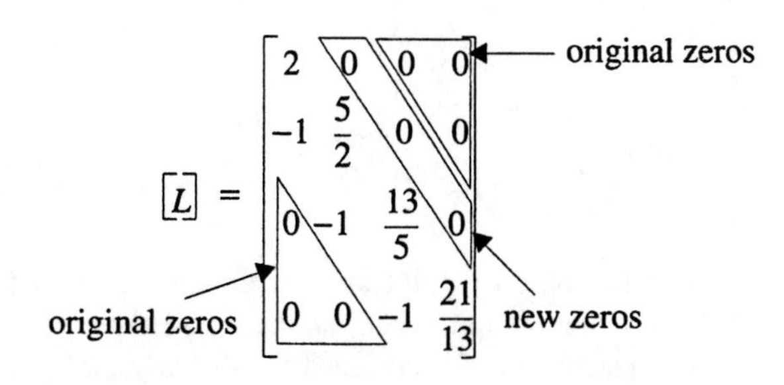

Figure 7.2 LU factorization of sparse matrices.

Now suppose that for the same circuit the nodes had been numbered differently, as shown in Figure 7.1. The nodal admittance matrix has the same number of zero-valued elements as before, so it is equally sparse:

$$\begin{bmatrix} 3 & \boxed{0} & -1 & -1 \\ \boxed{0} & 2 & \boxed{0} & -1 \\ -1 & \boxed{0} & 2 & \boxed{0} \\ -1 & -1 & \boxed{0} & 3 \end{bmatrix} \begin{bmatrix} v_1 \\ v_2 \\ v_3 \\ v_4 \end{bmatrix} = \begin{bmatrix} 0 \\ 1 \\ 1 \\ 0 \end{bmatrix} \qquad \textbf{(7.2.1)}$$

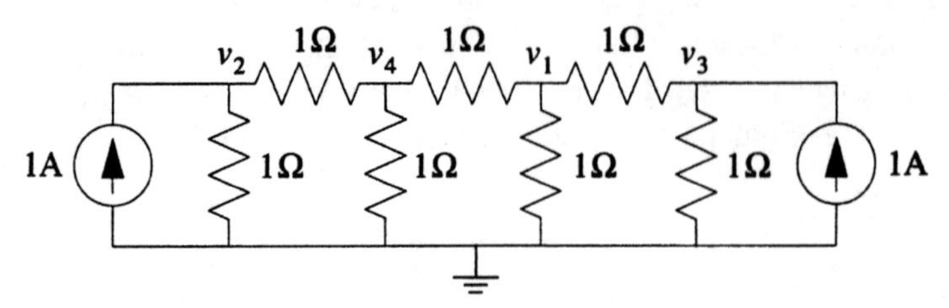

Figure 7.3 The circuit in Figure 7.1 with a different node number ordering.

Let us investigate how well LU factorization will maintain this sparsity with the new node ordering. The normalized first row must be applied to both the third and the fourth rows to proceed toward the upper triangular form.

$$\begin{bmatrix} 1 & \boxed{0} & -\frac{1}{3} & -\frac{1}{3} \\ \boxed{0} & 2 & \boxed{0} & -1 \\ 0 & \boxed{0} & \frac{5}{3} & \diamond\left(-\frac{1}{3}\right) \\ 0 & -1 & \diamond\left(-\frac{1}{3}\right) & \frac{8}{3} \end{bmatrix} \begin{bmatrix} v_1 \\ v_2 \\ v_3 \\ v_4 \end{bmatrix} = \begin{bmatrix} 0 \\ 1 \\ 1 \\ 0 \end{bmatrix} \tag{7.2.2}$$

A zero-element that turns into a nonzero during the solution procedure is called a *fill-in*. Such elements must be stored and computed, unlike the zero-elements that they replaced. Further, such elements may cause more fill-ins. Clearly, fill-ins must be avoided wherever possible. The fill-ins are denoted in (7.2.2) by $\diamond$.

Continuing with the Gaussian Elimination, we obtain

$$\begin{bmatrix} 1 & \boxed{0} & -\frac{1}{3} & -\frac{1}{3} \\ \boxed{0} & 1 & \boxed{0} & -\frac{1}{2} \\ 0 & \boxed{0} & \frac{5}{3} & \diamond\left(-\frac{1}{3}\right) \\ 0 & 0 & \diamond\left(-\frac{1}{3}\right) & \frac{13}{6} \end{bmatrix} \begin{bmatrix} v_1 \\ v_2 \\ v_3 \\ v_4 \end{bmatrix} = \begin{bmatrix} 0 \\ \frac{1}{2} \\ 1 \\ \frac{1}{2} \end{bmatrix} \tag{7.2.3}$$

and, finally,

$$\begin{bmatrix} 1 & \boxed{0} & -\frac{1}{3} & -\frac{1}{3} \\ \boxed{0} & 1 & \boxed{0} & -\frac{1}{2} \\ 0 & \boxed{0} & 1 & \diamond{-\frac{1}{5}} \\ 0 & 0 & \not{0} & \frac{21}{10} \end{bmatrix} \begin{bmatrix} v_1 \\ v_2 \\ v_3 \\ v_4 \end{bmatrix} = \begin{bmatrix} 0 \\ \frac{1}{2} \\ \frac{3}{5} \\ \frac{7}{10} \end{bmatrix} \tag{7.2.4}$$

When we apply Back Substitution we obtain $v_4 = \frac{1}{3}$, $v_3 = \frac{2}{3}$, $v_2 = \frac{2}{3}$, and $v_1 = \frac{1}{3}$, which is the same solution as before except for the node renumbering. Even though the results are the same as before, we had to do more work to obtain them. We created a fill-in in the $(4, 3)$ position in the lower triangle, and then we had to annihilate it later (shown by the $\times$ in (7.2.4)). We also created a fill-in, which required extra storage, in the upper triangle in the $(3, 4)$ position.

In terms of LU factorization, in this instance we obtain the following:

$$[U] = \begin{bmatrix} 1 & \boxed{0} & -\frac{1}{3} & -\frac{1}{3} \\ 0 & 1 & \boxed{0} & -\frac{1}{2} \\ 0 & 0 & 1 & \diamond{-\frac{1}{5}} \\ 0 & 0 & 0 & 1 \end{bmatrix} \qquad [L] = \begin{bmatrix} 3 & 0 & 0 & 0 \\ \boxed{0} & 2 & 0 & 0 \\ -1 & \boxed{0} & \frac{5}{3} & 0 \\ -1 & -1 & \diamond{-\frac{1}{3}} & \frac{21}{10} \end{bmatrix} \tag{7.2.5}$$

The second node numbering scheme requires more arithmetic operations and more storage than the first. So we ask the question: is there an optimal node ordering scheme to reduce the fill-ins and maintain the original sparsity as much as possible? If we had such a scheme, the simulation program, unbeknownst to the user, could renumber nodes (permute rows and columns of the nodal admittance matrix) to minimize the number of fill-ins that occur in $\boldsymbol{L}$ and $\boldsymbol{U}$. If we don't create nonzero elements, we don't have to store or annihilate them! Intuitively, we should suspect by now that the earliest processed rows and columns should have the fewest nonzero-valued elements. So, we might attempt to renumber the nodes so as to shape the nodal admittance matrix something like this:

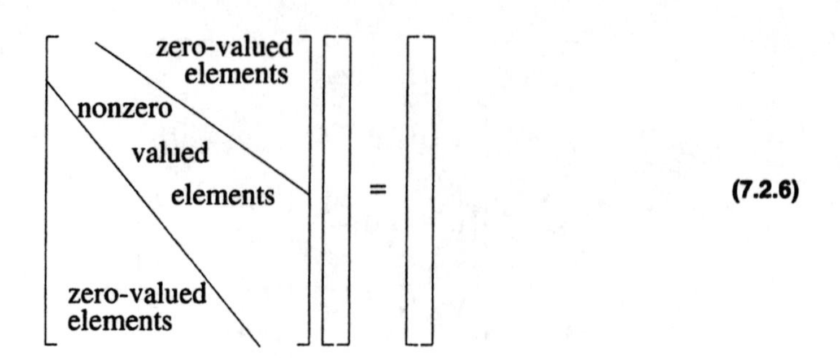

(7.2.6)

7.3 Ordering of Sparse Matrices

Since the MNA matrix Y is sparse, we don't build the full matrix. Rather, we compute the locations and values of the nonzeros. Certain sparse matrix algorithms are computationally efficient if the locations of the nonzeros do not change through the matrix processing. Given a (square) sparse matrix, it is possible to *symbolically* determine the locations of the potential fill-ins before the LU factorization is commenced. This section will explain how these locations are determined, and how sparse matrices can be reordered to minimize fill-ins. This approach applies to the LU factorization of any square nonsingular sparse matrix.

To begin to understand what must be done we consider three rows of a 7×7 sparse matrix, where x's indicate nonzero-valued elements:

$$\begin{matrix} x & 0 & 0 & x & 0 & 0 & x \\ 0 & x & 0 & 0 & x & 0 & x \\ x & 0 & x & x & 0 & x & 0 \end{matrix} \tag{7.3.1}$$

If we consider choosing the $(1, 1)$ position element to be the first pivot, we find the following:

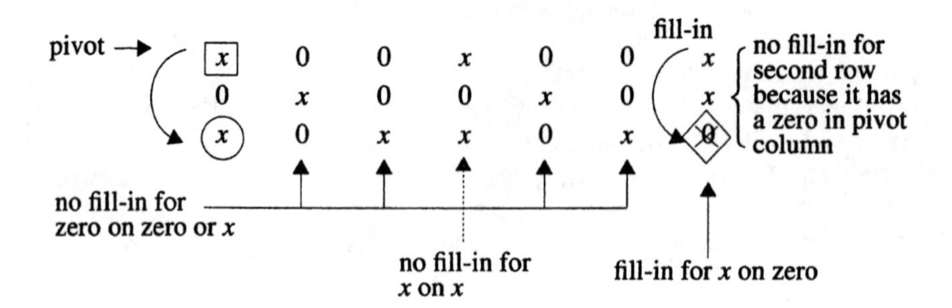

For this pivot choice there would only be one fill-in in the first three rows. Of course, we would not attempt to use any of the zeros as pivot elements, but there are two other non-zero possibilities, the choice of which would correspond to reordering the variables:

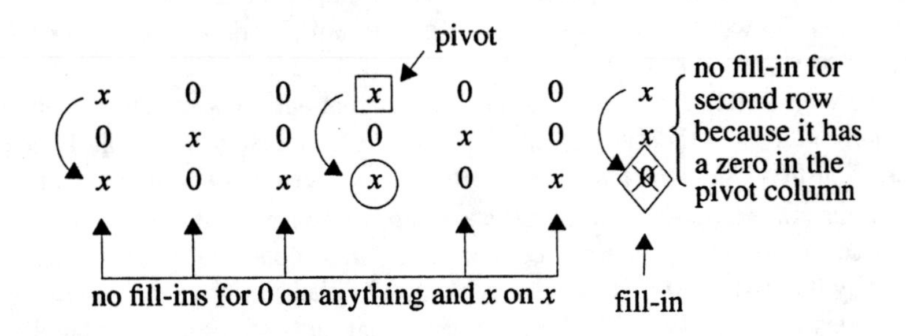

and

fill-in (first column), fill-in (fourth column), pivot (seventh column, first row), no fill-ins in this row (third row)

$$
\begin{array}{ccccccc}
x & 0 & 0 & x & 0 & 0 & \boxed{x} \\
\text{⊗}0 & x & 0 & \text{⊗}0 & x & 0 & (x) \\
x & 0 & x & x & 0 & x & 0
\end{array}
$$

Even in this limited illustration, we can see that different pivot choices in the first row (i.e., variable reordering) can cause different numbers of fill-ins for subsequent rows.

We can also consider the interchange of rows, which corresponds to reordering the equation numbers, and check what would happen if we were to take the first pivot from other rows. We will illustrate this without actually interchanging the rows so that the results can be easily compared with the above. We will first consider the two pivot selections shown below.

$$
\begin{array}{ccccccc}
x & 0 & 0 & x & 0 & 0 & x \\
0 & \boxed{x} & 0 & 0 & \boxed{x} & 0 & x \\
x & 0 & x & x & 0 & x & 0
\end{array}
$$

Neither of these two second row pivot choices would cause any fill-ins in rows one or three because both rows have zeros in their respective pivot columns. If we were to choose the $(2, 7)$ position element to be the first pivot element, we would obtain two fill-ins in the first row and none in the third.

$$
\begin{array}{ccccccc}
x & \text{⊗}0 & 0 & x & \text{⊗}0 & 0 & (x) \\
0 & x & 0 & 0 & x & 0 & \boxed{x} \leftarrow \text{pivot} \\
x & 0 & x & x & 0 & x & 0 \leftarrow \text{no fill-ins}
\end{array}
\qquad \textbf{(7.3.2)}
$$

For an $(n \times n)$ dense matrix there would be $[n^2 + (n-1)^2 + \ldots + 2^2 + 1^2]$ pivot choices for a full LU factorization and somewhat less than that for a sparse matrix. Every one of these possibilities would have to be evaluated to optimally order for fill-in minimization. In general the number of possible sets is too large to check in a computationally efficient manner. So we typically resort to some suboptimal ordering scheme, such as the one we will discuss in the next section.

But before we go on to that we should mention a bit about pivot selection as it has been applied traditionally to Modified Nodal Analysis. As enunciated originally by its initial advocates, Modified Nodal Analysis expects to find nonzero pivot elements on the diagonal of the $\boldsymbol{Y}$ matrix. Modified Nodal Analysis swaps those rows with zeros on their diagonals (because of a voltage source connected to that node) with the "excess row" contributed by the voltage source voltage constraint to place unity valued elements on the diagonal entries of both rows. So, the modified nodal equations are formulated so as to have nonzero-valued elements on the diagonal of the augmented $\boldsymbol{Y}$ matrix. Given that we are starting off with no zeros on the diagonal, we can resort to *diagonal pivoting*. In this simple-minded scheme, the "best" pivot (based on fill-in or numerical considerations) is chosen from among the diagonal elements only. Thus diagonal elements stay on the diagonal. Stated differently, whenever rows i and j are exchanged, columns i and j are exchanged, too. Without further explanation at this point, we will state that pivot selection schemes that are more sophisticated than diagonal pivoting typically work better and bring Modified Nodal Analysis into closer correspondence with the original compact Nodal Analysis described in Chapter 3.

Recall that Chapter 3 briefly introduced two types of pivoting schemes: *full pivoting*, which involves choice of a pivot from any of the rows and columns in the sub-matrix being LU factored and *partial pivoting*, which involves the choice of a pivot from just the first row or first column of the sub-matrix being LU factored.

7.4 Suboptimal Ordering

Several suboptimal ordering schemes have been proposed since that presented originally by Markowitz [Markowitz57], but we will discuss the method of Markowitz here since it captures the essence of suboptimal ordering. All such schemes boil down to estimating the fill-in potential in the sub-matrix being factored in order to make a pivot selection from the remaining rows and columns not yet processed. To explain the method of Markowitz, we first introduce some simple notation:

- NZUR is the number of **N**on **Z**ero-valued elements in the first **U**pper triangular **R**ow;
- NZLC is the number of **N**on **Z**ero-valued elements in the first **L**ower triangular **C**olumn as shown in Figure 7.4.

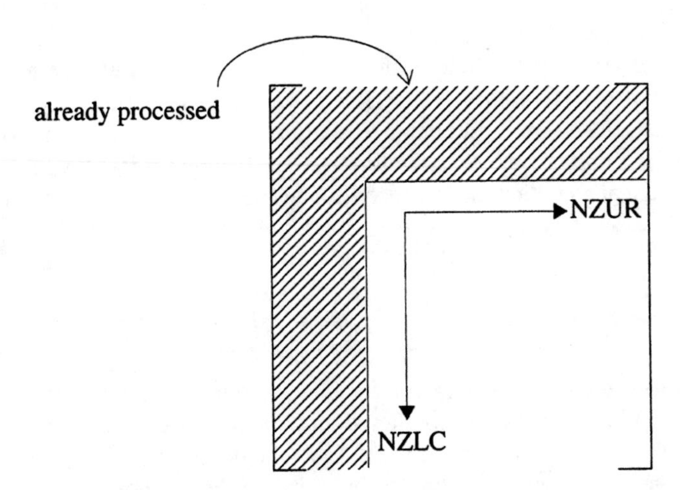

Figure 7.4 The NZUR and NZLC counts for a sparse matrix during LU factorization.

The consequences of the rows and columns already processed are included (as fill-ins) in NZUR and NZLC. So, as a pivot is selected, it can affect the values of NZUR and NZLC for subsequent rows and columns from which pivots are yet to be selected. Each nonzero potential pivot element is assigned the NZUR of its row and the NZLC of its column. Then, according to Markowitz, pivots should be selected as follows:

1. $Min\,[\,(\mathrm{NZUR}-1)\times(\mathrm{NZLC}-1)\,]$ and
2. $Min\,(\mathrm{NZLC})$.

The idea behind the Markowitz criterion is quite simple. Each of the NZURs except the pivot *potentially* causes one fill-in in its column in each of the $(\mathrm{NZLC}-1)$ rows below the pivot. The product $[\,(\mathrm{NZUR}-1)\times(\mathrm{NZLC}-1)\,]$ is not the actual number of fill-ins, but the worst-case number of fill-ins. The second criterion is a tie-breaker. If the product is the same for two pivots, we choose the one with the smaller NZLC (equivalently, the larger NZUR). The lower value of NZLC guarantees that one or more row(s) will have no fill-ins whereas a lower NZUR guarantees that one of more column(s) will be free of fill-ins. Obviously, we could equally well consider $Min\,(\mathrm{NZUR})$ as a tie-breaker, but experience has shown that not much is gained by such elaborations. Note that the Markowitz criterion only considers fill-in potential. Some sparse matrix manipulation schemes also reject any potential pivot that is too small in magnitude even if it is most desirable based on the Markowitz criteria.

In essence, the challenge of suboptimal ordering is to find row/column interchanges such that the ones with the fewest nonzero-valued elements are first. Fill-in potentials are used to find the pivots. Even for a simple pivot selection scheme, such as that advanced by Markowitz, the cost of setting up the appropriate data structures for subsequent LU factorization can be large relative to that of a simple LU factorization and Forward and Back Substitution. So, conventional wisdom says that such a symbolic analysis must be amortized over several subsequent solutions of a circuit. But picking a pivot merely on the basis of fill-in potential ignores entirely numerical problems that may arise because of small pivot size. SPICE, for example, attempts to apply the same pivot selection to all LU factorizations of the same circuit. As a nonlinear circuit may go through various regions of operation, numerical conditioning can become a problem, as described in the next section.

7.5 Numerical Conditioning and Partial Pivoting

During a nonlinear transient analysis or a frequency-domain analysis, matrix element values change from one Newton-Raphson iteration to the next and/or from one time point to the next or from one frequency to the next. Any scheme that considers numerical conditioning would have a dynamic pivoting scheme, unlike the symbolic scheme proposed in the previous section. In essence, a symbolic analysis of fill-ins would have to be carried out repeatedly for complete dynamic pivoting. Further, a dynamic scheme would lead to an unpredictable fill-in pattern, which is both cumbersome and inefficient to handle in a program. Hence, most general purpose circuit simulators do not consider numerical conditioning during pivot selection. However, this section will explore methods for and the consequences of numerical conditioning.

We have already discussed pivot conditioning in some detail at the end of Chapter 3. The essence of what we wish to avoid is the following:

$$\left.\begin{array}{l} \dfrac{big}{small} \rightarrow too\ big \\ \\ big - big \rightarrow\ too\ small \end{array}\right\}\ bad\ numbers$$

One way to take numerical conditioning into account is by means of partial pivoting, which is the interchange of columns (or rows) only, but not both, to obtain the largest (best) pivot element. With active devices in a circuit, the largest element in a row or column of the MNA $\boldsymbol{Y}$ matrix may not be on the diagonal. So it might be advantageous not to be bound exclusively by diagonal pivoting.

Under the assumption that we may be able to do it fast enough (perhaps with special purpose hardware), we can consider the following suboptimal ordering, which incorporates partial pivoting for numerical accuracy:

1. Pick the "pivot row" as that which has the lowest NZUR (break ties with the lowest NZLC).

2. Pick as the pivot element in that row its largest magnitude nonzero-valued element.
3. Carry out one step of LU factorization with that pivot; update NZURs and NZLCs as and when annihilation and/or fill-ins occur.
4. Repeat this procedure for the remaining rows/columns.

With regard to the above, note the following:

1. Processing a row with NZUR $= 1$ actually reduces by one the NZURs of the succeeding rows upon which its pivot impinges.
2. Processing a row with NZUR $= 2$ cannot increase the NZURs of succeeding rows upon which its pivot impinges, and it may soon reduce some of them.
3. Processing a row with NZUR $= 3$ increases the NZURs of subsequent rows by no more than 1.
4. Processing a row with NZUR $= 4$ is probably better than processing a row with NZUR $= 5$, etc.; this conclusion actually depends on the value of NZLC, too.

So a simple nested reduction scheme based only on minimum NZURs allows partial pivoting, and it may even go a long way toward the actual LU factorization of a very sparse matrix before causing any fill-in.

To illustrate this point, recall the first Nodal Analysis example with the nodes numbered consecutively from left to right. Its Y matrix structure was as follows:

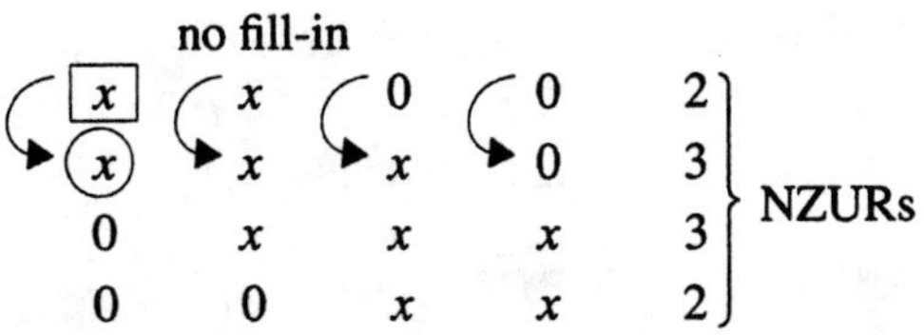

$$\begin{array}{cccc} & \text{no fill-in} & & \\ \boxed{x} & x & 0 & 0 \\ \textcircled{x} & x & x & 0 \\ 0 & x & x & x \\ 0 & 0 & x & x \end{array} \quad \left.\begin{array}{c} 2 \\ 3 \\ 3 \\ 2 \end{array}\right\} \text{NZURs} \tag{7.5.1}$$

So after one step of Gaussian Elimination we have

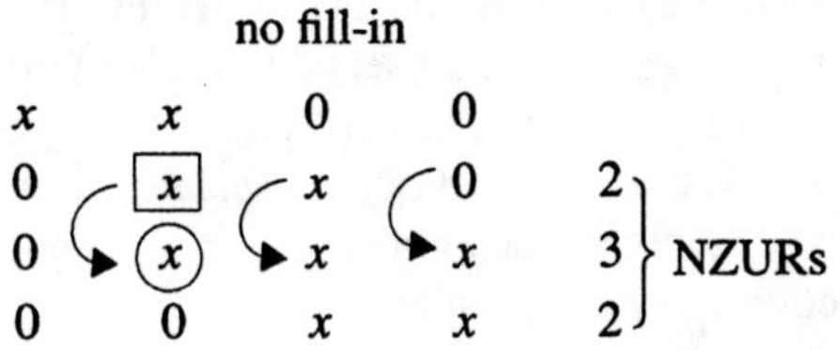

$$\begin{array}{cccc} & & \text{no fill-in} & \\ x & x & 0 & 0 \\ 0 & \boxed{x} & x & 0 \\ 0 & \textcircled{x} & x & x \\ 0 & 0 & x & x \end{array} \quad \left.\begin{array}{c} \\ 2 \\ 3 \\ 2 \end{array}\right\} \text{NZURs} \tag{7.5.2}$$

And after another step of Gaussian Elimination we have

$$\begin{matrix} x & x & 0 & 0 & \\ 0 & x & x & 0 & \\ 0 & 0 & \boxed{x} & x & 2 \\ 0 & 0 & \textcircled{x} & x & 2 \end{matrix} \left.\vphantom{\begin{matrix}a\\a\end{matrix}}\right\} \text{NZURs} \quad \text{(no fill-in)} \tag{7.5.3}$$

Recall that the Modified Nodal Analysis excess elements (e.g., zero-valued sources introduced to model controlled sources) produce rows and columns with NZURs and NZLCs that are typically one or two and sometimes three. So we can process many of them first with no fear of fill-in, and then proceed to the remainder of the $\boldsymbol{Y}$ matrix. That is precisely what was done in the name of (compact) Nodal Analysis.

Now that we understand sparse matrix considerations, a valid question to ask is whether (Modified) Nodal Analysis provides the best formulation of circuit equations with which to take advantage of sparsity. The next section will address this question.

7.6 Sparse Tableau Analysis

Sparse Tableau Analysis [Hachtel71] is described in Section A.4 of Appendix A. A quick review is provided below. For an $(n+1)$ node b branch circuit:

$$\boldsymbol{A}\boldsymbol{i}_b = 0 \qquad (n \text{ KCL equations}) \tag{7.6.1}$$

$$\boldsymbol{v}_b = \boldsymbol{A}^T\boldsymbol{v}_n \qquad (b \text{ KVL equations}) \tag{7.6.2}$$

$$\alpha\boldsymbol{v}_b + \beta\boldsymbol{i}_b = \gamma \qquad (b \text{ BCR equations}) \tag{7.6.3}$$

Thus we have a total of $(2b+n)$ equations in $(2b+n)$ unknowns. In Chapter 2 and Chapter 3 we discussed smaller, more manageable forms of circuit equations like Nodal Analysis and Modified Nodal Analysis. But now that we understand sparse matrix considerations, we might ask, "Why not address the larger set of $(2b+n)$ equations in $(2b+n)$ unknowns directly? Perhaps the larger set of equations will be compensated for by gains in sparsity." And that is exactly what Sparse Tableau Analysis attempts to do.

Before getting into sparsity considerations in Sparse Tableau Analysis, it helps to have some insight to the generalized BCRs in equation (7.6.3). Consider, for example, the circuit in Figure 7.5. The branch-node incidence matrix is

$$A = \begin{bmatrix} 1 & 1 & 0 & 0 & 0 & 0 & 0 \\ 0 & -1 & 1 & 1 & 1 & 0 & 0 \\ 0 & 0 & 0 & 0 & -1 & 1 & 1 \end{bmatrix} \tag{7.6.4}$$

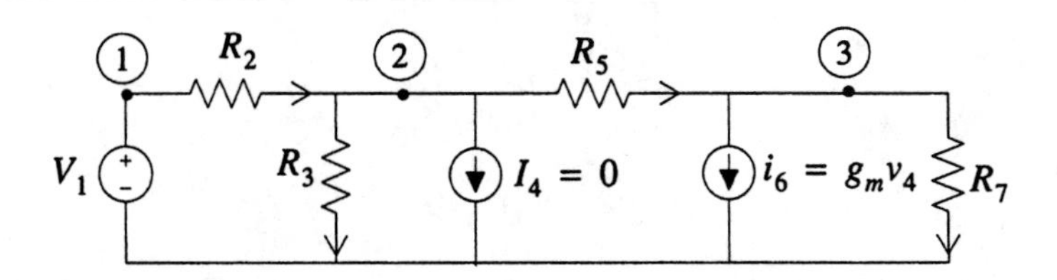

Figure 7.5 A simple linear circuit with a voltage controlled current source.

The seven Branch Constitutive Relations are the following:

$$\begin{aligned}
v_1 &= V_1 \\
v_2 - R_2 i_2 &= 0 \qquad or \qquad \frac{1}{R_2} v_2 - i_2 = 0 \\
v_3 - R_3 i_3 &= 0 \qquad or \qquad \frac{1}{R_3} v_3 - i_3 = 0 \\
i_4 &= 0 \\
v_5 - R_5 i_5 &= 0 \qquad or \qquad \frac{1}{R_5} v_5 - i_5 = 0 \\
g_m v_4 - i_6 &= 0 \qquad or \qquad v_4 - \frac{1}{g_m} i_6 = 0 \\
v_7 - R_7 i_7 &= 0 \qquad or \qquad \frac{1}{R_7} v_7 - i_7 = 0
\end{aligned} \tag{7.6.5}$$

We will consider only the left hand set of equations for the purpose of this example. But in actual circuit simulation we may want to renormalize some of them as on the right to aid in numerical conditioning. The various matrices of the generalized BCRs follow from the set of seven equations on the left:

$$\alpha = \begin{bmatrix} 1 & 0 & 0 & 0 & 0 & 0 & 0 \\ 0 & 1 & 0 & 0 & 0 & 0 & 0 \\ 0 & 0 & 1 & 0 & 0 & 0 & 0 \\ 0 & 0 & 0 & 0 & 0 & 0 & 0 \\ 0 & 0 & 0 & 0 & 1 & 0 & 0 \\ 0 & 0 & 0 & g_m & 0 & 0 & 0 \\ 0 & 0 & 0 & 0 & 0 & 0 & 1 \end{bmatrix}$$

$$\beta = \begin{bmatrix} 0 & 0 & 0 & 0 & 0 & 0 & 0 \\ 0 & -R_2 & 0 & 0 & 0 & 0 & 0 \\ 0 & 0 & -R_3 & 0 & 0 & 0 & 0 \\ 0 & 0 & 0 & 1 & 0 & 0 & 0 \\ 0 & 0 & 0 & 0 & -R_5 & 0 & 0 \\ 0 & 0 & 0 & 0 & 0 & -1 & 0 \\ 0 & 0 & 0 & 0 & 0 & 0 & -R_7 \end{bmatrix} \quad (7.6.6)$$

$$\gamma^T = \begin{bmatrix} V_1 & 0 & 0 & 0 & 0 & 0 & 0 \end{bmatrix}$$

Both the α and β matrices are very sparse. They are also singular (both the matrices have a row full of zeros), but we can invert them "partially" as in going from the left to the right hand branch relations, as shown in (7.6.5) above. We note too that it is very straightforward in this approach to characterize any kind of dependent source as well as so called "singular elements," such as the limiting case ideal operational amplifier shown in Figure 7.6. The operational amplifier provides three BCR equations in six unknown terminal variables, which is consistent with the observation that the BCRs in general must provide b equations in $2b$ unknowns.

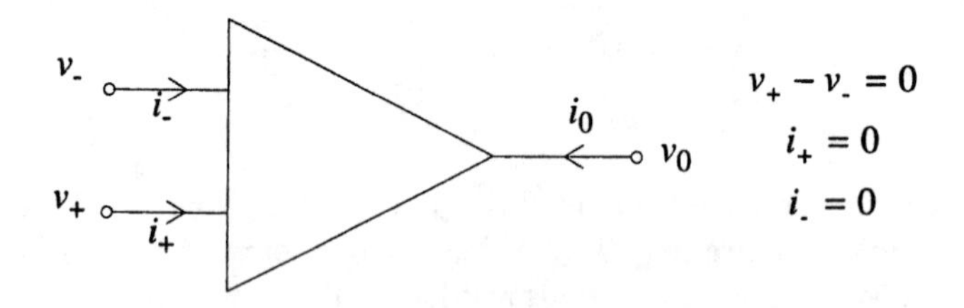

Figure 7.6 An ideal operational amplifier.

In general, the BCRs of the form in (7.6.3) can be formed for all of the following elements:

- the linear independent and dependent elements

$$\left.\begin{aligned} v - Ri &= 0 \\ Gv - i &= 0 \end{aligned}\right\} \quad \text{resistance/conductance}$$

$$\left.\begin{aligned} v &= V \\ i &= I \end{aligned}\right\} \quad \text{independent sources} \tag{7.6.7}$$

$$\left.\begin{aligned} g_m v_x - i_y &= 0 \\ i_x - \beta i_y &= 0 \\ v_x - \mu v_y &= 0 \\ v_x - r_m i_y &= 0 \end{aligned}\right\} \quad \text{dependent sources}$$

- the companion models for the energy storage elements

$$\begin{aligned}
v(t+\Delta t) &= v(t) + \frac{\Delta t}{C} i(t^+) && \text{FE Capacitance} \\
v(t+\Delta t) &= v(t) + \frac{\Delta t}{C} i(t+\Delta t) && \text{BE Capacitance} \\
v(t+\Delta t) &= v(t) + \frac{\Delta t}{2C} \{ i(t^+) + i(t+\Delta t) \} && \text{TR Capacitance} \\
i(t+\Delta t) &= i(t) + \frac{\Delta t}{L} v(t^+) && \text{FE Inductance} \\
i(t+\Delta t) &= i(t) + \frac{\Delta t}{L} v(t+\Delta t) && \text{BE Inductance} \\
i(t+\Delta t) &= i(t) + \frac{\Delta t}{2L} \{ v(t^+) + v(t+\Delta t) \} && \text{TR Inductance}
\end{aligned} \tag{7.6.8}$$

The notation t^+ refers to situations involving step voltages or currents, as described in Chapter 4.

- the Newton-Raphson linearization model for a nonlinear element characterized by $f(v, i) = 0$

$$\frac{\partial f}{\partial v} \hat{v} + \frac{\partial f}{\partial i} \hat{i} = -f(v, i) \tag{7.6.9}$$

In terms of equations (7.6.1), (7.6.2), and (7.6.3), we can now formulate the Sparse Tableau Analysis equations:

$$\begin{bmatrix} 1 & 0 & -A^T \\ 0 & A & 0 \\ \alpha & \beta & 0 \end{bmatrix} \begin{bmatrix} v_b \\ i_b \\ v_n \end{bmatrix} = \begin{bmatrix} 0 \\ 0 \\ \gamma \end{bmatrix} \begin{matrix} \text{KVL} \\ \text{KCL} \\ \text{BCRs} \end{matrix} \tag{7.6.10}$$

Not only are there four zero-matrix entries in the Sparse Tableau Matrix, but the remaining five nonzero matrix partitions are sparse as well. So we set out to solve this very large, sparse set of $(2b + n)$ equations in $(2b + n)$ unknowns. The advantages are

- sparsity of equations.
- all circuit variables are directly obtained by the solution procedure.
- BCRs are straightforward.

Of course, we have more equations, but the increased sparsity may compensate for that.

7.7 *Qualitative Attributes of the Sparse Tableau*

Recall the first Markowitz criterion for suboptimal pivot selection to attempt to reduce potential fill-ins:

$$Min\,[\,(\text{NZUR} - 1) \times (\text{NZLC} - 1)\,] \tag{7.7.1}$$

Tacit to this criterion is the observation that an NZUR $= 1$ row or an NZLC $= 1$ column can create no further fill-ins. In fact, an NZUR $= 1$ row will reduce by one the NZUR of succeeding rows with nonzeros in their pivot columns. Recall from our previous discussion that rows with NZUR $= 2$ cannot increase the NZUR of subsequent rows. Hence we could conceive of a strategy by which all the NZUR $= 1$ rows would be processed first, and then the NZUR $= 2$ rows, and so on. Let us see how such a $Min\,(\text{NZUR})$ scheme would fare on a Sparse Tableau matrix.

- The KCL rows have NZURs equal to the number of branches incident on the corresponding node (2 for the node between branches in series, 3 or more otherwise).
- The KVLs have NZURs of two for all branches with one end grounded and NZURs of three for all floating branches. (Note that branches in parallel can be combined on a local basis without generating fill-in by eliminating one of the two branches' voltage variables in favor of the other's.)
- The BCRs have NZUR of one for all independent sources, including those that arise from Forward Euler approximated energy storage elements; and NZURs of two for

all other elements, including dependent sources and Newton-Raphson linearized equivalents of nonlinear elements. (Some multi-terminal elements may have BCRs with NZURs of three or greater.)

On the basis of immediate fill-in potential, the following ordering would be followed:

1. Independent source BCRs.
2. Remaining BCRs with NZUR = 2.
3. KVLs for grounded branches.
4. KCLs for series connections.
5. Remaining BCRs with NZUR = 3.
6. Combination of parallel branches (this is a special NZUR = 3 operation).
7. Remaining KVLs for floating branches.
8. KCLs with NZUR = 3 (which are typical!).
9. KCLs or any remaining BCRs with NZUR ≥ 4.

Remember that selecting a pivot element from an NZUR = 1 row can only reduce the NZURs of some rows to be processed subsequently. And selecting a pivot element from an NZUR = 2 row cannot increase the NZURs of any row to be processed subsequently. So further fill-in economies can be generated by the row processing in categories 1 to 4, above.

We note that "processing a row" is equivalent to eliminating its pivot element as an unknown variable in subsequent rows of the matrix. It is recovered later in the course of Back Substitution. Since the nontrivial KCLs tend to be processed last, or so it appears qualitatively, we would expect some branch currents to remain as essential variables in the final flow columns. This approach tends to be opposite what occurs in Nodal Analysis, which eliminates branch currents in favor of node voltages to the extent possible.

Much of the above discussion is merely qualitative. We need only consider

$$\min (\text{NZUR});\ \max (\text{Pivot Magnitude})$$

to implement a reasonable STA LU factorization scheme. We should note though that we would require dynamic storage allocation and expect a new ordering with each new analysis. Traditional circuit simulators, such as SPICE, seek to use a single suboptimal ordering over many analyses so as to amortize its cost of creation. But the scheme discussed above would be more numerically robust, and the simplicity of the scheme would render it an excellent candidate for hardware acceleration. If we are careful in working with properly normalized rows, we can even consider replacing underflows with zero values and proceeding to process the rows in which they occur with appropriately reduced NZURs.

7.8 Relation of the Sparse Tableau to Other Solution Schemes

Reduced Tableau equations are derived in Appendix A (see equation (A.4.3)). In essence, Reduced Tableau equations are generated as follows. Substitute the KVL equations

$$v_b = A^T v_n \tag{7.8.1}$$

into the BCRs

$$\alpha v_b + \beta i_b = \gamma \tag{7.8.2}$$

to get

$$\alpha A^T v_n + \beta i_b = \gamma \tag{7.8.3}$$

Add in the KCL equations

$$A i_b = 0 \tag{7.8.4}$$

to get the Reduced Tableau equations

$$\begin{array}{c} \\ n \\ b \end{array}\begin{array}{c} \begin{array}{cc} b & n \end{array} \\ \begin{bmatrix} A & 0 \\ \beta & \alpha A^T \end{bmatrix} \end{array} \begin{bmatrix} i_b \\ v_n \end{bmatrix}\begin{array}{c} b \\ n \end{array} = \begin{bmatrix} 0 \\ \gamma \end{bmatrix}\begin{array}{c} n \\ b \end{array} \tag{7.8.5}$$

The above procedure can be thought of as a partial LU factorization. Starting with the Sparse Tableau equations,

$$\begin{array}{c} \\ b \\ n \\ b \end{array}\begin{array}{c} \begin{array}{ccc} b & b & n \end{array} \\ \begin{bmatrix} 1 & 0 & -A^T \\ 0 & A & 0 \\ \alpha & \beta & 0 \end{bmatrix} \end{array} \begin{bmatrix} v_b \\ i_b \\ v_n \end{bmatrix}\begin{array}{c} b \\ b \\ n \end{array} = \begin{bmatrix} 0 \\ 0 \\ \gamma \end{bmatrix}\begin{array}{c} b \\ n \\ b \end{array} \tag{7.8.6}$$

We can write the following partial LU factorization because the upper left partition is an identity matrix:

$$\begin{bmatrix} 1 & 0 & 0 \\ 0 & 1 & 0 \\ \alpha & 0 & 1 \end{bmatrix}\begin{bmatrix} 1 & 0 & -A^T \\ 0 & A & 0 \\ 0 & \beta & \alpha A^T \end{bmatrix}\begin{bmatrix} v_b \\ i_b \\ v_n \end{bmatrix} = \begin{bmatrix} 0 \\ 0 \\ \gamma \end{bmatrix} \tag{7.8.7}$$

The overall LU factorization of a $(2b+n) \times (2b+n)$ matrix has been reduced to that of the following $(b+n) \times (b+n)$ matrix:

$$\begin{array}{c c} & \begin{array}{cc} b & n \end{array} \\ \begin{array}{c} n \\ b \end{array} & \begin{bmatrix} A & 0 \\ \beta & \alpha A^T \end{bmatrix} \end{array} \tag{7.8.8}$$

Once we have solved the Reduced Tableau equations (7.8.5), we can easily recover the branch voltages v_b via KVL.

To get from the Reduced Tableau to (compact) Nodal Analysis, if possible, we invert β:

$$\begin{aligned} \beta i_b + \alpha A^T v_n &= \gamma \\ i_b + \beta^{-1} \alpha A^T v_n &= \beta^{-1} \gamma \end{aligned} \tag{7.8.9}$$

and then

$$\begin{aligned} A\,(i_b + \beta^{-1} \alpha A^T v_n) &= A \beta^{-1} \gamma \\ A i_b + A \beta^{-1} \alpha A^T v_n &= A \beta^{-1} \gamma \\ \underbrace{A \beta^{-1} \alpha A^T}_{Y}\ v_n &= \underbrace{A \beta^{-1} \gamma}_{J} \end{aligned} \tag{7.8.10}$$

The incidence matrix A is comprised of "connection vectors" (columns with just a +1 and −1 in them), so $A\beta^{-1}\alpha A^T$ is another way of characterizing the composition of the Y matrix via stamps. But β does not always have an inverse, as in the earlier example, in which case we would have to resort to Modified Nodal Analysis. But MNA equations -- and even pure nodal equations -- are best formulated directly, and not via the circuitous route sketched above. The above discussion was to show the relation between MNA and STA for intuition only.

We can repeat the qualitative row-ordering analysis of the previous section on the Reduced Tableau formulation of equation (7.8.5). Prior to any row processing we have the following:

1. NZUR = 1 for the BCRs of independent current sources and grounded independent voltage sources.
2. NZUR = 2 for KCL of nodes with incidence 2 and for the BCRs of floating independent voltage sources, grounded two-terminal elements.

3. NZUR = 3 for parallel connections of floating two-terminal elements and for BCRs of floating two-terminal elements including singly-controlled dependent sources, and for most KCL equations.
4. NZUR ≥ 4 for remaining KCL relations and some multi-terminal element BCRs.

Qualitatively, this ordering appears to be opposite that of Nodal Analysis.

7.9 The Original Sparse Tableau Approach

In the overview in Chapter 1 we observed that nonlinear transient analysis involves repeated nonlinear iterations performed within a time advancement loop. Either to account for manufacturing variations or to understand the sensitivity of the circuit to various parameters, there may be an outer loop performed with perturbed element values. The design process, in turn, consists of repeatedly evaluating multiple circuit configurations. These analysis procedures can be viewed as being comprised as a set of nested loops, shown here in Figure 7.7. Hachtel et al. [Hachtel71] advocated the ordering of the equations to maximize reuse of LU operations across successive iterations. To accomplish that, equations would be ordered such that topological constraints like KVL and KCL would be processed first since they change least often.

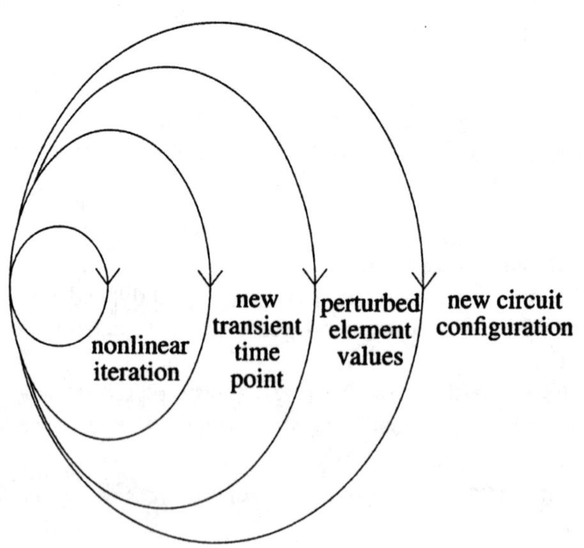

Figure 7.7 The nested loops for circuit simulation.

Given the Sparse Tableau equations in (7.8.6), the first $b+n$ rows are processed *a priori,* since only α, β, and γ may change through the course of the analysis. And we need not worry about the size of the pivots in processing the first $b+n$ rows since they are comprised only of ± 1 s or 0s. In fact, we can use integer arithmetic, or even purely logical operations, in processing the effects of those rows upon themselves. Nor are floating point multiplies entailed in the processing of the first $b+n$ rows, but only floating point additions and/or subtractions in the final b BCR rows. For example, we need only add and subtract appropriately to go from the Sparse Tableau to the Reduced Tableau shown in (7.8.5).

Similarly, in processing the next n rows no floating point multiples are entailed. But if we proceed in that order, we are trying to preserve the node voltages as essential variables. And that is what is done for Modified Nodal Analysis. So it is not surprising that the original form of Sparse Tableau Analysis has been largely abandoned in favor of Modified Nodal Analysis, which arrives at roughly the same set of reduced equations in a more straightforward fashion. The generally accepted conclusion is that MNA is computationally more efficient than STA, but the modeling ease and numerical robustness of STA are superior.

Sparse Tableau Analysis might be a good approach for AWE (see Chapter 6), if frequency shifting can be avoided, because with AWE, energy storage elements are replaced by $(\text{NZUR} = 1)$ independent sources. But such an approach coincides more closely with the

$$Min\,(\text{NZUR})\,;Max\,(\text{Pivot Magnitude})$$

scheme than does the originally advocated Sparse Tableau Analysis.

The original advocates of Sparse Tableau Analysis only considered pivot size in the processing of the final b BCR rows if it dropped below a predetermined threshold magnitude. Their LU factorization scheme was based first on element variability, and then on fill-in potential, and only finally on pivot magnitude:

1. First, process topological constraints as described above.
2. Next, process linear time-varying elements (linear resistance/conductance elements and fixed time step, Backward Euler or Trapezoidal capacitance and inductance elements).
3. Next, process linear time-varying elements (variable time step, Backward Euler or Trapezoidal capacitance and inductance elements).
4. Finally, process nonlinear elements.

The rationale here is that iteration is entailed in a nonlinear transient analysis, so the elements are ordered to render the inner loops most efficient. Of course, within each category, fill-in potential is minimized.

7.10 Some Sparse Tableau Modeling Considerations

For the bipolar junction transistor (BJT) shown in Figure 7.8 the most simplified models for the cutoff, forward active and saturation regions of operation are shown in Figures 7.9, 7.10, and 7.11 respectively.,

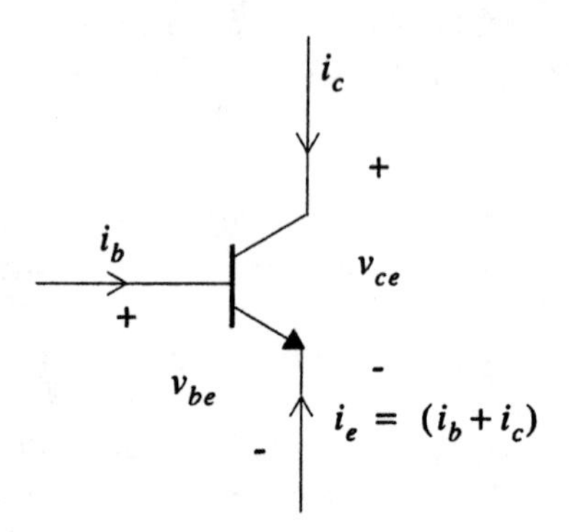

Figure 7.8 An NPN bipolar junction transistor (BJT).

Cutoff: $v_b < V_{be}(on) \sim 0.65\text{V} \Rightarrow i_b = 0 \;\&\; i_c = 0 \;\&\; i_e = 0$

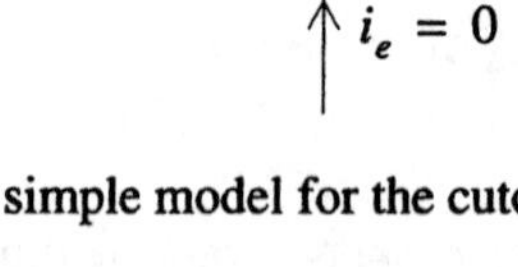

$i_c = 0$

$i_b = 0$

$i_e = 0$

Figure 7.9 A simple model for the cutoff region of operation.

There is also a reverse active mode of operation, which mirrors forward active but usually with a much smaller value of current gain, β_R, which we need not consider for the purpose of the present discussion.

The above model works very well for a "back-of-the-envelope" analysis of bipolar digital circuits. It is even reasonable for a "first cut" analysis of analog bipolar dc circuit behavior. (We would need to supplement such a simple model with the bipolar junction current-voltage relation of 60mV/decade at room temperature, etc., to begin to approach a

Forward Active: $v_{be} = V_{be}(on) \sim 0.65\text{V}$ & $v_{ce} < V_{ce}(sat) \sim 0.15\text{V} \Rightarrow i_c = \beta_f i_b$

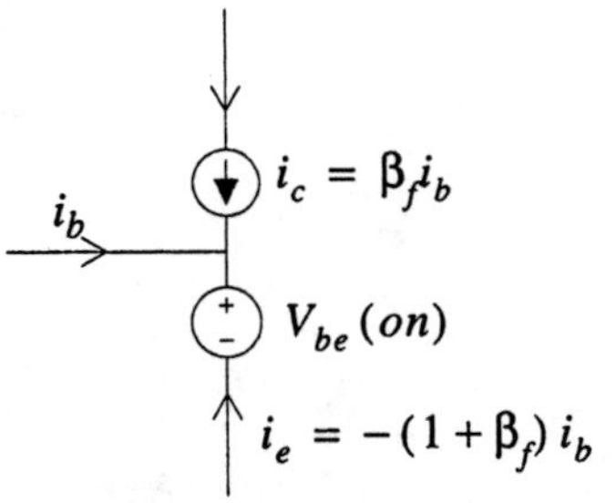

$i_c = \beta_f i_b$

i_b

$V_{be}(on)$

$i_e = -(1 + \beta_f)\, i_b$

Figure 7.10 A simple model for the forward active region of operation.

Saturation: $v_{be} = V_{be}(on) \sim 0.65\text{V}$ & $v_{ce} = V_{ce}(sat) \sim 0.15\text{V}$

i_c

$V_{ce}(sat)$

i_b

$V_{be}(on)$

$i_e = -(i_b + i_c)$

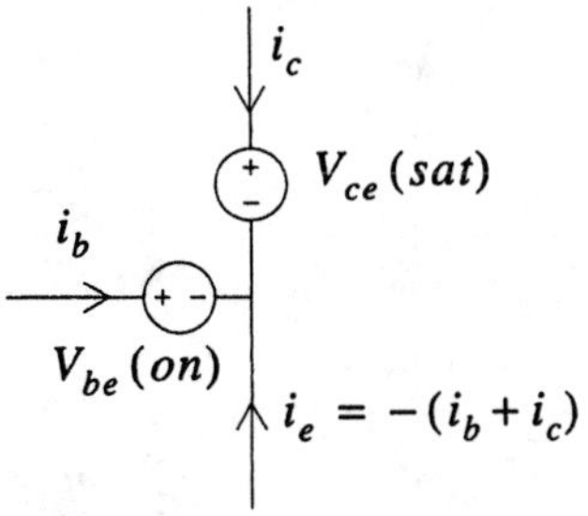

Figure 7.11 A simple model for the saturation region of operation.

more exact analysis.) But such a simple model would not work well in (Modified) Nodal Analysis, because its changes from one region of operation to another are tantamount to topological changes. It would work well with a

$$Min\,(\text{NZUR})\,;Max\,(\text{Pivot Magnitude})$$

Sparse Tableau Analysis scheme, however, which would reformulate the circuit equations for each analysis iteration.

Consider voltage sources to be generalized short circuits and current sources to be generalized open circuits, resulting in Figure 7.12. It is apparent that moving from one model to the next during the course of a simulation would be equivalent to changing the circuit topology. It is, therefore, not only for accuracy that more complete transistor models such as the one shown in Figure 7.13 are used, but also because they are more suitable to simulation algorithms.

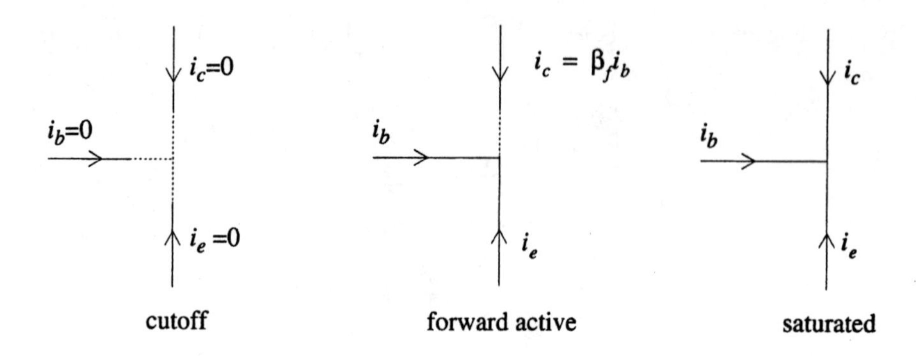

Figure 7.12 Regions of operation with the voltage sources modeled as short circuits.

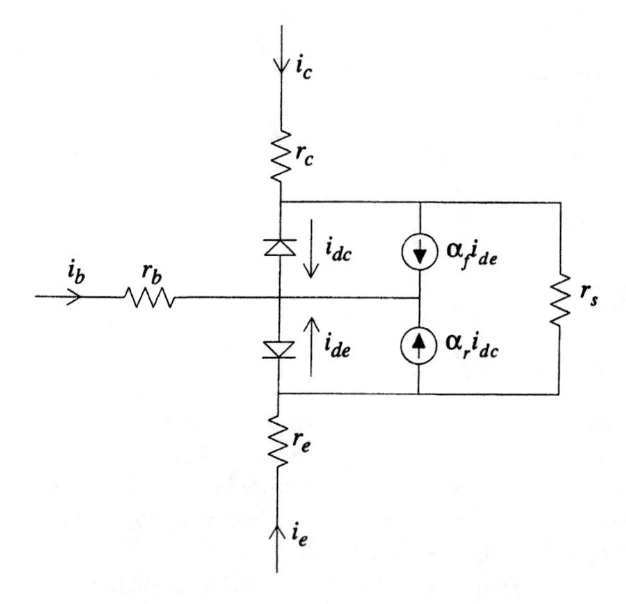

Figure 7.13 A more detailed BJT model.

In the model in Figure 7.13,

$$i_d = I_s\left(e^{\frac{qv}{nkT}} - 1\right) \tag{7.10.1}$$

for each junction. There are even more elaborate models with even more parasitic elements, which are said to provide even greater accuracy. For a "quick and dirty" analysis we may attempt to zero or ignore some of the parasitic element values. But a (Modified) Nodal Analysis program may rely on them to be finite to avoid producing a singular matrix that cannot be LU factored. Note that there are approximately four branches per node inside the dc device model. There are probably fewer branches impinging on those nodes where the devices actually interconnect! So it may require several additional nodes and branches in a circuit to formulate "reliable and accurate" BJT device models. But the simpler model introduced earlier often provides (more than) adequate answers with much better computational efficiency if we can handle it. In Chapter 9 we'll see that we may be able to use adjoint sensitivity to elaborate a simple model efficiently, if necessary. And a simple model may be preferable for statistical analysis, where we must make several (efficient) runs based on variations in environmental and processing parameters.

We have seen that Modified Nodal Analysis cannot cope with simple BJT models. When bipolar transistors are used for analog design, more accurate models are needed given the nature of the design, so this failing is not an issue. Simple MOS devices however, such as that in Figure 7.14, are well suited to Nodal Analysis.

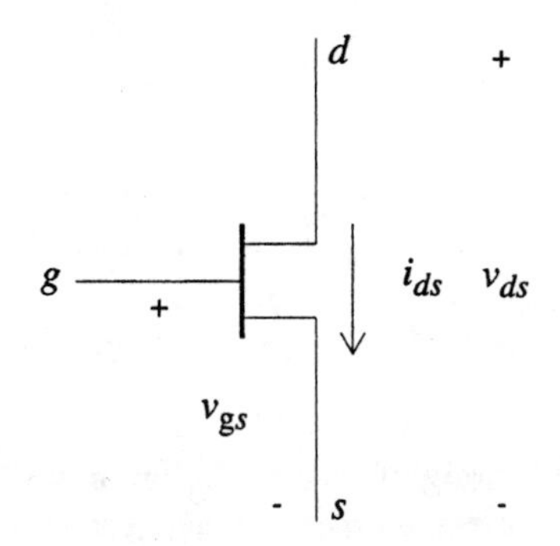

Figure 7.14 An N type MOS transistor.

For dc analysis the MOS gate terminal is always open. For small-signal analysis, when $v_{gs} > V_T$ and for large values of v_{ds} the source-drain channel can be modeled in terms of a voltage-controlled current source (a generalized open) as shown in Figure 7.15. This is, of course, a simplification of the small-signal MOSFET model shown in Chapter 2. Note that g_m is a nonlinear function of v_{ds}. Nodal Analysis works well for models such as this one.

Switch level simulators model the source-drain channel as a voltage-controlled switch that is on or off depending on whether v_{gs} is greater or less than V_T, as shown in Figure 7.16. Such a simple model is not handled well by Nodal Analysis, so it is difficult to mix switch and circuit level simulators. The simulator, RSIM [Terman83] adopts the compromise shown in Figure 7.17.

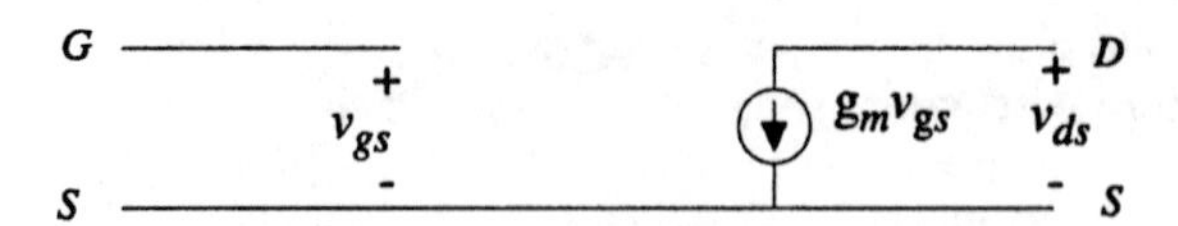

Figure 7.15 A simple small-signal model for an NMOSFET in saturation.

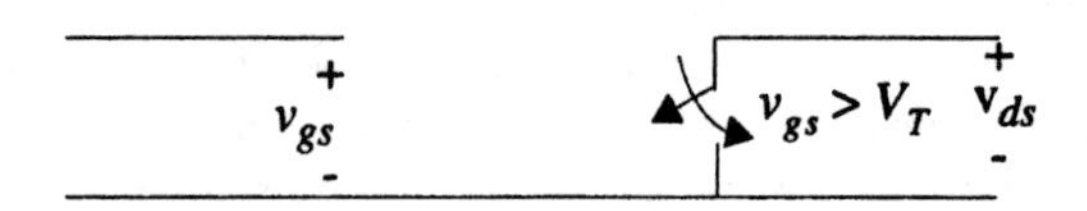

Figure 7.16 A switch-level model for an NMOSFET.

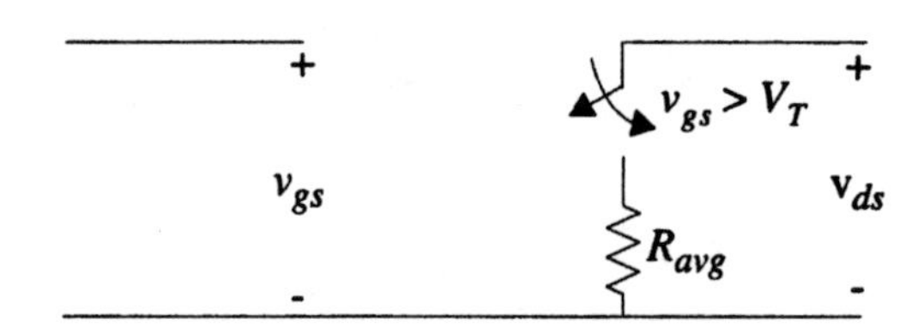

Figure 7.17 A switch-resistance model for an NMOSFET.

This model can be used in a straightforward fashion in Nodal Analysis and in conjunction with capacitance energy storage elements it can provide crude timing information. It is not always easy, though, to find a good value of R_{avg}, which may differ between turn-on and turn-off. "Model tuning" is usually employed to make such crude models reasonably accurate.

7.11 References

[Markowitz57] H. M. Markowitz. The Elimination Form of the Inverse and Its Application to Linear Programming. *Management Science*, vol. 3, pp. 255-269, April 1957.

[Desoer69] C. A. Desoer and E. S. Kuh. *Basic Circuit Theory.* McGraw-Hill, 1969.

[Hachtel71] G. D. Hachtel, R. K. Brayton, and F. G. Gustavson. The Sparse Tableau Approach to Network Analysis and Design. *IEEE Transactions on Circuit Theory,* vol. CT-18, pp. 101-118, January 1971.

[Terman83] C. J. Terman. Simulation Tools for Digital LSI Design. *Ph.D. Thesis, Massachusetts Institute of Technology,* September 1983.

Chapter 8 Circuit Partitioning and Large Change Sensitivity

So far, we have considered the analysis of a single circuit with a given topology and a given set of component values. In practical applications, however, there are many reasons for analyzing variants of the basic circuit design. We may, for example, need to ensure that the circuit works as specified in the context of manufacturing variations or operating temperature variations. We may want to assess the behavior of the circuit for a range of values of a certain element. Computing the change in circuit response with respect to circuit changes is called *sensitivity analysis*. The goal is to find the sensitivity of the behavior of a circuit to variations in the underlying, or *nominal*, circuit. Sensitivity computation has many applications in circuit tuning, optimization, reliability analysis, periodic steady state analysis, critical path analysis, and so on.

A change in the nominal circuit can be a *large change* (or *large scale change*) or a *small change*. Examples of large changes are the addition of a finite resistance between an existing pair of nodes, the removal ("cutting") of an existing circuit branch, or the "splitting" of an existing node. They usually involve either topological changes or large variations in component values. Understanding how to analyze these situations leads to the ability to analyze circuit partitions separately and then combine the subcircuit analyses to obtain the solution of a larger circuit. An example of a small change is the increase or decrease of a resistance value or the width of a transistor by a fraction of 1 percent. Sensitivity values give us the direction and magnitude of the circuit response change with respect to a change in the nominal circuit. Small change sensitivity is a special case of large change sensitivity and special, more efficient, techniques exist for computing small change sensitivities.

Note that one method of computing sensitivities is by using *finite differences*, whereby multiple simulations are performed to compute the sensitivity. Finite difference methods are often inefficient and numerically sensitive. In contrast, this chapter and the next will address incremental methods of sensitivity analysis. An incremental method assumes that the nominal analysis has been performed, and asks the question, "Is it possible to re-use the nominal solution and determine the sensitivity with a small incremental overhead?" Note that the nominal analysis might have to be performed in a particular manner so as to be prepared for the incremental sensitivity analysis. This chapter will address large change sensitivity and the next chapter will address small change sensitivity, both on incremental bases.

8.1 Adding a Resistance Between Two Nodes

Given an original circuit with nodes k and l shown in Figure 8.1, suppose that we have already solved this circuit in terms of Nodal Analysis equations, so that we know

$$Yv = J \qquad \text{or} \qquad v = Y^{-1}J \tag{8.1.1}$$

The latter is, of course, our notation to indicate the solution. We would not actually invert the Y matrix, but rather LU factor it. Next, we pose the following question: "What is the effect on this solution of adding a resistance of value R between nodes k and l (as shown in Figure 8.2)?"

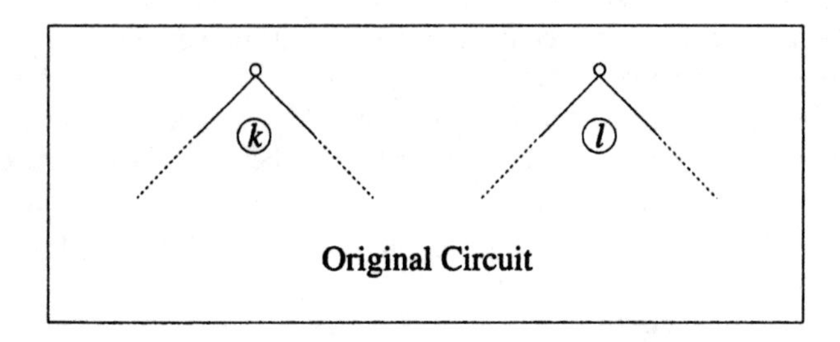

Figure 8.1 A linear(ized) circuit with nodes k and l.

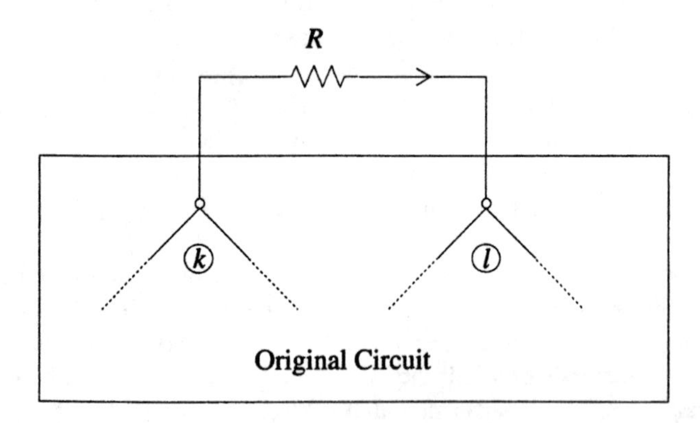

Figure 8.2 Adding a resistance between nodes k and l.

Note that we have provided this added resistance branch with an arbitrary orientation, from node k to node l, which we will need for convenience in subsequent discussions. To solve this problem formally, we first define a connection vector, ξ_{kl}:

$$\xi_{kl} = \begin{bmatrix} 0 \\ \vdots \\ 0 \\ +1 \\ 0 \\ \vdots \\ 0 \\ -1 \\ 0 \\ \vdots \\ 0 \end{bmatrix} \begin{matrix} \\ \\ \\ \leftarrow k^{th} \text{ row} \\ \\ \\ \\ \leftarrow l^{th} \text{ row} \\ \\ \\ \\ \end{matrix} \left. \begin{matrix} \\ \\ \\ \\ \\ \\ \\ \\ \\ \\ \\ \end{matrix} \right\} \text{all other rows are zero}$$

In terms of this connection vector we can write the open circuit voltage between nodes k and l as (see Figure 8.3):

$$v_{oc} = v_k - v_l = \xi_{kl}^T v = \xi_{kl}^T Y^{-1} J \tag{8.1.2}$$

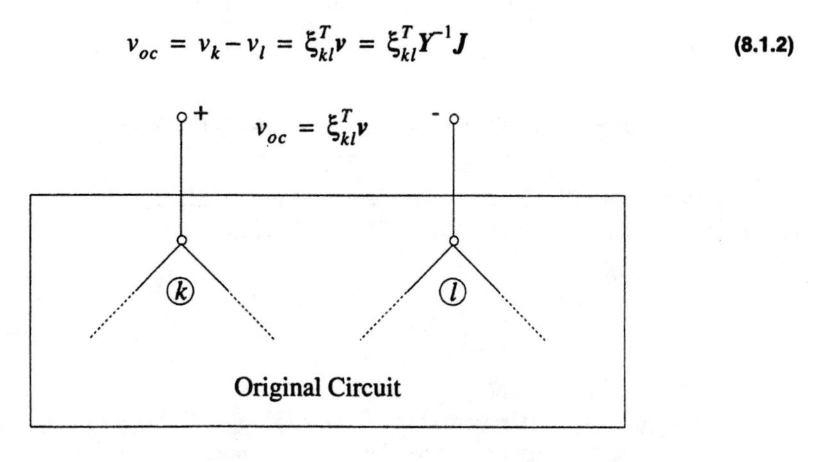

Figure 8.3 The open circuit voltage between nodes k and l.

Next, we note that the solution $Y^{-1}\xi_{kl}$ would provide the vector of voltages that would prevail if all independent sources in the original circuit were zeroed (independent voltage sources replaced by short circuits and independent current sources replaced by open circuits) and an independent current source of unit value were connected from node l to node k. The voltage that would appear across nodes k and l would then be $\xi_{kl}^T Y^{-1} \xi_{kl}$.

Therefore, the Thevenin equivalent resistance across that $\{k, l\}$ node pair is

$$R_{TH} = \xi_{kl}^T Y^{-1} \xi_{ki} \tag{8.1.3}$$

as shown in Figure 8.4. Again, we emphasize that we would not invert Y; to obtain R_{TH} is a straightforward computation in terms of the LU factors of Y.

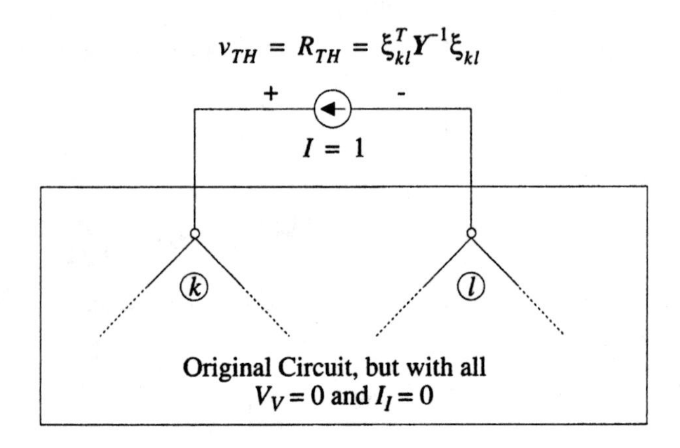

Figure 8.4 The Thevenin equivalent resistance between nodes k and l.

Using (8.1.2) and (8.1.3), from the viewpoint of the added resistance R, the situation appears to be that shown in Figure 8.5. The resulting (loop) current that flows in the resistance is

$$i_R = \frac{v_{oc}}{R + R_{TH}} \tag{8.1.4}$$

The substitution theorem [Desoer69] says that, so long as the circuit solution is unique, we can replace any element by an independent voltage source that constrains the same value of voltage or an independent current source that constrains the same value of current and not change the values of any voltages or currents throughout the circuit (refer to Figure 8.6). Under these circumstance *all* circuit voltages and currents are the same as they would be if the resistance R were connected between nodes k and l.

We will now use superposition to derive the required solution of the circuit in Figure 8.6. First, we will only apply I_R by zeroing out all other independent sources in the circuit, to obtain the set of voltages

$$v' = Y^{-1} \xi_{kl} (-I_R) \tag{8.1.5}$$

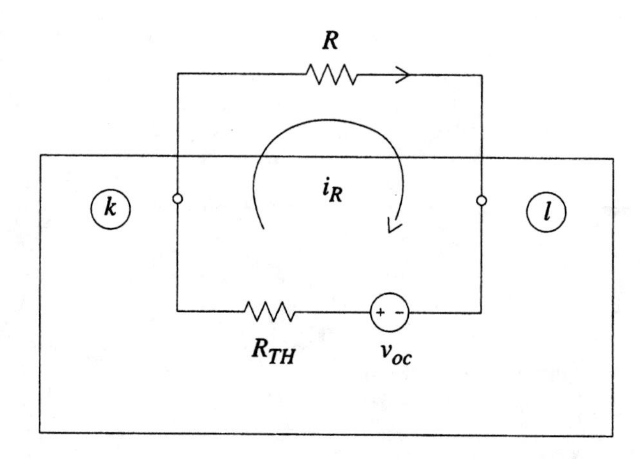

Figure 8.5 The equivalent circuit between nodes k and l.

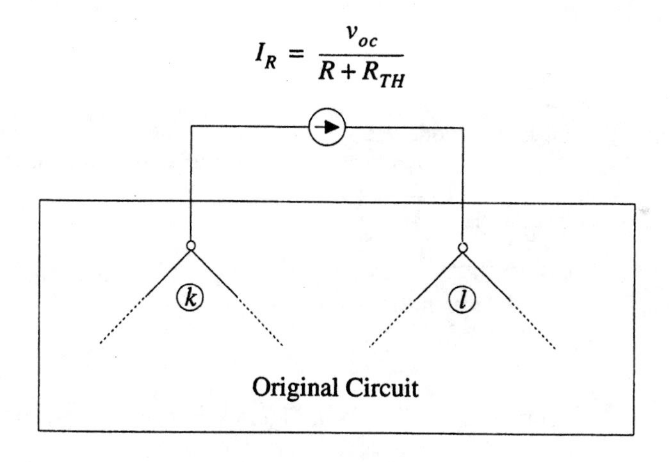

Figure 8.6 Applying the substitution theorem for the resistor between nodes k and l.

by linearity, as shown in Figure 8.7. Next, we will zero I_R and apply all the independent sources in the original circuit. We know the original solution is $v = Y^{-1}J$, so we add these two solutions together to get

$$\hat{v} = v + v' = v - I_R Y^{-1} \xi_{kl} \tag{8.1.6}$$

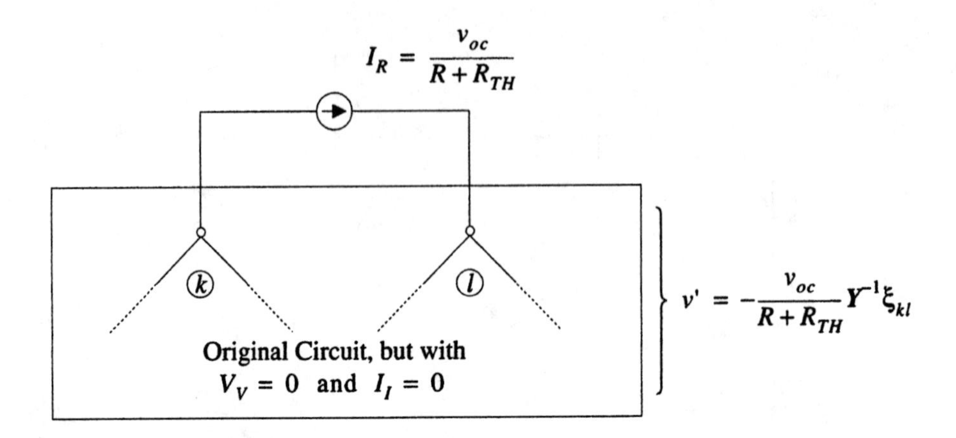

Figure 8.7 Applying the substitution theorem and calculating the change in node voltages.

or

node voltage contribution due to +1A independent current source from node l to node k

$$\hat{v} = v - \frac{v_{oc}}{R+R_{TH}} \overbrace{Y^{-1}\xi_{kl}}$$

original source contribution

$$I_R \text{ normalization} = \frac{\xi_{kl}^T v}{R + \xi_{kl}^T Y^{-1} \xi_{kl}}$$

So, finally, we have the complete solution from Figure 8.8 which is equivalent to that in Figure 8.9.

$$\begin{aligned}\hat{v} &= v - \frac{v_{oc}}{R+R_{TH}} Y^{-1}\xi_{kl} \\ &= Y^{-1}J - \frac{\xi_{kl}^T Y^{-1} J}{R + \xi_{kl}^T Y^{-1}\xi_{kl}} Y^{-1}\xi_{kl}\end{aligned} \tag{8.1.7}$$

Note that $\xi_{kl} Y^{-1} J$ is a scalar, v_{oc}, hence

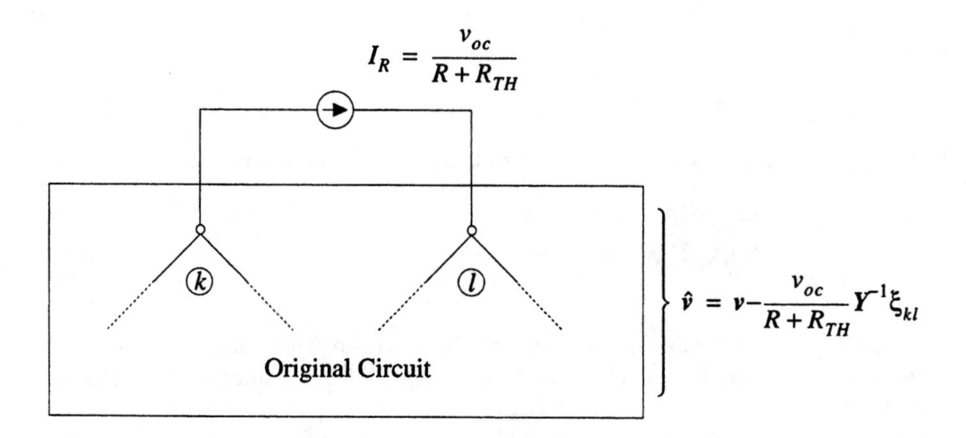

Figure 8.8 Complete solution for the addition of a resistor between nodes k and l.

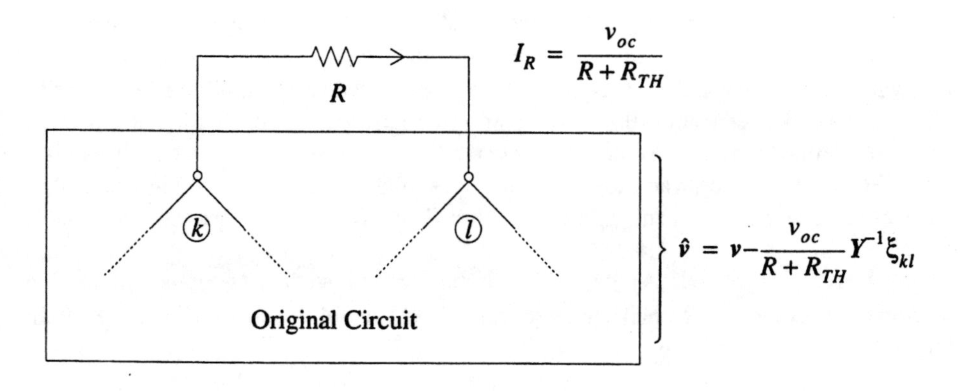

Figure 8.9 The complete solution for the actual circuit.

$$\begin{aligned}\hat{v} &= Y^{-1}J - \frac{1}{R+\xi_{kl}^T Y^{-1}\xi_{kl}} Y^{-1}\xi_{kl}\xi_{kl}^T Y^{-1}J \\ &= \left(Y^{-1} - \frac{1}{R+\xi_{kl}^T Y^{-1}\xi_{kl}} Y^{-1}\xi_{kl}\xi_{kl}^T Y^{-1}\right)J\end{aligned} \quad (8.1.8)$$

The revised circuit can be described conveniently by

$$(\boldsymbol{Y}+\frac{1}{R}\xi_{kl}\xi_{kl}^T)\,\hat{\boldsymbol{v}} = \boldsymbol{J} \tag{8.1.9}$$

where $\frac{1}{R}\xi_{kl}\xi_{kl}^T$ is the dyad (outer product) that shows explicitly the stamp contributed by the resistance R to the nodal admittance formulation. So we have obtained

$$(\boldsymbol{Y}+\frac{1}{R}\xi_{kl}\xi_{kl}^T)^{-1} = \boldsymbol{Y}^{-1} - \frac{1}{R+\xi_{kl}^T\boldsymbol{Y}^{-1}\xi_{kl}}\boldsymbol{Y}^{-1}\xi_{kl}\xi_{kl}^T\boldsymbol{Y}^{-1} \tag{8.1.10}$$

an explicit expression for the inverse of a matrix plus a dyad. This is known as Kron's formula to electrical engineers, or Householder's formula to mathematicians [Kron39, Householder57].

Seldom, if ever, would we use this formula to find the explicit form of an inverse. Even if $\boldsymbol{Y}$ and $\frac{1}{R}\xi_{kl}\xi_{kl}^T$ were sparse matrices, neither $\boldsymbol{Y}^{-1}$ nor the dyad $\boldsymbol{Y}^{-1}\xi_{kl}\xi_{kl}^T\boldsymbol{Y}^{-1}$ are guaranteed to be sparse. Rather we recognize that we can efficiently compute the new node voltage vector $\hat{\boldsymbol{v}}$ with no new LU factorizations required and only one new Forward and Back Substitution in terms of the original LU factors of $\boldsymbol{Y}$. Because the value of resistance R is explicit, we can use these formulas as design equations to study the variation of node voltages with R. And we can then consider with no more difficulty the addition of another single resistance between any other node pair. Alternatively, by considering the parallel addition of either a positive or negative resistance to an existing one, we can investigate the effect of resistance variation. Simultaneous multiple resistance additions and/or changes can be similarly handled, but with more difficulty as we will see later in this section.

As an example, we consider the ladder circuit we have studied previously but with a new resistance of value R connected between nodes 1 and 4 as shown in Figure 8.10. In this symmetric case we have

$$\xi_{14}v^T = v_1 - v_4 = \frac{2}{3}\mathrm{V} - \frac{2}{3}\mathrm{V} = 0\mathrm{V} \tag{8.1.11}$$

which implies no change regardless of the value of the resistance. This is not surprising since nodes 1 and 4 are at the same potential and the added resistance would not upset the symmetry of the circuit or its solution. So we consider the more interesting case in which the arbitrary resistance is connected between nodes 2 and 4, as shown in Figure 8.11. Now we have

$$v_{oc} = \xi_{24}^T\boldsymbol{v} = v_2 - v_4 = \frac{1}{3}\mathrm{V} - \frac{2}{3}\mathrm{V} = -\frac{1}{3}\mathrm{V} \tag{8.1.12}$$

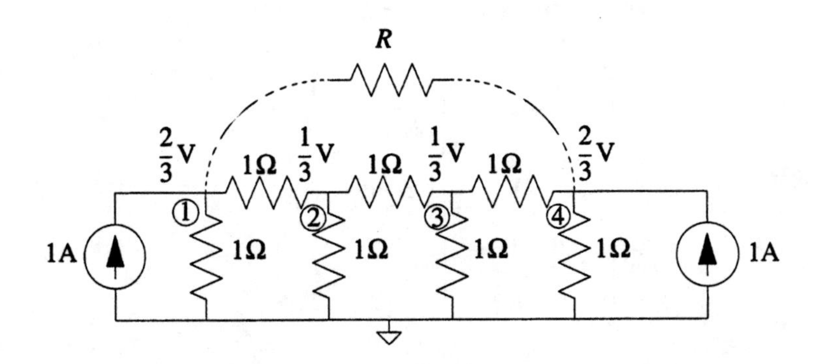

Figure 8.10 Example of adding a resistance between nodes 1 and 4 for the resistor ladder circuit from Chapter 1.

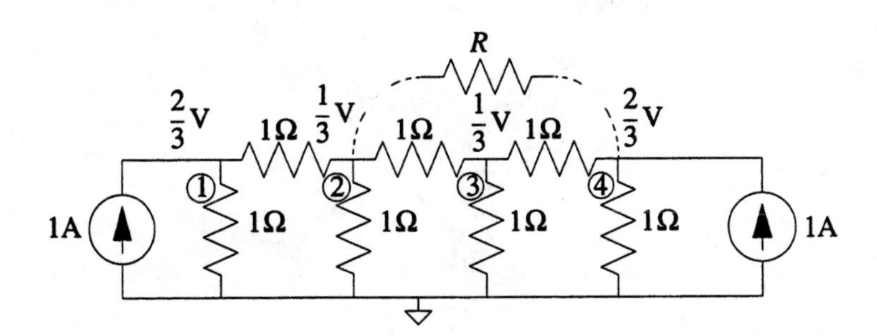

Figure 8.11 Example of adding a resistance between nodes 2 and 4 for the resistor ladder circuit from Chapter 1.

To find $\boldsymbol{Y}^{-1}\xi_{24}$ and $\xi_{24}^T\boldsymbol{Y}^{-1}\xi_{24}$ we use the original LU factors of $\boldsymbol{Y}$; $\boldsymbol{Lx} = \xi_{24}$:

$$\begin{bmatrix} 2 & 0 & 0 & 0 \\ -1 & \frac{5}{2} & 0 & 0 \\ 0 & -1 & \frac{13}{5} & 0 \\ 0 & 0 & -1 & \frac{21}{13} \end{bmatrix} \begin{bmatrix} x_1 \\ x_2 \\ x_3 \\ x_4 \end{bmatrix} = \begin{bmatrix} 0 \\ 1 \\ 0 \\ -1 \end{bmatrix} \tag{8.1.13}$$

which yields $x_1 = 0$, $x_2 = \frac{2}{5}$, $x_3 = \frac{2}{13}$ and $x_4 = -\frac{11}{21}$. Following that Forward Substitution, we can apply Back Substitution to $Uv' = x$:

$$\begin{bmatrix} 1 & -\frac{1}{2} & 0 & 0 \\ 0 & 1 & -\frac{2}{5} & 0 \\ 0 & 0 & 1 & -\frac{5}{13} \\ 0 & 0 & 0 & 1 \end{bmatrix} \begin{bmatrix} v_1' \\ v_2' \\ v_3' \\ v_4' \end{bmatrix} = \begin{bmatrix} 0 \\ \frac{2}{5} \\ \frac{2}{13} \\ -\frac{11}{21} \end{bmatrix} \tag{8.1.14}$$

which yields $v_4' = -\frac{11}{21}$, $v_3' = -\frac{1}{21}$, $v_2' = \frac{8}{21}$, and $v_1' = \frac{4}{21}$.

So,

$$R_{TH} = \xi_{24}^T Y^{-1} \xi_{24} = v_2' - v_4' = \frac{8}{21} - (-\frac{11}{21}) = \frac{19}{21} \tag{8.1.15}$$

And finally,

$$\hat{v} = v - \frac{\xi_{24} v}{R + R_{TH}} v' \tag{8.1.16}$$

or

$$\hat{v} = \begin{bmatrix} \frac{2}{3} \\ \frac{1}{3} \\ \frac{1}{3} \\ \frac{2}{3} \end{bmatrix} - \left(\frac{-\frac{1}{3}}{R + \frac{19}{21}} \right) \begin{bmatrix} \frac{4}{21} \\ \frac{8}{21} \\ -\frac{1}{21} \\ -\frac{11}{21} \end{bmatrix} \tag{8.1.17}$$

Now we can vary the value of R arbitrarily and easily compute the results. Even $R = 0$ and $R = \infty$ are possibilities.

Using Kron's method, we could also consider the hypothetical cutting of a circuit branch that might partition a circuit into two smaller subcircuits (refer to Figure 8.12), and then build up the overall solution by piecing the subcircuit solutions together. This approach was called *branch tearing* by Kron, and it forms the basis for his *link-at-a-time* algorithm. In the extreme, we could consider cutting enough (link) branches that the remaining circuit would be reduced to merely a tree. Refer to Appendix A for a discussion of trees and links. Then we could add back in a link at a time to ultimately solve the original circuit. This solution method is seldom used because it is difficult to exploit the matrix sparsity of the original circuit. *Node tearing* is more popular and is the topic of the next section.

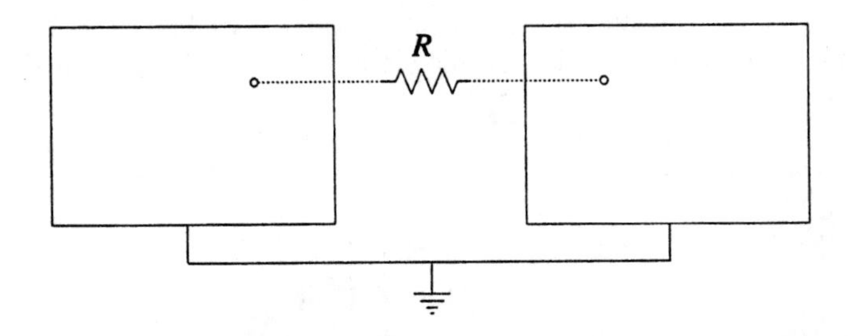

Figure 8.12 Representation of branch tearing between two circuits.

8.2 Node Tearing

Recall

$$\hat{v} = v - \frac{\xi_{kl}^T v}{R + \xi_{kl}^T Y^{-1} \xi_{kl}} Y^{-1} \xi_{kl} \tag{8.2.1}$$

or

$$\hat{v} = v - \frac{v_{oc}}{R + R_{TH}} Y^{-1} \xi_{kl} \tag{8.2.2}$$

These formulas hold in the limit as $R \to 0$:

$$\hat{v} = v - \frac{\xi_{kl}^T v}{\xi_{kl}^T Y^{-1} \xi_{kl}} Y^{-1} \xi_{kl} \tag{8.2.3}$$

or

$$\hat{v} = v - \frac{v_{oc}}{R_{TH}} Y^{-1} \xi_{kl} \tag{8.2.4}$$

So we can consider even the addition of a short circuit between a node pair.

Equivalently, we can consider the *tearing* (splitting) of a node to create two smaller circuits as shown in Figure 8.13. Then we could solve the two smaller subcircuits separately, as shown in Figure 8.14, to obtain v_{oc}. Next, we could consider obtaining R_{TH} as the voltage that would appear across a 1A independent current source placed across the torn node(s) with all other independent source values set to zero (see Figure 8.15). But in terms of current "source transportation," this situation is equivalent to the analysis shown in Figure 8.16.

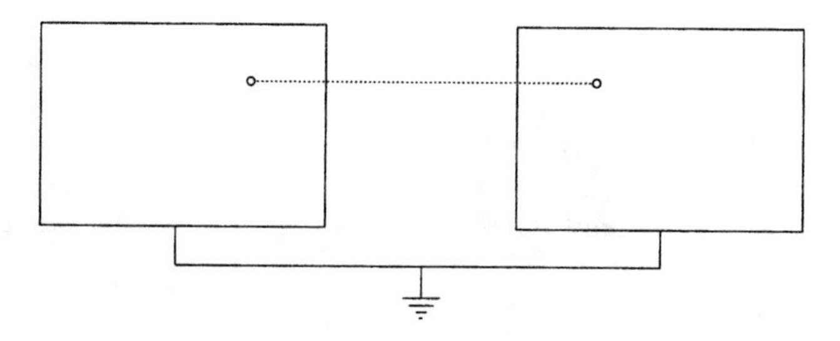

Figure 8.13 Node tearing.

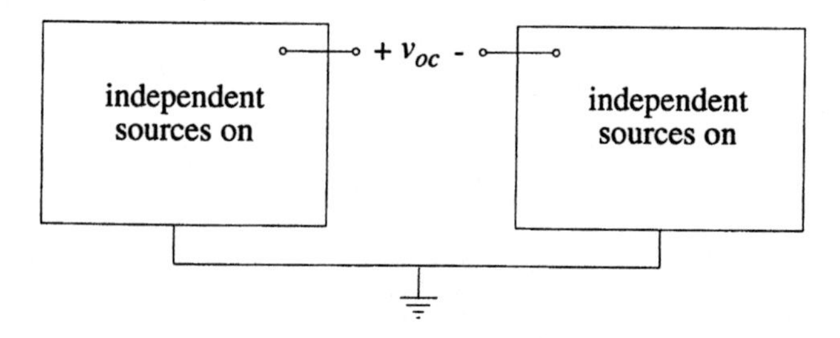

Figure 8.14 Solving two "torn" subcircuits separately.

So if we save the LU factors from the two separate solutions of the partitioned subcircuits (that we use separately to obtain v_{oc}), we can apply them separately once again (to unit independent current source vectors) to obtain $v = R_{TH}$. In the course of that computation, we of course find $Y_1^{-1}\xi_1$ and $Y_2^{-1}\xi_2$ as well. Note that since these unit vectors representing the independent current sources have one end grounded, we can drop the double subscript notation that we had to use earlier. If "tearing" a node splits the circuit into two smaller subcircuits, we can solve (LU factor) the subcircuits independently and then piece the overall solution together as shown in the previous section.

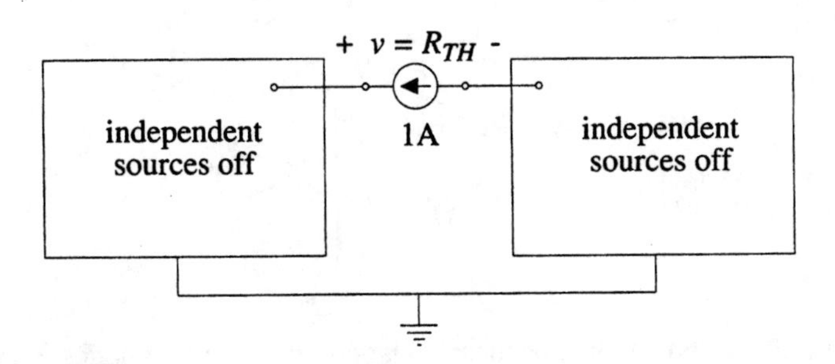

Figure 8.15 Evaluating the Thevenin resistance between the torn subcircuits.

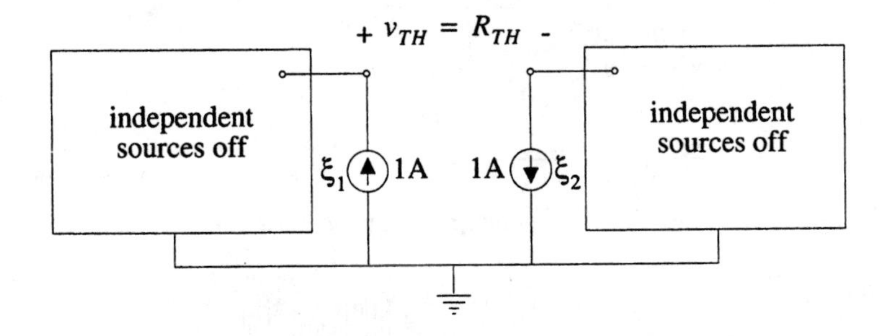

Figure 8.16 Equivalent representation of Figure 8.15.

$$\hat{v} = \begin{bmatrix} Y_1^{-1} J_1 \\ Y_2^{-1} J_2 \end{bmatrix} - \frac{v_{oc}}{R_{TH}} \begin{bmatrix} Y_1^{-1} \xi_1 \\ Y_2^{-1} \xi_2 \end{bmatrix} \tag{8.2.5}$$

or

$$\hat{v} = \begin{bmatrix} Y_1^{-1} J_1 \\ Y_2^{-1} J_2 \end{bmatrix} - \frac{\xi_1^T Y_1^{-1} J_1 + \xi_2^T Y_2^{-1} J_2}{\xi_1^T Y_1^{-1} \xi_1 + \xi_2^T Y_2^{-1} \xi_2} \begin{bmatrix} Y_1^{-1} \xi_1 \\ Y_2^{-1} \xi_2 \end{bmatrix} \tag{8.2.6}$$

Before we proceed to consider the simultaneous splitting of several nodes, we generalize the above short circuit case to include the addition of an independent voltage source between a pair of nodes (Figure 8.17). As usual, we first obtain the Thevenin equivalent circuit that faces the added branch, as shown in Figures 8.18 and 8.19, where v is the node voltage vector for the original circuit.

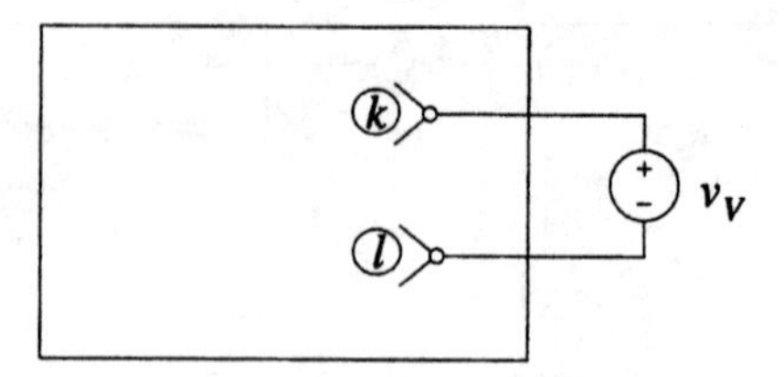

Figure 8.17 Adding an independent voltage source between nodes k and l.

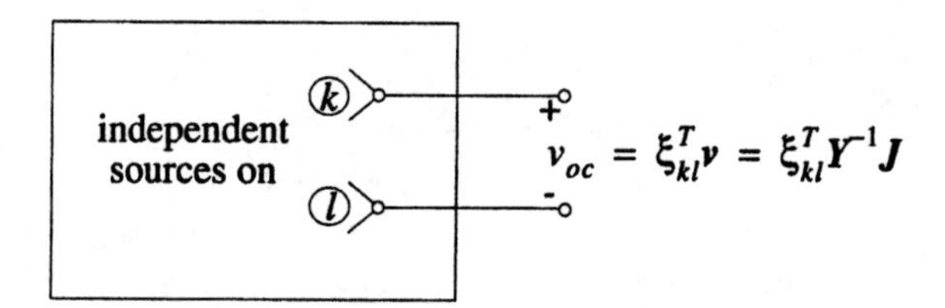

Figure 8.18 Measuring the open circuit voltage between nodes k and l.

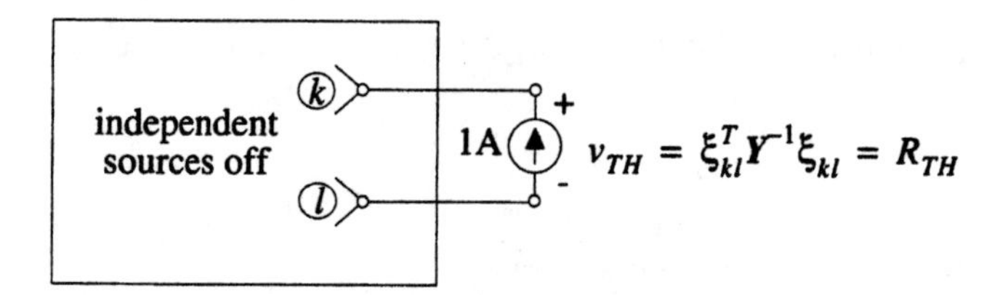

Figure 8.19 Calculating the Thevenin resistance between nodes k and l.

Again, we invoke the substitution theorem and we must find the value of the independent current source I (Figure 8.20) that causes the desired source voltage v_V to appear across it:

$$v_{oc} - R_{TH} I = v_V \Rightarrow I = \frac{v_{oc} - v_V}{R_{TH}} \tag{8.2.7}$$

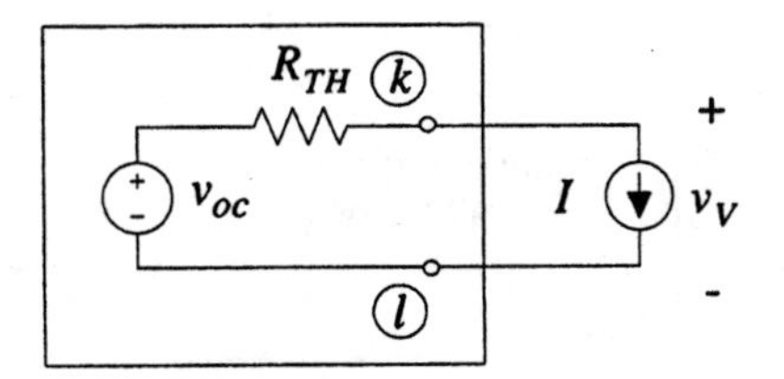

Figure 8.20 Applying source substitution.

Then, by superposition

$$\begin{aligned}\hat{v} &= v - IY^{-1}\xi_{kl} \\ &= v - \frac{v_{oc} - v_V}{R_{TH}} Y^{-1}\xi_{kl} \\ &= v + \frac{v_V - \xi_{kl}^T v}{\xi_{kl}^T Y^{-1}\xi_{kl}} Y^{-1}\xi_{kl}\end{aligned} \tag{8.2.8}$$

This result reduces to what we obtained earlier for a short circuit ($R \to 0$) when $v_V \to 0$, as we would expect. So we see that we can compensate for the addition of even an independent voltage source in terms of the LU factors of the original $\boldsymbol{Y}$.

8.3 Multiple Voltage Source Additions

We will now consider a multiplicity of split nodes. This generalization is as easily derived, poses some interest in its own right and is more useful. We proceed exactly as we did above, only now for a multi-port representation of the problem. First, find the open circuit voltages for the 2-port circuit in Figure 8.21:

$$\begin{aligned}Yv &= J \Rightarrow v = Y^{-1}J \\ v_{oc1} &= v_{kl} = v_k - v_l = \xi_{kl}^T v \\ v_{oc2} &= v_{pq} = v_p - v_q = \xi_{pq}^T v\end{aligned} \tag{8.3.1}$$

Next, excite the circuit separately with unit valued independent current sources at each port as shown in Figures 8.22 and 8.23:

k
+
v_{oc1}
-
l
independent
sources on
p
+
v_{oc2}
-
q

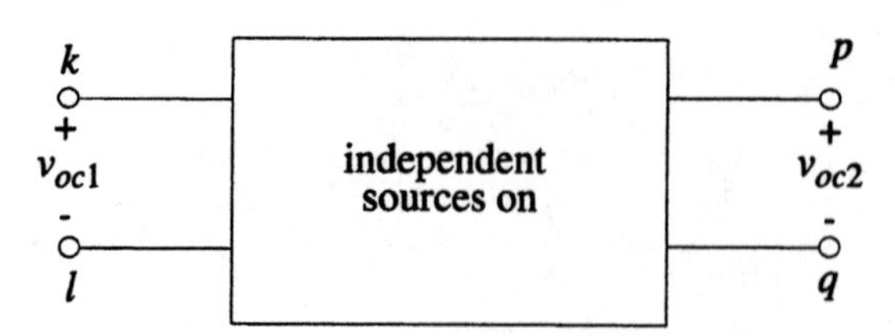

Figure 8.21 Measuring the open circuit voltages at two ports.

$$\begin{aligned} Yv' &= \xi_{kl} \Rightarrow v' = Y^{-1}\xi_{kl} \\ z_{11} &= v'_{kl} = v'_k - v'_l = \xi_{kl}^T v' = \xi_{kl}^T Y^{-1}\xi_{kl} \\ z_{21} &= v'_{pq} = v'_p - v'_q = \xi_{pq}^T v' = \xi_{pq}^T Y^{-1}\xi_{kl} \end{aligned} \tag{8.3.2}$$

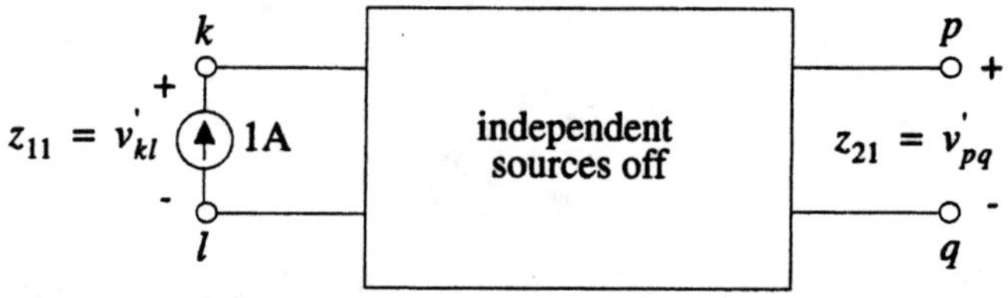

k
+
$z_{11} = v'_{kl}$
1A
-
l
independent
sources off
p
+
$z_{21} = v'_{pq}$
-
q

Figure 8.22 Calculating the two-port z parameters, z_{11} and z_{21}.

$$\begin{aligned} Yv'' &= \xi_{pq} \Rightarrow v'' = Y^{-1}\xi_{pq} \\ z_{22} &= v''_{pq} = v''_p - v''_q = \xi_{pq}^T v'' = \xi_{pq}^T Y^{-1}\xi_{pq} \\ z_{21} &= v''_{kl} = v''_k - v''_l = \xi_{kl}^T v'' = \xi_{kl}^T Y^{-1}\xi_{pq} \end{aligned} \tag{8.3.3}$$

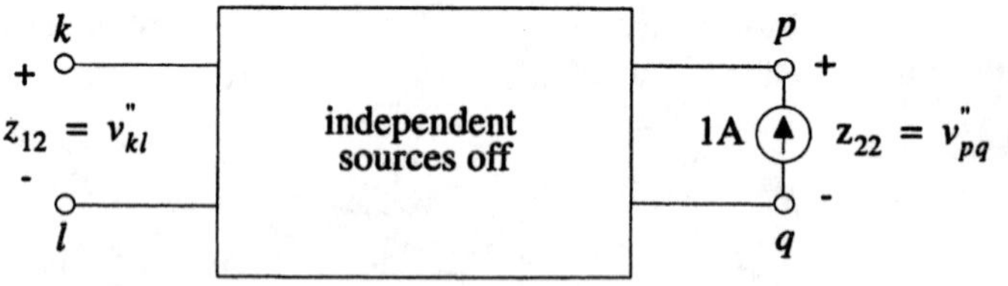

k
+
$z_{12} = v''_{kl}$
-
l
independent
sources off
p
+
1A
$z_{22} = v''_{pq}$
-
q

Figure 8.23 Calculating the two-port z parameters, z_{22} and z_{12}.

So, entirely in terms of $\boldsymbol{Y}^{-1}$ (the LU factors of $\boldsymbol{Y}$), we have found the following two-port impedance parameters:

$$[\boldsymbol{Z}] = \begin{bmatrix} z_{11} & z_{12} \\ z_{21} & z_{22} \end{bmatrix} = \begin{bmatrix} \xi_{kl}^T \boldsymbol{Y}^{-1} \xi_{kl} & \xi_{kl}^T \boldsymbol{Y}^{-1} \xi_{pq} \\ \xi_{pq}^T \boldsymbol{Y}^{-1} \xi_{kl} & \xi_{pq}^T \boldsymbol{Y}^{-1} \xi_{pq} \end{bmatrix} \tag{8.3.4}$$

The open circuit voltages were also obtained in terms of $\boldsymbol{Y}^{-1}$ (the LU factors of $\boldsymbol{Y}$), and overall we have the two-port equivalent circuit in Figure 8.24.

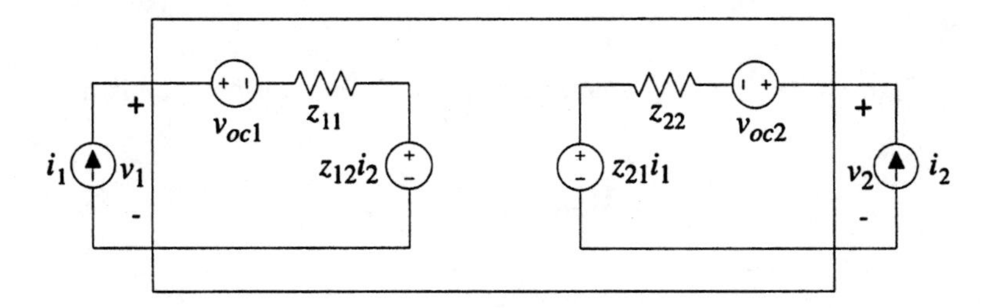

Figure 8.24 The equivalent circuit for the two-port z parameters.

To apply the substitution theorem, we must simultaneously adjust the independent current source values i_1 and i_2 so as to obtain the desired independent voltage source voltage values (refer to Figure 8.25):

$$v_1 = v_{V1} \quad \text{and} \quad v_2 = v_{V2} \tag{8.3.5}$$

So

$$\begin{aligned} v_{V1} &= v_{oc1} + z_{11}i_1 + z_{12}i_2 \\ v_{V2} &= v_{oc2} + z_{21}i_1 + z_{22}i_2 \end{aligned} \tag{8.3.6}$$

or

$$\begin{bmatrix} z_{11} & z_{12} \\ z_{21} & z_{22} \end{bmatrix} \begin{bmatrix} i_1 \\ i_2 \end{bmatrix} = \begin{bmatrix} v_{V1} - v_{oc1} \\ v_{V2} - v_{oc2} \end{bmatrix} \tag{8.3.7}$$

or

$$\begin{bmatrix} i_1 \\ i_2 \end{bmatrix} = \begin{bmatrix} Z^{-1} \end{bmatrix} \begin{bmatrix} v_{V1} - v_{oc1} \\ v_{V2} - v_{oc2} \end{bmatrix} \tag{8.3.8}$$

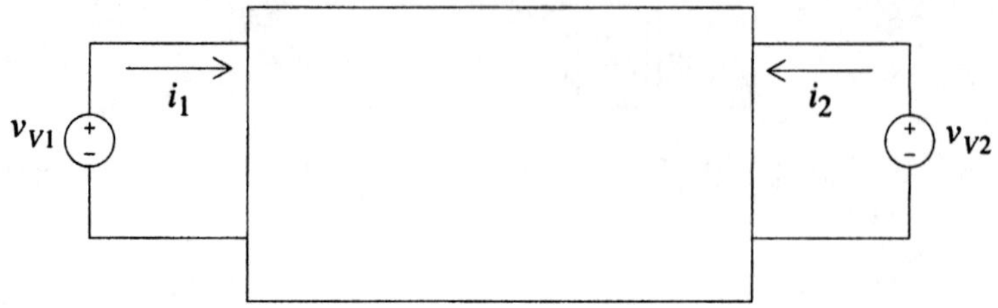

Figure 8.25 Applying source substitution for the two-port problem.

Now, since

$$i_1 = 1 \Rightarrow v' = Y^{-1}\xi_{kl} \tag{8.3.9}$$

and

$$i_2 = 1 \Rightarrow v'' = Y^{-1}\xi_{pq} \tag{8.3.10}$$

by superposition we have

$$\hat{v} = v + \begin{bmatrix} Y^{-1}\xi_{kl} & Y^{-1}\xi_{pq} \end{bmatrix} \begin{bmatrix} Z^{-1} \end{bmatrix} \begin{bmatrix} v_{V1} - v_{oc1} \\ v_{V2} - v_{oc2} \end{bmatrix} \tag{8.3.11}$$

or

$$\hat{v} = v + \begin{bmatrix} Y^{-1}\xi_{kl} & Y^{-1}\xi_{pq} \end{bmatrix} Z^{-1} \begin{bmatrix} v_{V1} - \xi_{kl}^T v \\ v_{V2} - \xi_{pq}^T v \end{bmatrix} \tag{8.3.12}$$

$$\hat{v} = Y^{-1}J + \begin{bmatrix} Y^{-1}\xi_{kl} & Y^{-1}\xi_{pq} \end{bmatrix} Z^{-1} \begin{bmatrix} v_{V1} - \xi_{kl}^T Y^{-1} J \\ v_{V2} - \xi_{pq}^T Y^{-1} J \end{bmatrix} \tag{8.3.13}$$

Except for the 2×2 matrix Z to be inverted, all constituents of this result are in terms of

$\boldsymbol{Y}^{-1}$ (the LU factors of $\boldsymbol{Y}$).

In the special case where the added source voltages are zero -- the reconstitution of split nodes with short circuits -- we have

$$\hat{\boldsymbol{v}} = \boldsymbol{Y}^{-1}\boldsymbol{J} - \begin{bmatrix} \boldsymbol{Y}^{-1}\xi_{kl} & \boldsymbol{Y}^{-1}\xi_{pq} \end{bmatrix} \boldsymbol{Z}^{-1} \begin{bmatrix} \xi_{kl}^{T}\boldsymbol{Y}^{-1}\boldsymbol{J} \\ \xi_{pq}^{T}\boldsymbol{Y}^{-1}\boldsymbol{J} \end{bmatrix} \tag{8.3.14}$$

or

$$\hat{\boldsymbol{v}} = \left\{ \boldsymbol{1} - \boldsymbol{Y}^{-1} [\xi_{kl} \quad \xi_{pq}] \boldsymbol{Z}^{-1} \begin{bmatrix} \xi_{kl}^{T} \\ \xi_{pq}^{T} \end{bmatrix} \right\} \boldsymbol{Y}^{-1}\boldsymbol{J} \tag{8.3.15}$$

In other words

$$\hat{\boldsymbol{Y}}^{-1} = \boldsymbol{Y}^{-1} - \boldsymbol{Y}^{-1} [\xi_{kl} \quad \xi_{pq}] \boldsymbol{Z}^{-1} \begin{bmatrix} \xi_{kl}^{T} \\ \xi_{pq}^{T} \end{bmatrix} \boldsymbol{Y}^{-1} \tag{8.3.16}$$

is the generalization to the addition of two dyads of Householder's -- or Kron's -- formula.

The extension of these results to the case of many added independent voltage sources, say γ of them, is straightforward:

$$\hat{\boldsymbol{v}} = \boldsymbol{v} + \boldsymbol{Y}^{-1} [\xi_1 \quad \xi_2 \quad \dots \quad \xi_\gamma] \boldsymbol{Z}^{-1} \begin{bmatrix} v_{V1} - \xi_1^{T}\boldsymbol{v} \\ v_{V2} - \xi_2^{T}\boldsymbol{v} \\ \vdots \\ v_{V\gamma} - \xi_\gamma^{T}\boldsymbol{v} \end{bmatrix} \tag{8.3.17}$$

Here the ξ_j vector is the appropriate ξ_{kl} connection vector that designates the node pair $k-l$ between which the jth independent voltage source is (to be) connected, $\boldsymbol{v} = \boldsymbol{Y}^{-1}\boldsymbol{J}$ as usual, and $\boldsymbol{Z}$ is a $\gamma \times \gamma$ matrix with the following components:

$$z_{ij} = \xi_i^{T}\boldsymbol{Y}^{-1}\xi_j \tag{8.3.18}$$

Such a partitioning can always be effected provided that $\boldsymbol{Y}^{-1}$ and $\boldsymbol{Z}^{-1}$ exist. For $\boldsymbol{Z}^{-1}$ to exist the ξ_j's must be linearly independent. That is, there can be no loops of voltage sources as we would expect. Common sense conditions also prevail for Y^{-1} to exist. For the node-splitting case, if the sources are zero (short circuits) the above result particularizes to

$$\hat{v} = v - Y^{-1}[\xi_1 \;\; \xi_2 \;\; \dots \;\; \xi_\gamma] Z^{-1} \begin{bmatrix} \xi_1^T \\ \xi_2^T \\ \vdots \\ \xi_\gamma^T \end{bmatrix} v \tag{8.3.19}$$

And we can easily derive from it the γ dyad extension of the Kron-Householder formula.

So we can consider tearing a large circuit into several subcircuits by means of node-splitting, and then piecing the resulting smaller analyses together via the above formula. If the above partitions are cleverly defined, we may get Y and/or Z to be of the form shown in Figure 8.26.

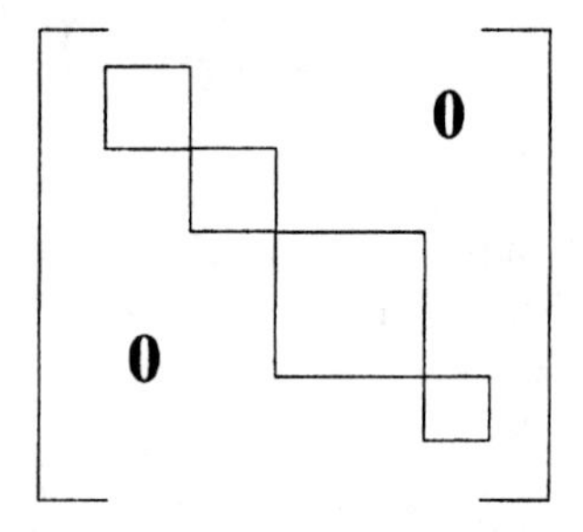

Figure 8.26 The overall appearance of a circuit matrix with tearing.

How to effect "optional partitionings" has been the subject of much research [Branin75, Rabbat76, Rabbat79]. Of those (usually digital) circuit simulators that use partitioning at all, most merely take it along the lines of user-defined (library cell) building blocks. Partitioning may be used to exploit subcircuit latency. If the inputs to a circuit have not changed since the last analysis we need not re-solve it but rather can reuse the old solution.

We note that the above results for addition of independent voltage sources are what we could also obtain upon appropriate reduction of MNA equations to first eliminate excess currents.

Another possible application of the above is to analyze a circuit full of switches: first with all switches open and then with them closed as appropriate. The method can be applied to diode circuits, too. Let us say we wish to use a simple model for the diode, whereby a diode is open (OFF) when its voltage is less than say 0.6V, and behaves like a voltage source of 0.6V otherwise. Then the circuit can be initially analyzed with all diodes

OFF. Then, those diodes with 0.6V or more across them can be replaced with independent voltage sources and the solution iteratively recomputed efficiently using the formulas derived in this chapter. Another possible application of partitioning is switched capacitor filter circuit analysis.

We note finally, that all of the results of this chapter can be generalized beyond Nodal Analysis. A word of warning, though: one should be careful with notation, since it can easily get out of hand.

8.4 References

[Kron39] G. Kron, *Tensor Analysis of Networks*. Wiley, 1939.

[Householder57] A. S. Householder. A Survey of Some Closed Methods of Inverting Matrices. *SIAM Journal of Applied Mathematics*, vol. 5, pp. 153-169, 1957.

[Desoer69] C. A. Desoer and E. S. Kuh. *Basic Circuit Theory*. McGraw-Hill, 1969.

[Branin75] F. H. Branin. A Sparse Matrix Modification of Kron's Method of Piecewise Analysis. *Proc. IEEE International Symposium on Circuits and Systems (ISCAS)*, pp. 383-386, 1975.

[Rabbat76] N. B. Rabbat and H. Y. Hsieh. A Latent Macromodular Approach to Large-scale Sparse Networks. *IEEE Trans. on Circuits and Systems*, vol. CAS-23, pp. 745-752, December 1976.

[Rabbat79] N. B. Rabbat, A. L. Sangiovanni-Vincentelli, and H. Y. Hsieh. A Multilevel Newton Algorithm with Macromodeling and Latency for the Analysis of Large-scale Nonlinear Circuits in the Time Domain. *IEEE Transactions on Circuits and Systems*, vol. CAS-26, pp. 733-741, September 1979.

Chapter 9

Incremental Sensitivity

In the last chapter we saw that the "large change sensitivity" of a circuit's response with respect to a single element value can be computed relatively efficiently. For example, the node voltage changes that result from variation of a single parameter can be computed in terms of a new Forward and Back Substitution by reusing the LU factors of the original nodal $\boldsymbol{Y}$ matrix. There are a few other computations required to obtain the final results, but they are relatively inexpensive compared to LU factorization and Forward and Back Substitution.

We saw too in the previous chapter that we could consider the simultaneous variation of more than one parameter, but at a higher run-time cost. Each parameter requires a new Forward and Back Substitution; and a new square matrix $\boldsymbol{Z}$, of size equal to the number of parameters, must be LU factored as well. The situation remains workable provided there are not too many parameters. By these methods, we can efficiently find the *large change* variation of *all* circuit responses with respect to *arbitrary* variations of a *few* circuit parameters.

This chapter will focus on *incremental sensitivity* or *small change sensitivity*. Incremental sensitivity is defined as the partial derivative of a circuit response with respect to a parameter of interest,

$$\frac{\partial\,(response)}{\partial\,(parameter)}$$

It is valid in a small range around the nominal value of the parameter of interest. We ask the question, "Can the incremental sensitivity of all the circuit responses be computed with respect to single parameter efficiently?" Further, "Can the incremental sensitivity of a single circuit response be computed with respect to multiple parameters efficiently?" This chapter will describe *direct sensitivity* and *adjoint sensitivity*, which are methods that answer these two questions, respectively. Both of these methods involve construction of a new circuit with the same topology as the original one, whose solution yields the required set of sensitivities. In the case of direct sensitivity, the auxiliary circuit is called the *sensitivity circuit* and in the case of adjoint sensitivity it is called the *adjoint circuit*. In both methods, LU factors from the original circuit solution are reused with new Forward and Back Substitutions to compute sensitivities. We alluded to this process in Chapter 3.

9.1 Direct Circuit Sensitivities

We begin the discussion of direct sensitivity by considering a simple example. The circuit in Figure 9.1 has the following voltage at node 2:

$$v_2 = V_S\left[\frac{R_2}{R_1 + R_2}\right] \tag{9.1.1}$$

the following voltage at node 1:

$$v_1 = V_S \tag{9.1.2}$$

and the following current through the independent voltage source:

$$i_S = -\frac{V_S}{R_1 + R_2} \tag{9.1.3}$$

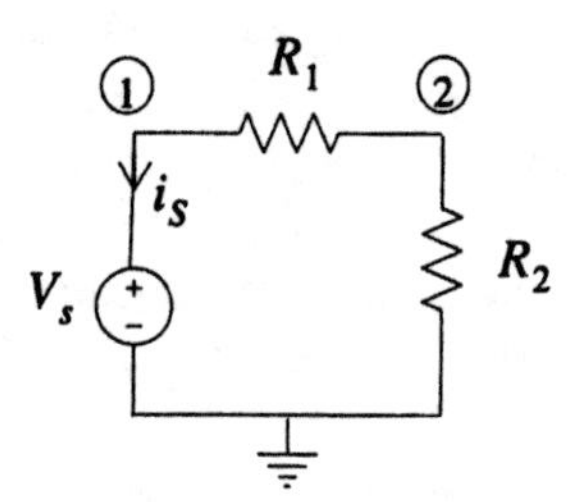

Figure 9.1 A simple voltage divider circuit.

The sensitivity of this voltage response with respect to some circuit parameter is easily obtained from (9.1.1) by partial differentiation. For example, the sensitivity of v_2 with respect to R_1 is

$$\frac{\partial v_2}{\partial R_1} = -V_s\left[\frac{R_2}{(R_1 + R_2)^2}\right] \tag{9.1.4}$$

The sensitivity of v_1 with respect to R_1 is 0 and the sensitivity of i_S with respect to R_1 is

$$\frac{\partial i_S}{\partial R_1} = \frac{V_S}{(R_1 + R_2)^2} \tag{9.1.5}$$

We don't usually have the luxury of analytical expressions for the response voltages. Instead, sensitivity circuits are used to obtain this information in general. These sensitivity circuits are obtained by *direct differentiation* of the original circuit equations.

The easiest way to think of direct sensitivity is in terms of differentiation of the BCRs of the original circuit. We will first explain direct sensitivity in this manner, and then discuss a matrix interpretation of the same method in section 9.2. Consider a circuit consisting of conductors, resistors, independent current sources, and independent voltage sources only. Assume that we are interested in the sensitivity of the circuit response to some parameter x, which can be any quantity like temperature (which might effect many element values), value of a resistance, value of the voltage supply to a circuit, and so on. Differentiating the BCRs, we find

$$\begin{aligned}
&\text{Conductors:} \quad i_G = G v_G \Rightarrow \frac{\partial i_G}{\partial x} = G\frac{\partial v_G}{\partial x} + \frac{\partial G}{\partial x} v_G \\
&\text{Resistors:} \quad v_R = R i_R \Rightarrow \frac{\partial v_R}{\partial x} = R\frac{\partial i_R}{\partial x} + \frac{\partial R}{\partial x} i_R \\
&\text{Independent current sources:} \quad i_I = I_S \Rightarrow \frac{\partial i_I}{\partial x} = \frac{\partial I_S}{\partial x} \\
&\text{Independent voltage sources:} \quad v_V = V_S \Rightarrow \frac{\partial v_V}{\partial x} = \frac{\partial V_S}{\partial x}
\end{aligned} \tag{9.1.6}$$

Postulate a new circuit η called the *sensitivity circuit* that has the same topology as the original circuit. The sensitivity circuit η has branch voltages, branch currents, and node voltages ψ_b, ϕ_b, and ψ_n, respectively, as shown in Figure 9.6. In tabular form, we can summarize the relationship between the two circuits as follows:

	original circuit	sensitivity circuit
branch voltages	v_b	ψ_b
branch currents	i_b	φ_b
node voltages	v_n	ψ_n

topologically identical

N	η
v_b, i_b, v_n	$\psi_b, \varphi_b, \psi_n$

Figure 9.2 A circuit N and its sensitivity circuit η.

We would like the currents and voltages of the sensitivity circuit to be the sensitivities we seek. Hence, we can write the BCRs for the sensitivity circuit as follows:

$$
\begin{aligned}
&\text{Conductors:} \quad \varphi_G = G\psi_G + \frac{\partial G}{\partial x} v_G \\
&\text{Resistors:} \quad \psi_R = R\varphi_R + \frac{\partial R}{\partial x} i_R \\
&\text{Independent current sources:} \quad \Phi_I = \frac{\partial I_S}{\partial x} \\
&\text{Independent voltage sources:} \quad \Psi_V = \frac{\partial V_S}{\partial x}
\end{aligned}
\tag{9.1.7}
$$

Thus each branch voltage is the sensitivity of the same branch voltage of the original circuit and likewise with branch currents. Each conductor in the original circuit is replaced by a conductor of the same value in parallel with an independent current source whose value is known from the original circuit solution. Thus each conductor becomes a Norton model in the sensitivity circuit. Likewise, resistors become Thevenin models in the sensitivity circuit (which can easily be converted to Norton models if necessary). Each independent source is replaced by the partial derivative of its value with respect to the parameter of interest, which could possibly be zero.

Next, we solve the sensitivity circuit. The sensitivity circuit has the same topology as the original circuit and the same nodal admittance matrix $\boldsymbol{Y}$. The excitation vector $\boldsymbol{J}$, however, is different. Hence the LU factors of the nominal solution can be reused by Forward and Back Substitution with a new $\boldsymbol{J}$ vector to determine all the sensitivities at once. Node voltages are simple linear combinations of branch voltages, so their sensitivities are automatically determined by the same process.

For example, let us apply the above method to our simple example in Figure 9.1. First we construct the sensitivity circuit by directly differentiating the BCRs:

$$\Psi_S = \frac{\partial V_S}{\partial R_1} = 0$$

$$\psi_{R1} = R_1\varphi_{R1} + i_{R1} = R_1\varphi_{R1} + \frac{V_S}{R_1 + R_2} \qquad \textbf{(9.1.8)}$$

$$\psi_{R2} = R_2\varphi_{R2}$$

The sensitivity circuit is shown in Figure 9.1. Solving this circuit, we obtain the necessary sensitivities:

$$\frac{\partial v_1}{\partial R_1} = \Psi_S = 0 \qquad \textbf{(9.1.9)}$$

$$\psi_{R2} = \frac{\partial v_2}{\partial R_1} = -i_{R1}\left[\frac{R_2}{R_1 + R_2}\right] = -V_S\left[\frac{R_2}{(R_1 + R_2)^2}\right] \qquad \textbf{(9.1.10)}$$

and

$$\varphi_S = \frac{\partial i_S}{\partial R_1} = \frac{i_{R1}}{R_1 + R_2} = \frac{V_S}{(R_1 + R_2)^2} \qquad \textbf{(9.1.11)}$$

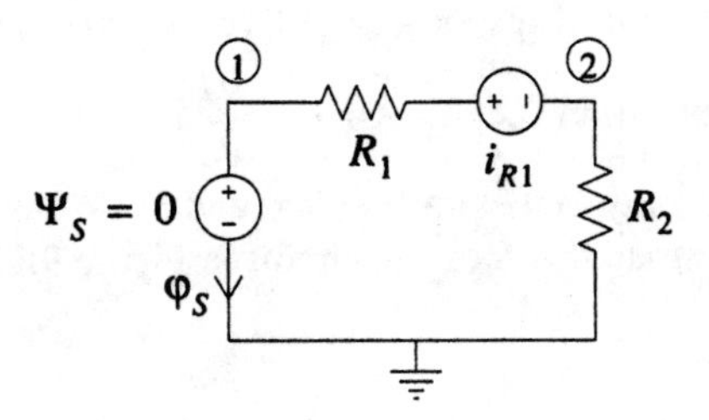

Figure 9.3 Sensitivity circuit for example in Figure 9.1.

Thus the sensitivities of *all* of a linear dc circuit's responses with respect to *one* parameter of interest can be computed simultaneously with just one extra Forward and Back Substitution. Each new parameter of interest will entail one more Forward and Back Substitution.

9.2 Matrix Interpretation of Direct Sensitivity

Consider the MNA matrix equation for a linear time-invariant circuit

$$Yv = J \tag{9.2.1}$$

which has a solution

$$v = Y^{-1}J \tag{9.2.2}$$

Of course, as before, v is obtained by LU factorization, not by matrix inversion. Assuming we have this nominal solution, we would now like to calculate the sensitivities of the response vector v, to changes in a single circuit parameter x.

Differentiating (9.2.1),

$$\frac{\partial}{\partial x}(Yv) = \frac{\partial}{\partial x}(J) \tag{9.2.3}$$

results in

$$\frac{\partial v}{\partial x} = -Y^{-1}\left[\frac{\partial Y}{\partial x}v - \frac{\partial J}{\partial x}\right] \tag{9.2.4}$$

So, we can calculate the vector of response sensitivities, $\partial v/\partial x$, using the original circuit LU factors and the new excitation vector $-\left[\frac{\partial Y}{\partial x}v - \frac{\partial J}{\partial x}\right]$.

Solving (9.2.4) requires only one extra Forward and Back Substitution using the original LU factors. To demonstrate this fact, the circuit in Figure 9.1 has the following MNA equations

$$\begin{bmatrix} 1 & 0 & 0 \\ -\frac{1}{R_1} & (\frac{1}{R_1}+\frac{1}{R_2}) & 0 \\ \frac{1}{R_1} & -\frac{1}{R_1} & 1 \end{bmatrix} \begin{bmatrix} v_1 \\ v_2 \\ i_S \end{bmatrix} = \begin{bmatrix} V_s \\ 0 \\ 0 \end{bmatrix} \tag{9.2.5}$$

This Y matrix is lower-triangular, therefore, the LU factors are simply $L = Y$ and $U = 1$. To solve the sensitivity equations, (9.2.4), we build the appropriate RHS vector. For our example above, $x = R_1$, therefore

$$\frac{\partial \boldsymbol{Y}}{\partial x} = \begin{bmatrix} 0 & 0 & 0 \\ \frac{1}{R_1^2} & -\frac{1}{R_1^2} & 0 \\ -\frac{1}{R_1^2} & \frac{1}{R_1^2} & 0 \end{bmatrix} \tag{9.2.6}$$

$$\boldsymbol{v} = \begin{bmatrix} V_s \\ V_s\left(\frac{R_2}{R_1+R_2}\right) \\ \frac{-V_s}{R_1+R_2} \end{bmatrix} \tag{9.2.7}$$

and

$$\frac{\partial \boldsymbol{J}}{\partial x} = \begin{bmatrix} 0 \\ 0 \\ 0 \end{bmatrix} \tag{9.2.8}$$

From (9.2.4), the new RHS term for the sensitivity circuit analysis is

$$\text{RHS} = -\begin{bmatrix} 0 \\ \frac{V_s}{R_1(R_1+R_2)} \\ \frac{-V_s}{R_1(R_1+R_2)} \end{bmatrix} \tag{9.2.9}$$

Applying the original LU factors (the lower triangular $\boldsymbol{Y}$ matrix for this example) to the new RHS vector,

$$\begin{bmatrix} 1 & 0 & 0 \\ -\frac{1}{R_1} & (\frac{1}{R_1}+\frac{1}{R_2}) & 0 \\ \frac{1}{R_1} & -\frac{1}{R_1} & 1 \end{bmatrix} \begin{bmatrix} \frac{\partial v_1}{\partial x} \\ \frac{\partial v_2}{\partial x} \\ \frac{\partial i_S}{\partial x} \end{bmatrix} = -\begin{bmatrix} 0 \\ \frac{V_s}{R_1(R_1+R_2)} \\ \frac{-V_s}{R_1(R_1+R_2)} \end{bmatrix} \tag{9.2.10}$$

Hence the voltage sensitivities are

$$\frac{\partial v_1}{\partial x} = 0 \tag{9.2.11}$$

and

$$\frac{\partial v_2}{\partial x} = -V_s\left[\frac{R_2}{(R_1+R_2)^2}\right] \tag{9.2.12}$$

and

$$\frac{\partial i_S}{\partial x} = \frac{V_S}{(R_1+R_2)^2} \tag{9.2.13}$$

These expressions match our earlier sensitivity calculations from (9.1.4) and (9.1.5), as well as from the sensitivity circuit in (9.1.10) and (9.1.11).

The matrix interpretation produces the same results as directly differentiating BCRs and creating a sensitivity circuit. The excitation vector for the sensitivity circuit (from the right hand side of (9.2.4)) contains two terms. The first is $\partial J/\partial x$, which means that independent sources are replaced by their derivatives. The ith element of the other term is $-\sum_{j=1}^{N}\frac{\partial Y_{ij}}{\partial x}v_j$. So, for example, if there is a conductance G connected from node i to node j, then a current source of value $\partial G/\partial x\,(v_i - v_j)$ or $(\partial G/\partial x)\,v_G$ is connected from node i to node j in the sensitivity circuit. We arrived at the same result in the previous section by directly differentiating the BCRs.

9.3 *Controlled Sources and Nonlinear Elements*

This section will extend direct sensitivity to circuits containing linear controlled sources and nonlinear elements. To begin, we differentiate the BCRs of the four types of linear controlled sources to obtain the corresponding sensitivity circuit elements as follows.

$$\text{Voltage-controlled current source:}\quad i_2 = g_m v_1 \rightarrow \frac{\partial i_2}{\partial x} = g_m\frac{\partial v_1}{\partial x} + \frac{\partial g_m}{\partial x}v_1$$

$$\therefore\ \phi_2 = g_m\psi_1 + \frac{\partial g_m}{\partial x}v_1 \tag{9.3.1}$$

So a voltage-controlled current source is replaced in the sensitivity circuit by an identical voltage-controlled current source in parallel with an independent current source whose value is known from the original circuit solution.

$$\text{Voltage-controlled voltage source:} \quad v_2 = \mu v_1 \rightarrow \frac{\partial v_2}{\partial x} = \mu \frac{\partial v_1}{\partial x} + \frac{\partial \mu}{\partial x} v_1$$
$$\therefore \quad \psi_2 = \mu \psi_1 + \frac{\partial \mu}{\partial x} v_1 \tag{9.3.2}$$

Hence, a voltage-controlled voltage source will manifest itself as an identical element in the sensitivity circuit, in series with an independent voltage source.

$$\text{Current-controlled current source:} \quad i_2 = \alpha i_1 \rightarrow \frac{\partial i_2}{\partial x} = \alpha \frac{\partial i_1}{\partial x} + \frac{\partial \alpha}{\partial x} i_1$$
$$\therefore \quad \varphi_2 = \alpha \varphi_1 + \frac{\partial \alpha}{\partial x} i_1 \tag{9.3.3}$$

Thus a current-controlled current source is replaced by an identical source in parallel with an independent current source in the sensitivity circuit. Finally,

$$\text{Current-controlled voltage source:} \quad v_2 = r_m i_1 \rightarrow \frac{\partial v_2}{\partial x} = r_m \frac{\partial i_1}{\partial x} + \frac{\partial r_m}{\partial x} i_1$$
$$\therefore \quad \psi_2 = r_m \varphi_1 + \frac{\partial r_m}{\partial x} i_1 \tag{9.3.4}$$

whereby a current-controlled voltage source is substituted by a similar source in series with an independent voltage source of known value in the sensitivity circuit.

As before, the linear sensitivity circuit is built, and then solved using the same LU factors as for the original circuit. All the required sensitivities are obtained by this procedure.

In the case of nonlinear circuits, since we are considering small change sensitivities, we can only discuss sensitivity about an operating point. Once the original nonlinear circuit has been solved, we know from Chapter 1 that a linearized companion model has been determined for each element at the operating point. Figure 9.4 depicts a linearized model consisting of a Norton equivalent with a conductance G_{eq} in parallel with a current source I_{eq}. The BCR for this linearized element,

$$i_b = G_{eq} v_b + I_{eq} \tag{9.3.5}$$

can be directly differentiated to obtain

$$\frac{\partial i_b}{\partial x} = G_{eq}\frac{\partial v_b}{\partial x} + \frac{\partial G_{eq}}{\partial x}v_b + \frac{\partial I_{eq}}{\partial x} \tag{9.3.6}$$

or

$$\phi_b = G_{eq}\psi_b + (\frac{\partial G_{eq}}{\partial x}v_b + \frac{\partial I_{eq}}{\partial x}) \tag{9.3.7}$$

Thus the element in the sensitivity circuit is also a linearized Norton equivalent as shown in Figure 9.4.

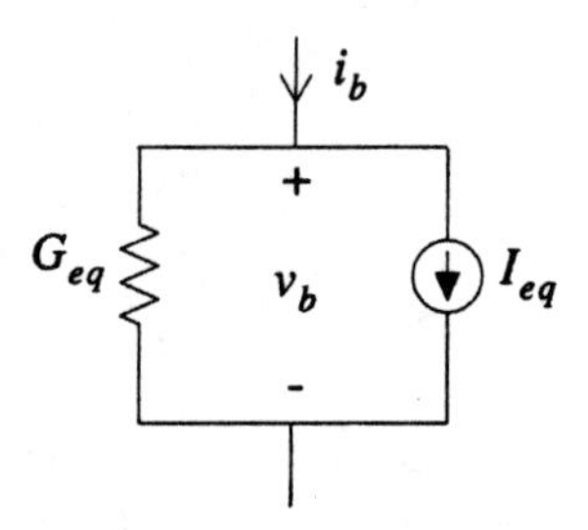

Figure 9.4 Companion model for a linearized element.

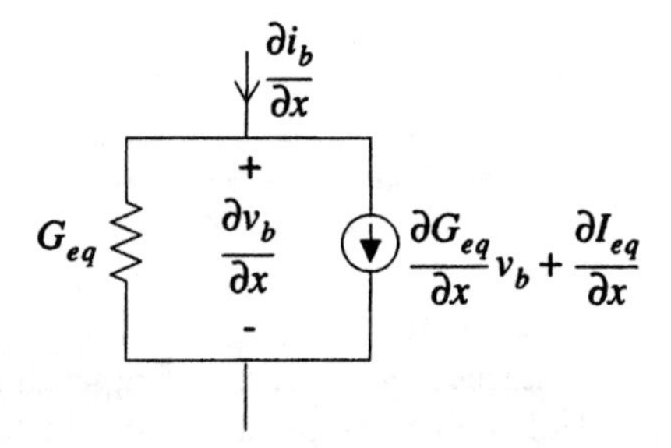

Figure 9.5 Companion model for a linearized element in the sensitivity circuit.

In our simple example in Figure 9.1, we considered the sensitivities of the response variables due to a change in a single resistance value. It is more likely that we would have occasion to calculate the sensitivities with respect to some complex parameter. In such cases, we can use the chain rule of differentiation to obtain the sensitivity circuit element.

As an example, consider the case of calculating the sensitivities of the node voltages with respect to the change in the saturation current of a diode. The diode branch relation is

$$i_d = f(v_d, x) = I_{SAT}\left(e^{\frac{v_d}{V_T}} - 1\right) \tag{9.3.8}$$

where $x = I_{SAT}$. Differentiating,

$$\frac{\partial i_d}{\partial x} = \frac{\partial f}{\partial v_d}\frac{\partial v_d}{\partial x} + \frac{\partial f}{\partial x} \tag{9.3.9}$$

we recognize that $G_{eq} = \partial f / \partial v_d$ and $\partial f / \partial x$ computed at the operating point give us a companion model for the sensitivity circuit.

In summary, the direct sensitivity analyses in Sections 9.1, 9.2, and 9.3 compute the sensitivities of all responses with respect to one parameter using the LU factors of the original circuit.

9.4 Adjoint Sensitivity Analysis

The approach described in the previous section for calculating sensitivities directly yields the sensitivities of all of the responses to changes in one parameter. Conversely, during circuit optimization, we are sometimes interested in the sensitivity of one output, with respect to many parameter values (perhaps all element values). In Chapter 3 we introduced adjoint sensitivity analysis and showed how to evaluate the adjoint matrix equations to obtain such sensitivities.

Consider once again the example shown in Figure 9.1. Recall the steps for adjoint sensitivity computation from section 3.6. To obtain the sensitivities of v_2 with respect to all parameters in $\boldsymbol{Y}$ and $\boldsymbol{J}$ (which are shown in (9.1.7)), we define

$$\boldsymbol{e}_2 = \begin{bmatrix} 0 \\ 1 \\ 0 \end{bmatrix} \tag{9.4.1}$$

where a "1" entry appears in the v_2 location for the solution vector. We solve for an intermediate vector $\boldsymbol{g}_2$ (the subscript 2 refers to the second variable in the solution vector) in

$$\boldsymbol{U}^T \boldsymbol{g}_2 = \boldsymbol{e}_2 \tag{9.4.2}$$

by Forward Substitution using the original LU factors. For our example in Figure 9.1, $\boldsymbol{U} = \boldsymbol{1}$, therefore $\boldsymbol{g}_2 = \boldsymbol{e}_2$. By Back Substitution we then solve

$$L^T \xi_2 = g_2 \tag{9.4.3}$$

or

$$\begin{bmatrix} 1 & \frac{-1}{R_1} & \frac{1}{R_1} \\ 0 & \left(\frac{1}{R_1} + \frac{1}{R_2}\right) & \frac{-1}{R_1} \\ 0 & 0 & 1 \end{bmatrix} [\xi_2] = \begin{bmatrix} 0 \\ 1 \\ 0 \end{bmatrix} \tag{9.4.4}$$

to obtain the vector ξ_2 from which we can specify the sensitivities. For our example,

$$\xi_2 = \begin{bmatrix} \frac{R_2}{R_1 + R_2} \\ \frac{R_1 R_2}{R_1 + R_2} \\ 0 \end{bmatrix} \tag{9.4.5}$$

From section 3.6, we know

$$\frac{\partial v_2}{\partial J_j} = \xi_{2j} \tag{9.4.6}$$

where ξ_{2j} is the jth entry of ξ_2 and J_j is the jth entry of $\boldsymbol{J}$. Or, we can calculate the sensitivities with respect to the $\boldsymbol{Y}$ matrix using

$$\frac{\partial v_2}{\partial Y_{kl}} = -\xi_{2k} v_l \tag{9.4.7}$$

where Y_{kl} is the (k, l) component of $\boldsymbol{Y}$ and v_l is the lth solution variable.

The $\partial v_2 / \partial R_1$ term, for example, is now obtained by applying the chain rule of differentiation. We first find all the locations in $\boldsymbol{Y}$ where R_1 appears and combine that stamp with the sensitivity with respect to those stamp locations to obtain the composite sensitivity that we seek.

Now that we have recalled the material from section 3.6 and applied it to our simple example, we are ready to proceed further. Like the sensitivity circuit for direct sensitivity, this matrix-based adjoint analysis has a physical circuit interpretation. We did not describe the adjoint circuit in Section 3.6, but we do so here in the remainder of this chapter.

9.5 The Adjoint Sensitivity Relation

Before introducing the adjoint circuit, we must review and extend Tellegen's theorem (which is mentioned in (A.5.11)), where we demonstrated that conservation of energy is a consequence of the KCL and KVL topological constraints.

To obtain the basic sensitivity relation, first suppose that we have two topologically identical circuits. That is, they have the same graph, with the same branch numbering, but their branch constitutive relations (BCRs) may be different. As with direct sensitivity, the original circuit N has branch voltages $\boldsymbol{v}_b$, $\boldsymbol{i}_b$ and $\boldsymbol{v}_n$, respectively and the related "adjoint circuit" η has branch voltages, branch currents, and node voltages ψ_b, ϕ_b, and ψ_n, respectively, as shown in Figure 9.6.

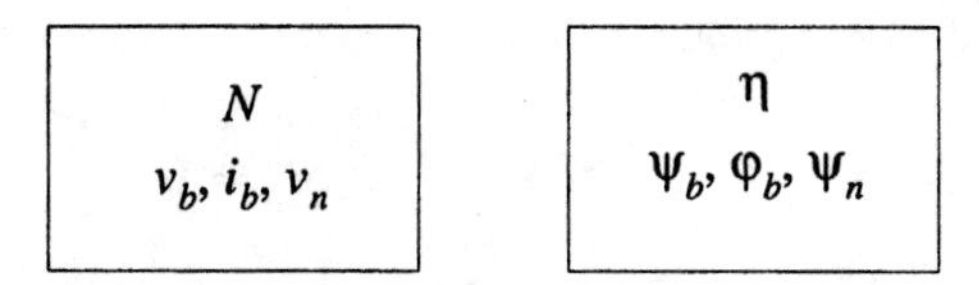

Figure 9.6 A circuit N and its adjoint circuit η.

Because the circuits are topologically identical, their topological constraints (KCL and KVL) are the same (see (A.3.2) and (A.3.3)):

$$\begin{aligned} &\text{KCL:} \quad \boldsymbol{A}\boldsymbol{i}_b = \boldsymbol{0} \quad \text{and} \quad \boldsymbol{A}\varphi_b = \boldsymbol{0} \\ &\text{KVL:} \quad \boldsymbol{v}_b = \boldsymbol{A}^T\boldsymbol{v}_n \quad \text{and} \quad \psi_b = \boldsymbol{A}^T\psi_n \end{aligned} \tag{9.5.1}$$

We begin with Tellegen's theorem [Desoer69] in its most general form:

$$\boldsymbol{v}_b^T\varphi_b = 0 \quad \text{and} \quad \psi_b^T\boldsymbol{i}_b = 0 \tag{9.5.2}$$

Equation (9.5.2) can easily be derived from the above topological constraints [Desoer69] since

$$\boldsymbol{v}_b^T\varphi_b = (\boldsymbol{A}^T\boldsymbol{v}_n)^T\varphi_b = \boldsymbol{v}_n^T\boldsymbol{A}\varphi_b = 0 \tag{9.5.3}$$

and

$$\psi_b^T\boldsymbol{i}_b = (\boldsymbol{A}^T\psi_n)^T\boldsymbol{i}_b = \psi_n^T\boldsymbol{A}\boldsymbol{i}_b = 0 \tag{9.5.4}$$

We note that the energy conservation relations

$$v_b^T i_b = 0 \qquad \text{and} \qquad \psi_b^T \varphi_b = 0 \tag{9.5.5}$$

are special cases of Tellegen's theorem.

Suppose that we alter the BCRs, but not the topology of the original circuit. Because of the variation of some parameters of interest, let us assume that the nominal branch voltages and currents show a variation:

$$v_b \to v_b + \delta v_b \qquad \text{and} \qquad i_b \to i_b + \delta i_b \tag{9.5.6}$$

Again, by Tellegen's theorem

$$(v_b^T + \delta v_b^T)\,\varphi_b = 0 \qquad \text{and} \qquad \psi_b^T (i_b + \delta i_b) = 0 \tag{9.5.7}$$

Subtracting the original pair of relations (9.5.2) from this pair, we obtain

$$\delta v_b^T \varphi_b = 0 \qquad \text{and} \qquad \psi_b^T \delta i_b = 0 \tag{9.5.8}$$

Finally, we can subtract the second of these relations from the first to obtain

$$\delta v_b^T \varphi_b - \psi_b^T \delta i_b = 0 \tag{9.5.9}$$

which is the basic sensitivity relation. So far, we have made no assumptions regarding the Branch Constitutive Relations in either circuit; this result is general. The only assumption made so far is that the two circuits share a common incidence matrix A. We will see that it is more convenient for subsequent manipulations to consider the basic sensitivity relation in scalar form:

$$\sum_{\text{all branches}} (\varphi \delta v - \psi \delta i) = 0 \tag{9.5.10}$$

9.6 Simple Reciprocal (R, G, V, I) Linear dc Circuits

Consider a circuit that has only independent voltage and current sources, resistances, and conductances. We expand the basic sensitivity relation as follows:

$$\begin{aligned}\sum_{\text{all branches}} (\varphi\delta v - \psi\delta i) = &\sum_V (\varphi_V\delta v_V - \psi_V\delta i_V) \\ &+\sum_I (\varphi_I\delta v_I - \psi_I\delta i_I) \\ &+\sum_R (\varphi_R\delta v_R - \psi_R\delta i_R) \\ &+\sum_G (\varphi_G\delta v_G - \psi_G\delta i_G) = 0\end{aligned} \tag{9.6.1}$$

Here the subscripts, V, I, R, and G, on the original and adjoint circuit branch voltages and currents indicate the generic branch types. And these same subscripts on the summation symbols ($\sum$s) indicate that the summation is to be taken over all such branches.

Next, we consider the varied BCRs for each generic branch type:

Independent voltage sources:

$$\begin{aligned}v_V &= V_V \\ v_V + \delta v_V &= V_V + \delta V_V \\ \delta v_V &= \delta V_V\end{aligned} \tag{9.6.2}$$

Independent current sources:

$$\begin{aligned}i_I &= I_I \\ i_I + \delta i_I &= I_I + \delta I_I \\ \delta i_I &= \delta I_I\end{aligned} \tag{9.6.3}$$

Resistors:

$$\begin{aligned}v_R &= R i_R \\ v_R + \delta v_R &= (R + \delta R)\,(i_R + \delta i_R) \\ \delta v_R &\approx R\delta i_R + \delta R i_R + \cancel{\delta R\delta i_R}\end{aligned} \tag{9.6.4}$$

Conductors:

$$\begin{aligned}i_G &= G v_G \\ i_G + \delta i_G &= (G + \delta G)\,(v_G + \delta v_G) \\ \delta i_G &\approx G\delta v_G + \delta G v_G + \cancel{\delta G\delta v_G}\end{aligned} \tag{9.6.5}$$

The final relations in (9.6.4) and (9.6.5) for the resistances and conductances are only approximate since we have crossed out the second-order variations. It is this assumption that limits the applicability of the following results to *incremental* sensitivity studies.

Upon substitution of the BCR sensitivity relations (9.6.2) to (9.6.5) into the basic sensitivity relation (9.6.1), we obtain

$$\begin{aligned}&\sum_V (\varphi_V \delta V_V - \psi_V \delta i_V) + \sum_I (\varphi_I \delta v_I - \psi_I \delta I_I)\\ &+\sum_R [\varphi_R (R\delta i_R + \delta R i_R) - \psi_R \delta i_R]\\ &+\sum_G [\varphi_G \delta v_G - \psi_G (G\delta v_G + \delta G v_G)] = 0\end{aligned} \tag{9.6.6}$$

We are free to choose any BCRs for the adjoint circuit elements. So we define the resistance and conductance BCRs for the adjoint circuit so as to remove their voltage and current variations from (9.6.6):

$$\text{Resistances:} \qquad \psi_R = R\varphi_R \tag{9.6.7}$$

$$\text{Conductances:} \qquad \varphi_G = G\psi_G \tag{9.6.8}$$

In other words, resistance and conductance branches in the original circuit correspond to resistance and conductance branches, with the same respective values, in the adjoint circuit. Substituting these adjoint element BCRs into (9.6.6),

$$\begin{aligned}&\sum_V (\varphi_V \delta V_V - \psi_V \delta i_V) + \sum_I (\varphi_I \delta v_I - \psi_I \delta I_I)\\ &+\sum_R \varphi_R i_R \delta R + \sum_G (-\psi_G v_G \delta G) = 0\end{aligned} \tag{9.6.9}$$

Separating independent source variations on the left hand side, we obtain the following useful result:

$$\begin{aligned}\sum_V \psi_V \delta i_V + \sum_I (-\varphi_I)\, \delta v_I = &\sum_V \varphi_V \delta V_V + \sum_I (-\psi_I)\, \delta I_I\\ &+\sum_R \varphi_R i_R \delta R + \sum_G (-\psi_G v_G)\, \delta G\end{aligned} \tag{9.6.10}$$

Here we have isolated the variations of the response variables, δi_V and δv_I, on the left hand side of the equation, and the variable parameters, δV_V, δI_I, δR, and δG, on the right. Without loss of generality, we can always take independent voltage source currents and independent current source voltages to be response variables, since we can introduce zero valued sources appropriately into a circuit wherever we want to define responses.

To complete the interpretation of the sensitivity relation (9.6.10), we choose BCRs for the independent sources in the adjoint circuit. We choose to replace independent sources in the original circuit with the same type of elements in the adjoint circuit. Thus,

$$\psi_V = \Psi_V \tag{9.6.11}$$

and

$$-\varphi_I = \Phi_I \tag{9.6.12}$$

can be taken to be independent voltage and current source branches, respectively, in the adjoint circuit.

Let us assume we have just one response of interest. First, we set all independent sources in the adjoint circuit to zero, except for the one in which we are interested, which we set to unity (+1 if the response is the current of an independent voltage source, -1 if its is the voltage of an independent current source). The next step is to solve the adjoint circuit by reusing the LU factors of the original circuit. Finally, we can pick off all the required sensitivities from the right hand side of (9.6.10).

More generally, given a scalar performance function

$$f(\boldsymbol{i}_V, \boldsymbol{v}_I) \tag{9.6.13}$$

then

$$\delta f = \sum_V \frac{\partial f}{\partial i_V}\delta i_V + \sum_I \frac{\partial f}{\partial v_I}\delta v_I \tag{9.6.14}$$

So, we can take the following as adjoint circuit independent source excitations:

$$\left.\begin{aligned} \Psi_V &= \frac{\partial f}{\partial i_V} \\ \Phi_I &= \frac{\partial f}{\partial v_I} \end{aligned}\right\} \text{for each such response (branch)} \tag{9.6.15}$$

Then the variation of the scalar performance function (9.6.14) coincides with the left hand side of (9.6.10), and we can pick from the right hand side the variations of the performance function $f(\boldsymbol{i}_V, \boldsymbol{v}_I)$ with respect to each of the individual elements (types):

$$\frac{\partial f}{\partial V_V} = \varphi_V \tag{9.6.16}$$

the current(s) through the adjoint voltage source(s);

$$\frac{\partial f}{\partial I_I} = -\psi_I \tag{9.6.17}$$

(the negative of) the voltage(s) across the adjoint current source(s);

$$\frac{\partial f}{\partial G} = -\psi_G v_G \tag{9.6.18}$$

(the negative of) the product(s) of the voltages across the original and adjoint circuit conductances;

$$\frac{\partial f}{\partial R} = \varphi_R i_R \tag{9.6.19}$$

the product(s) of the currents through the original and adjoint resistances.

In this simple case the original and adjoint circuits are identical except for their independent source values. And the adjoint circuit independent source excitation values are derived from the original circuit responses via the scalar performance function $f(i_V, v_I)$ and its variation (9.6.14). So we see that with two simple circuit solutions, first the original and then the adjoint, we can obtain the incremental sensitivity of an arbitrary scalar performance function $f(i_V, v_I)$ with respect to *every* circuit parameter. Since the original and adjoint circuits are the same except for their independent source excitation values, most of the effort entailed in the adjoint circuit solution (the LU Factorization) has already been performed for the original circuit. The original LU factors need merely be reapplied via Forward and Back Substitution to a new excitation vector defined by (9.6.15).

Before we leave this topic, we note finally that the resistance and conductance sensitivity expressions

$$\frac{\partial f}{\partial R} = \varphi_R i_R \tag{9.6.20}$$

and

$$\frac{\partial f}{\partial G} = -\psi_G v_G \tag{9.6.21}$$

are consistent. Because

$$G = \frac{1}{R} \tag{9.6.22}$$

we expect

$$\frac{\partial f}{\partial R} = \frac{\partial f}{\partial G}\frac{dG}{dR} = -\frac{1}{R^2}\frac{\partial f}{\partial G} \tag{9.6.23}$$

Using

$$\varphi_R = \frac{1}{R}\psi_R \qquad \text{and} \qquad i_R = \frac{1}{R}v_R \tag{9.6.24}$$

we obtain

$$\frac{\partial f}{\partial R} = \varphi_R i_R = \frac{1}{R^2}\psi_R v_R = \frac{1}{R^2}(-\psi_G v_G) = -\frac{1}{R^2}\frac{\partial f}{\partial G} \tag{9.6.25}$$

as expected. Since these sensitivities are identical, it does not matter whether we treat such elements as conductances or resistances. But there is a good reason to distinguish between them. Suppose that we want to find the sensitivity of a scalar performance function $f(i_V, v_I)$ with respect to a non-existent parasitic resistance (in series with another element) or conductance (across a node pair). For example, we may omit a parasitic resistance in the nominal analysis and then desire to find the sensitivity with respect to the parasitic effect that we omitted. In such cases, we must distinguish between conductances and resistances, since we are finding the sensitivity with respect to zero-valued elements. The ability to obtain such "parasitic sensitivities" aids in device model refinement and in simulation validation. Sometimes for simulation efficiency we may wish to suppress parasitic elements that we do not expect to have an appreciable effect on the outcome of the analysis. Then we can find the sensitivities of the simulation results with respect to the suppressed parasitic elements to verify our assumption in excluding them from the original analysis. Similarly, we can use such parasitic sensitivities to determine initial wiring paths in conjunction with performance driven layout programs.

Consider the circuit example in Figure 9.7. Suppose that our performance function is merely the output voltage

$$f \equiv v_I \tag{9.6.26}$$

Let us try to find the sensitivity of f with respect to all element values by the adjoint method. For the original circuit analysis we have the following results:

$$\begin{aligned} v_V &= V_V & v_{R2} &= \frac{R_2}{R_2+R_3}V_V & v_{R3} &= \frac{R_3}{R_2+R_3}V_V \\ i_{R1} &= \frac{V_V}{R_1} & i_{R2} &= \frac{V_V}{R_2+R_3} & i_{R3} &= \frac{V_V}{R_2+R_3} \end{aligned} \tag{9.6.27}$$

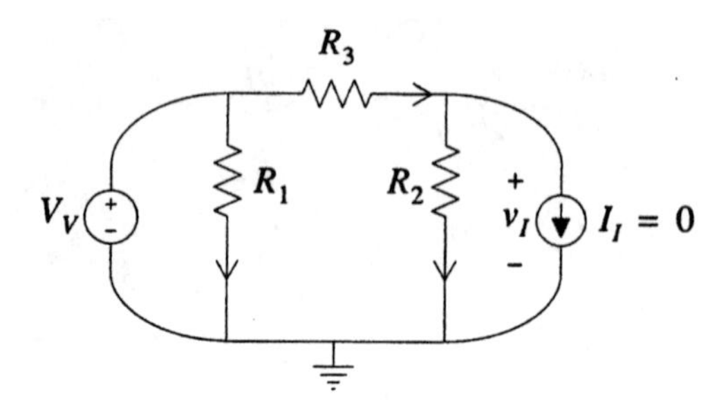

Figure 9.7 Circuit for sensitivity computation.

The adjoint circuit is shown in Figure 9.8. For the adjoint circuit analysis we have the following results:

$$\varphi_V = \frac{R_2}{R_2 + R_3} \qquad \varphi_{R1} = 0 \qquad \varphi_{R2} = \frac{R_3}{R_2 + R_3} \qquad \varphi_{R3} = \frac{-R_2}{R_2 + R_3} \tag{9.6.28}$$

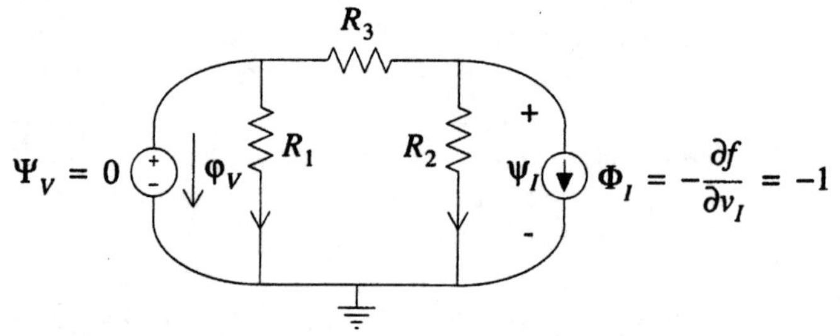

R_3

R_1 R_2

$\Psi_V = 0$ φ_V ψ_I $\Phi_I = -\frac{\partial f}{\partial v_I} = -1$

Figure 9.8 Sensitivity circuit for example in Figure 9.7.

So we obtain the sensitivity expression from (9.6.10)

$$\delta v_I = \underbrace{\frac{R_2}{R_2 + R_3}}_{\varphi_V} \delta V_V + \underbrace{0}_{\varphi_{R1} i_{R1}} (\delta R_1) + \underbrace{\frac{R_3 V_V}{(R_2 + R_3)^2}}_{\varphi_{R2} i_{R2}} \delta R_2 + \underbrace{\frac{-R_2 V_V}{(R_2 + R_3)^2}}_{\varphi_{R3} i_{R3}} \delta R_3 \tag{9.6.29}$$

From (9.6.29), we can pick off the required sensitivities

$$\frac{\partial f}{\partial V_V} = \frac{R_2}{R_2 + R_3} \qquad \frac{\partial f}{\partial R_1} = 0 \qquad \frac{\partial f}{\partial R_2} = \frac{R_3 V_V}{(R_2 + R_3)^2} \qquad \frac{\partial f}{\partial R_3} = \frac{-R_2 V_V}{(R_2 + R_3)^2} \tag{9.6.30}$$

For this simple circuit we can check these results analytically:

$$f = v_I = \frac{R_2}{R_2 + R_3} V_V$$

$$\frac{\partial f}{\partial V_V} = \frac{R_2}{R_2 + R_3} \qquad \frac{\partial f}{\partial R_1} = 0 \qquad \frac{\partial f}{\partial R_2} = \frac{R_3 V_V}{(R_2 + R_3)^2} \qquad \frac{\partial f}{\partial R_3} = \frac{-R_2 V_V}{(R_2 + R_3)^2} \quad \textbf{(9.6.31)}$$

In general we cannot compute the sensitivities analytically as we have been able to for this simple example. But, using the adjoint circuit, we can compute in general the required incremental sensitivities numerically and exactly.

9.7 *Sensitivities with Respect to Dependent Source Values*

The adjoint circuit formulation gets a little more complicated when the original circuit contains dependent sources and active devices. To begin to address this generalization we first consider the four basic dependent source relations and their perturbations in terms of the two branch models introduced first in Chapter 2.

Figure 9.9 shows a voltage-controlled current source and its sensitivity equations. Likewise, a voltage-controlled voltage source, current-controlled current source and current-controlled voltage source are shown along with their sensitivity equations in Figures 9.10, 9.11, and 9.12, respectively. In all of the above we have neglected second-order variations as indicated in the figures.

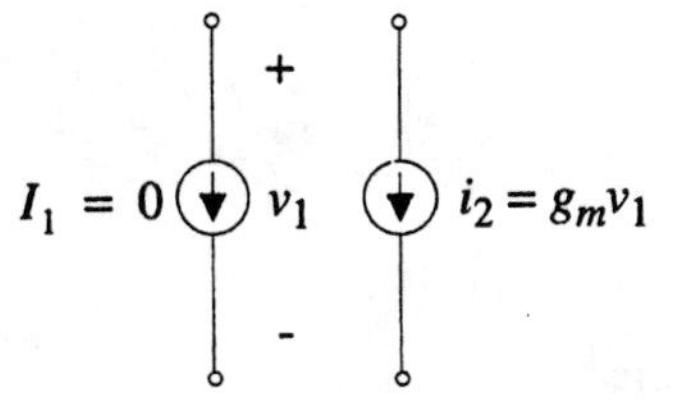

$I_1 = 0$ + v_1 - $\quad i_2 = g_m v_1$

$$\delta I_1 = 0$$

$$\delta i_2 \approx g_m \delta v_1 + \delta g_m v_1 + \delta g_m \delta v_1$$

Figure 9.9 Voltage-controlled current source.

To account for dependent sources in the basic sensitivity relation (9.6.10), we consider together both their controlling and controlled branches in a single composite term:

$$\sum_{\text{all branches}} (\varphi \delta v - \psi \delta i) = \ldots + (\varphi_1 \delta v_1 - \psi_1 \delta i_1 + \varphi_2 \delta v_2 - \psi_2 \delta i_2) + \ldots = 0 \quad \textbf{(9.7.1)}$$

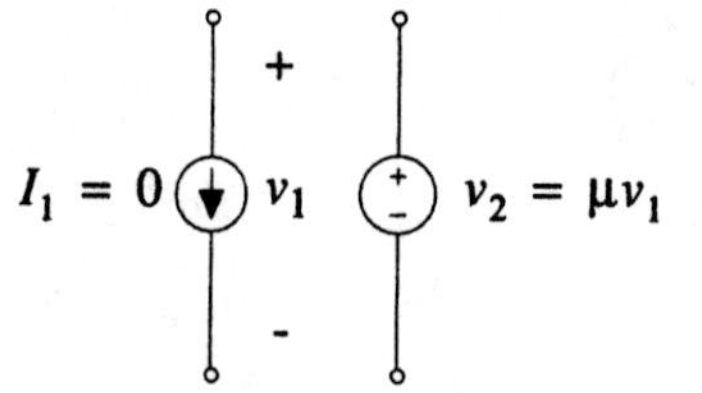

$$\delta I_1 = 0$$

$$\delta v_2 \approx \mu \delta v_1 + \delta\mu v_1 + \cancel{\delta\mu\delta v_1}$$

Figure 9.10 Voltage-controlled voltage source.

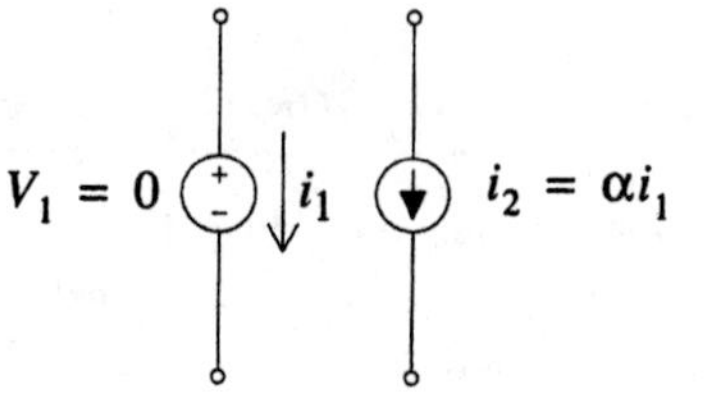

$$\delta V_1 = 0$$

$$\delta i_2 \approx \alpha \delta i_1 + \delta\alpha i_1 + \cancel{\delta\alpha\delta i_1}$$

Figure 9.11 Current-controlled current source.

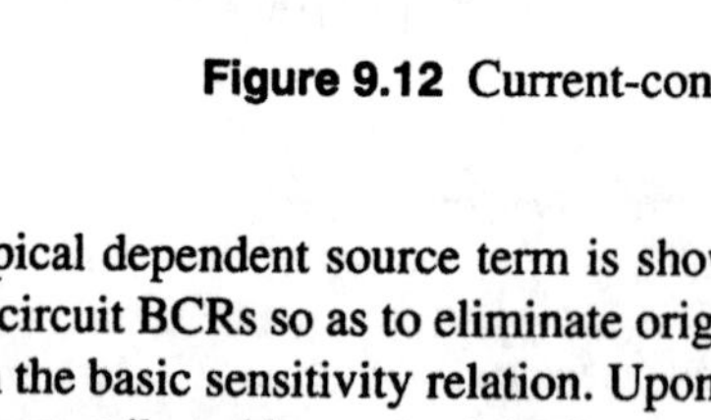

$$\delta V_1 = 0$$

$$\delta v_2 \approx r_m \delta i_1 + \delta r_m i_1 + \cancel{\delta r_m \delta i_1}$$

Figure 9.12 Current-controlled voltage source.

where a typical dependent source term is shown explicitly. As before, we seek to define the adjoint circuit BCRs so as to eliminate original circuit branch voltage and current variations from the basic sensitivity relation. Upon careful examination of the composite term in (9.7.1) as contributed by each of the four types of dependent sources, we obtain the following adjoint circuit relations.

For a voltage-controlled current source, the contribution to (9.7.1) is

$$\cancel{\varphi_1 \delta v_1} + \cancel{\varphi_2 \delta v_2} - \psi_2 (\cancel{g_m \delta v_1} + \delta g_m v_1) \tag{9.7.2}$$

We are free to choose the BCRs of the adjoint circuit elements. So we choose $\varphi_2 = 0$ and $\varphi_1 = g_m \psi_2$. Thus the adjoint circuit element is a voltage-controlled current source, too, but with the direction of control reversed, as shown in Figure 9.13. That leaves the following contribution to the left hand side of (9.7.1):

$$-\psi_2 v_1 \delta g_m \tag{9.7.3}$$

Thus the sensitivity with respect to g_m is the negative product of the original and adjoint controlling voltages.

Similarly for a voltage-controlled voltage source, with the appropriate choice of BCRs for the adjoint circuit,

$$\cancel{\varphi_1 \delta v_1} + \varphi_2 (\cancel{\mu \delta v_1} + \delta \mu v_1) - \cancel{\psi_2 \delta i_2} = \varphi_2 v_1 \delta \mu \tag{9.7.4}$$

and the corresponding sensitivity circuit is shown in Figure 9.14. For a current-controlled current source,

$$-\cancel{\psi_1 \delta i_1} + \cancel{\varphi_2 \delta v_2} - \psi_2 (\cancel{\alpha \delta i_1} + \delta \alpha i_1) = -\psi_2 i_1 \delta \alpha \tag{9.7.5}$$

and the corresponding sensitivity circuit is shown in Figure 9.15. Finally, for a current-controlled voltage source,

$$-\cancel{\psi_1 \delta i_1} + \varphi_2 (\cancel{r_m \delta i_1} + \delta r_m i_1) - \cancel{\psi_2 \delta i_2} = \varphi_2 i_1 \delta r_m \tag{9.7.6}$$

and the sensitivity circuit is shown in Figure 9.16.

$$\varphi_2 = 0 \qquad \text{and} \qquad \varphi_1 = g_m \psi_2$$

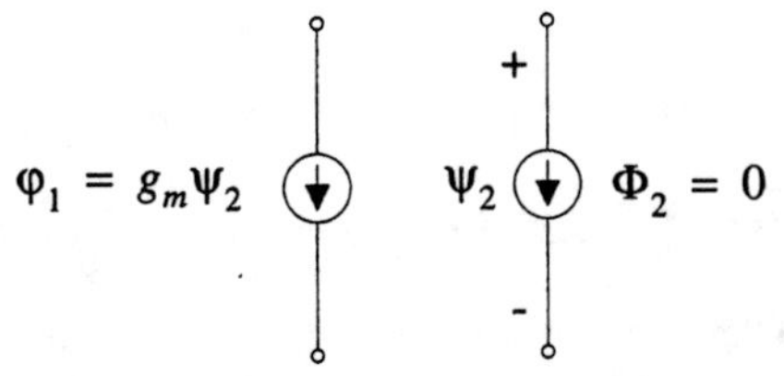

$\varphi_1 = g_m \psi_2$ + ψ_2 $\Phi_2 = 0$ -

Figure 9.13 Sensitivity circuit for a voltage-controlled current source.

$$\psi_2 = 0 \qquad \text{and} \qquad \varphi_1 = -\mu\varphi_2$$

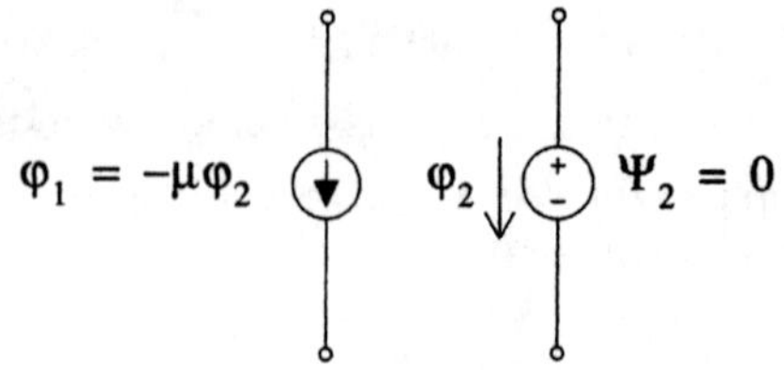

$$\varphi_1 = -\mu\varphi_2 \qquad \varphi_2 \qquad \Psi_2 = 0$$

Figure 9.14 Sensitivity circuit for a voltage-controlled voltage source.

$$\varphi_2 = 0 \qquad \text{and} \qquad \psi_1 = -\alpha\psi_2$$

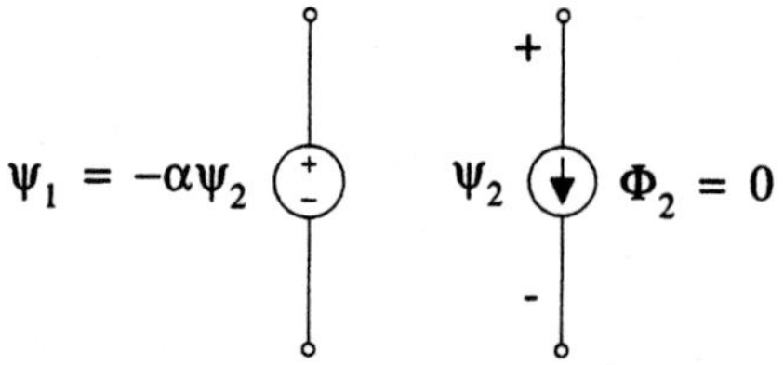

$$\psi_1 = -\alpha\psi_2 \qquad \psi_2 \qquad \Phi_2 = 0$$

Figure 9.15 Sensitivity circuit for a current-controlled current source.

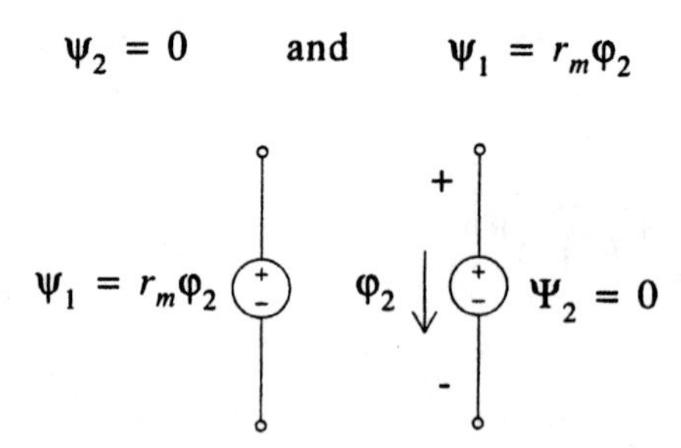

Figure 9.16 Sensitivity circuit for a current-controlled voltage source.

Whereas we could have derived equivalent results without them, the two-branch dependent source models have simplified considerably the manipulations. The adjoint circuit topology remains the same as that of the original circuit, only the direction of control in each dependent source is reversed. We could not have detected this effect with reciprocal circuits, by considering one element at a time as in the previous section.

If we form the adjoint circuit by replacing all of the original circuit elements with their appropriate adjoint elements as detailed above, we obtain the following generalization of the sensitivity relation shown in (9.6.10):

$$\begin{aligned}
\delta f(i_V, v_I) &= \sum_V \frac{\partial f}{\partial i_v}\delta i_v + \sum_I \frac{\partial f}{\partial v_I}\delta v_I \\
&= \sum_V \Psi_V \delta i_V + \sum_I (-\Phi_I)\,\delta v_I \\
&= \sum_V \varphi_V \delta V_V + \sum_I (-\psi_I)\,\delta I_I \\
&\quad + \sum_R \varphi_R i_R \delta R + \sum_I (-\psi_G v_G)\,\delta G \\
&\quad + \sum_{VCI} (-\psi_2 v_1)\,\delta g_m + \sum_{VCV} \varphi_2 v_1 \delta\mu \\
&\quad + \sum_{ICI} (-\psi_2 i_1)\,\delta\alpha + \sum_{ICV} \varphi_2 i_1 \delta r_m
\end{aligned} \tag{9.7.7}$$

As before, appropriate adjoint-circuit independent source excitations

$$\Psi_V = \frac{\partial f}{\partial i_V} \qquad \text{and} \qquad \Phi_I = -\frac{\partial f}{\partial v_I} \tag{9.7.8}$$

lead to adjoint circuit responses from which the desired sensitivities can be selected.

In the case of nonlinear circuits, the original circuit is first solved for an operating point. Each nonlinear element is represented by a linearized equivalent at its operating point. Then an adjoint sensitivity analysis can be carried out on the linearized circuit exactly as described above. The chain rule of differentiation can be applied to determine the sensitivity with respect to device model parameters.

We note that if the original linear(ized) circuit were solved in terms of a matrix $M = LU$, then the adjoint circuit can be solved in terms of $M^T = U^T L^T$. U^T and L^T are lower and upper triangular, respectively, so no new LU factorization needs to be performed. This relationship may not be obvious from the above dependent source relations, so we will discuss it in more detail later in the chapter.

In general, small change sensitivities in general can be determined by either the direct method or the adjoint method. In the direct method, we can consider many functions at once, but each parameter of interest requires a new Forward and Back Substitution. In the adjoint method, we can consider many parameters at once, but each response function requires a new Forward and Back Substitution. Hence, the adjoint method is used when there are more parameters than functions, and the direct method when there are more functions than parameters.

9.8 Adjoint Circuit Representation of Some Other Multi-Terminal Circuit Elements

In this section, we consider the adjoint circuit elements of some common linear multi-terminal circuit elements.

Perhaps the most interesting is the nullator/norator model of an ideal operational amplifier (see Figure 9.17) under appropriate stabilizing feedback. Incorporation of ideal operational amplifiers into Nodal Analysis is discussed in section 2.5. The circuit models for nullators and norators are shown in Figure 9.18. We can entertain such a model because two circuit branches should introduce two independent BCRs; both the relations, however, apply to the left hand nullator branch and the right hand norator branch is unconstrained. For the nullator we have

$$\delta v_1 = 0 \qquad \text{and} \qquad \delta i_1 = 0 \tag{9.8.1}$$

The composite term from the basic sensitivity expression (9.7.1) is

$$(\varphi_1 \delta v_1 - \psi_1 \delta i_1 + \varphi_2 \delta v_2 - \psi_2 \delta i_2) = 0 \tag{9.8.2}$$

provided we choose the adjoint circuit BCRs as

$$\varphi_2 = 0 \qquad \text{and} \qquad \psi_2 = 0 \tag{9.8.3}$$

The adjoint circuit models for a nullator and norator are shown in Figure 9.19. So, the adjoint circuit element for an ideal operational amplifier is another ideal operational amplifier, but reversed.

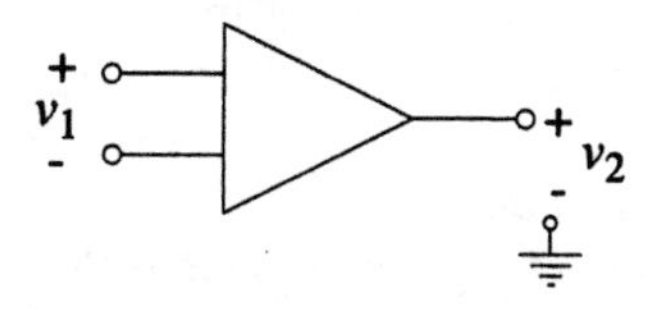

Figure 9.17 Ideal operational amplifier.

For an ideal transformer shown in Figure 9.20 we have

$$v_1 = nv_2 \Rightarrow \delta v_1 \approx n\delta v_2 + \delta n v_2 + \delta n \delta v_2 \tag{9.8.4}$$

and

$$i_2 = -ni_1 \Rightarrow \delta i_2 \approx -n\delta i_1 - \delta n i_1 - \delta n \delta i_1 \tag{9.8.5}$$

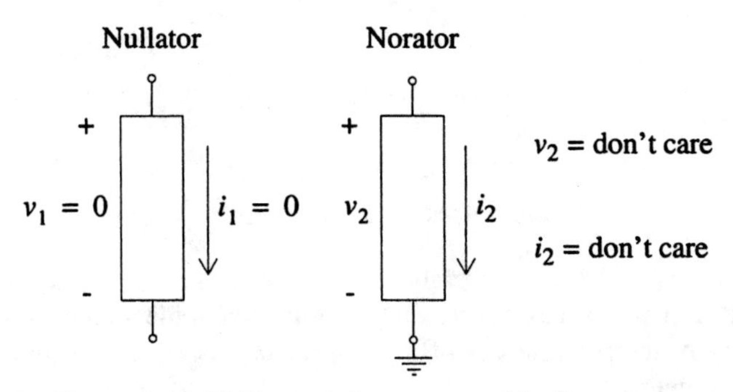

Figure 9.18 Nullator and norator model of an op-amp.

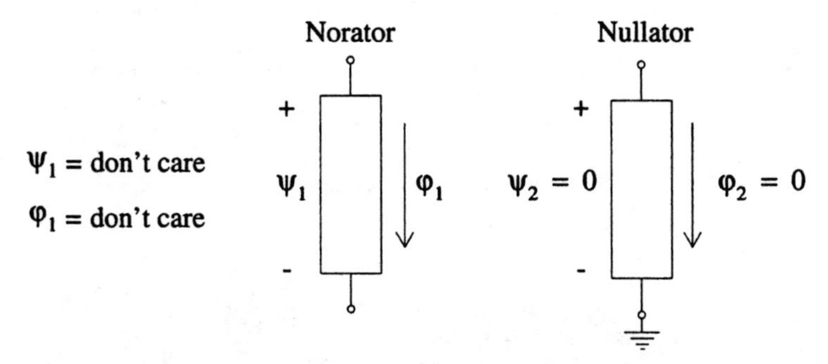

Figure 9.19 Adjoint circuit for an ideal op-amp.

So the composite term in the basic sensitivity expression (9.7.1) becomes

$$
\begin{aligned}
(\varphi_1 \delta v_1 - \psi_1 \delta i_1 + \varphi_2 \delta v_2 - \psi_2 \delta i_2) &= \varphi_1 (\cancel{n \delta v_2} + \delta n v_2) \\
&\quad - \cancel{\psi_1 \delta i_1} + \cancel{\varphi_2 \delta v_2} \\
&\quad - \psi_2 (-\cancel{n \delta i_1} - \delta n i_1) \\
&= (\varphi_1 v_2 + \psi_2 i_1)\, \delta n
\end{aligned}
\tag{9.8.6}
$$

provided that we choose the adjoint BCRs as

$$
\psi_1 = n \psi_2 \qquad \text{and} \qquad \varphi_2 = -n \varphi_1 \tag{9.8.7}
$$

Thus the adjoint circuit element for an ideal transformer element is the same as the original. Independent sources, resistances, conductances, and ideal transformers are *self-adjoint* (or reciprocal) circuit elements: their adjoint circuit models are the same as those of the original elements.

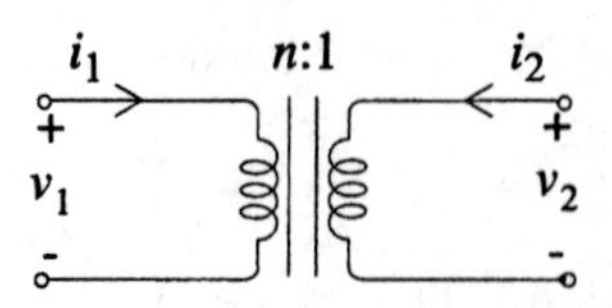

Figure 9.20 Ideal transformer.

Dependent sources and (ideal) operational amplifiers are not self-adjoint. We have observed this lack of reciprocity for dependent sources and ideal operational amplifiers above. A gyrator is another example of an element that is not self-adjoint. Consider a gyrator, shown in Figure 9.21.

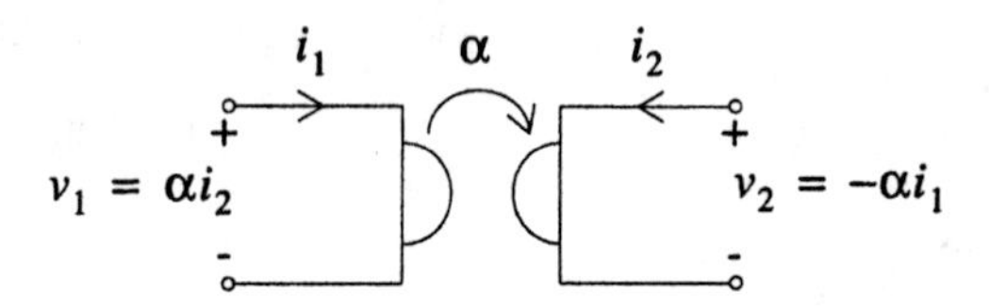

Figure 9.21 A gyrator.

For a gyrator, we have:

$$\begin{aligned} v_1 = \alpha i_2 &\Rightarrow \delta v_1 \approx \alpha \delta i_2 + \delta\alpha i_2 + \cancel{\delta\alpha\delta i_2} \\ v_2 = -\alpha i_1 &\Rightarrow \delta v_2 \approx -\alpha\delta i_1 - \delta\alpha i_1 - \cancel{\delta\alpha\delta i_1} \end{aligned} \tag{9.8.8}$$

and

$$\begin{aligned} (\varphi_1 \delta v_1 - \psi_1 \delta i_1 + \varphi_2 \delta v_2 - \psi_2 \delta i_2) &= \varphi_1 (\cancel{\alpha\delta i_2} + \delta\alpha i_2) - \cancel{\psi_1 \delta i_1} \\ &\quad + \varphi_2 (-\cancel{\alpha\delta i_1} - \delta\alpha i_1) - \cancel{\psi_2 \delta i_2} \\ &= (\varphi_1 i_2 - \varphi_2 i_1)\,\delta\alpha \end{aligned} \tag{9.8.9}$$

provided that

$$\psi_1 = -\alpha\varphi_2 \qquad \text{and} \qquad \psi_2 = \alpha\varphi_1 \tag{9.8.10}$$

The sensitivity circuit for a gyrator is shown in Figure 9.22. As expected, the gyration ratio is reversed. Hence the gyrator is not a self-adjoint element.

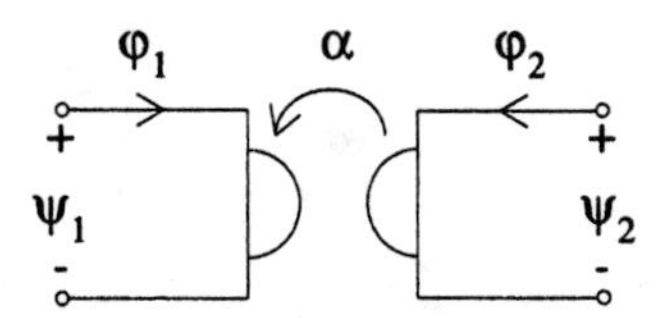

Figure 9.22 Sensitivity circuit for a gyrator.

9.9 The Sparse Tableau Interpretation of Adjoint Sensitivity

The only response variables that we have considered so far are currents through independent voltage sources and voltages across independent current sources. However, in the Sparse Tableau formulation of circuit equations (see section A.4 for details), all branch voltages, branch currents, and node voltages are explicitly determined. Hence, in sensitivity computation and optimization situations, it may be advantageous to work with the Sparse Tableau formulation of the circuit equations.

In this section, we will recall our matrix-based introduction to sensitivity in section 3.6. We will build on that foundation to derive more general sensitivity equations. Then we will interpret those equations in the context of Sparse Tableau formulation and tie them in with the sensitivity expressions we derived in earlier sections of this chapter based on Tellegen's theorem.

Given a set of linear(ized) equations

$$\boldsymbol{M}\boldsymbol{x} = \boldsymbol{b} \quad \textbf{(9.9.1)}$$

we can write

$$\boldsymbol{M}\delta\boldsymbol{x} + \delta\boldsymbol{M}\boldsymbol{x} = \delta\boldsymbol{b} \quad \textbf{(9.9.2)}$$

by neglecting second-order variations, and

$$\delta\boldsymbol{x} = \boldsymbol{M}^{-1}[\delta\boldsymbol{b} - \delta\boldsymbol{M}\boldsymbol{x}] \quad \textbf{(9.9.3)}$$

Suppose we have a scalar performance function $f(\boldsymbol{x})$. We can write

$$\delta f = \left[\frac{\partial f}{\partial \boldsymbol{x}}\right]^T \delta\boldsymbol{x} = \left[\frac{\partial f}{\partial \boldsymbol{x}}\right]^T \boldsymbol{M}^{-1}[\delta\boldsymbol{b} - \delta\boldsymbol{M}\boldsymbol{x}] \quad \textbf{(9.9.4)}$$

Postulate a matrix ξ such that

$$M^T\xi = \left[\frac{\partial f}{\partial x}\right] \tag{9.9.5}$$

Then

$$\xi^T M = \left[\frac{\partial f}{\partial x}\right]^T \tag{9.9.6}$$

and

$$\xi^T = \left[\frac{\partial f}{\partial x}\right]^T M^{-1} \tag{9.9.7}$$

From (9.9.7), ξ can easily be determined using the LU factors of M. Substituting (9.9.7) into (9.9.4),

$$\delta f = \xi^T [\delta b - \delta M x] \tag{9.9.8}$$

So the sensitivities of f with respect to the matrix entries of M and b can be computed.

Let us apply (9.9.13) and (9.9.16) to the Sparse Tableau equations (see section A.4):

$$\begin{bmatrix} \mathbf{1} & \mathbf{0} & -A^T \\ \mathbf{0} & A & \mathbf{0} \\ \alpha & \beta & \mathbf{0} \end{bmatrix} \begin{bmatrix} v_b \\ i_b \\ v_n \end{bmatrix} = \begin{bmatrix} o \\ o \\ \gamma \end{bmatrix} \tag{9.9.9}$$

and recognize that they are in the form $Mx = b$. Then to obtain the sensitivities of a scalar performance function

$$f(x) = f(v_b, i_b, v_n) \tag{9.9.10}$$

we must solve

$$\begin{bmatrix} \mathbf{1} & \mathbf{0} & \alpha^T \\ \mathbf{0} & A^T & \beta^T \\ -A & \mathbf{0} & \mathbf{0} \end{bmatrix} \begin{bmatrix} -\varphi_b \\ \psi_n \\ \eta_b \end{bmatrix} = \begin{bmatrix} \frac{\partial f}{\partial v_b} \\ \frac{\partial f}{\partial i_b} \\ \frac{\partial f}{\partial v_n} \end{bmatrix} \tag{9.9.11}$$

Note that we have given convenient names to the various adjoint variables

$$\xi \equiv \begin{bmatrix} -\varphi_b \\ \psi_n \\ \eta_b \end{bmatrix} \tag{9.9.12}$$

which will be determined when the adjoint circuit is solved.

Separated, these equations become

$$\left.\begin{aligned} -\varphi_b + \alpha^T \eta_b &= -\Phi_b \equiv \frac{\partial f}{\partial \boldsymbol{v}_b} \\ \boldsymbol{A}^T \psi_n + \beta^T \eta_b &= \Psi_b \equiv \frac{\partial f}{\partial \boldsymbol{i}_b} \\ \boldsymbol{A}\varphi_b &= \Phi_n \equiv \frac{\partial f}{\partial \boldsymbol{v}_n} \end{aligned}\right\} = \frac{\partial f}{\partial \boldsymbol{x}} \tag{9.9.13}$$

The subvectors $-\Phi_b$, Ψ_b, and Φ_n of $\partial f / \partial \boldsymbol{x}$ are adjoint circuit independent source excitations that provide the requisite right hand sides required for the desired sensitivity calculation. Note too that $\partial f / \partial \boldsymbol{x}$ could merely be the unit vector $\boldsymbol{e}_k$ if we were only interested in the sensitivities of the kth unknown variable in the original circuit.

Let us try to interpret (9.9.8) for a simple example. If a parameter of interest is a resistance value, one element of β is the only part of $\boldsymbol{M}$ that changes when the resistance value changes. From (9.9.16) we see for this example that $\delta \boldsymbol{b}$ is zero and $\delta \boldsymbol{M}\boldsymbol{x}$ picks out the current through that resistor in the original circuit, with all other values in the vector being zero. Hence the sensitivity is the product of the currents in the original and adjoint circuits, the same result as the one we obtained with Tellegen's theorem. This intuition will be formalized in the rest of this section.

Now, suppose that we define the adjoint circuit branch voltage vector

$$\psi_b \equiv \boldsymbol{A}^T \psi_n \tag{9.9.14}$$

then the above equations become

$$\begin{aligned} \varphi_b &= \alpha^T \eta_b + \Phi_b \\ \psi_b &= -\beta^T \eta_b + \Psi_b \\ \boldsymbol{A}\varphi_b &= \Phi_n \end{aligned} \tag{9.9.15}$$

And these are a description of a related adjoint circuit that is topologically identical to the original:

$$\text{KVL:} \qquad \psi_b = A^T \psi_n$$
$$\text{KCL:} \qquad A\varphi_b = \Phi_n \tag{9.9.16}$$

where the Φ_n represent current sources connected from ground to the nodes in the adjoint circuit which are nonzero only if the node voltages v_n appear explicitly in the performance function f.

$$\text{BCRs:} \qquad \varphi_b = \alpha^T \eta_b + \Phi_b \qquad \text{and} \qquad \psi_b = -\beta^T \eta_b + \Psi_b \tag{9.9.17}$$

Φ_b represents independent current sources in parallel and Ψ_b independent voltage sources in series with adjoint circuit branches, which are nonzero only if their respective voltages or currents appear explicitly in the performance function f.

To begin to interpret the BCRs, suppose that α and β have inverses, even though in general they probably do not. Then we can eliminate the dummy variable η_b:

$$(\alpha^T)^{-1} \varphi_b + (\beta^T)^{-1} \psi_b = (\alpha^T)^{-1} \Phi_b + (\beta^T)^{-1} \Psi_b \tag{9.9.18}$$

or

$$\psi_b + \beta^T (\alpha^T)^{-1} \varphi_b = \Psi_b + \beta^T (\alpha^T)^{-1} \Phi_b \tag{9.9.19}$$

or

$$\psi_b + (\alpha^{-1}\beta)^T \varphi_b = \Psi_b + (\alpha^{-1}\beta)^T \Phi_b \tag{9.9.20}$$

Upon reconsideration of the original BCRs

$$\alpha v_b + \beta i_b = \gamma \tag{9.9.21}$$

we see also that

$$v_b + \alpha^{-1}\beta i_b = \alpha^{-1}\gamma \tag{9.9.22}$$

So, with the exception of the adjoint circuit independent source excitations derived from the performance function f, we see that the BCRs for the adjoint circuit are transposes of the original BCRs, as expected.

From this point it is straightforward to obtain individual element sensitivities in terms of the expressions derived earlier via Tellegen's Theorem. In the Tellegen's Theorem formulation of the sensitivity expressions we only allowed the currents through independent voltage sources (or ammeters) or the voltages across independent current sources (or voltmeters) to be performance function variables. That formulation is more elegant than the results from the Sparse Tableau formulation, but the answers are identical. To corroborate this statement, we look at some familiar circuit elements and their corresponding entries in the α, β, and γ matrices.

	α	β	γ
Independent voltage source:	1	0	V_v
Independent current source:	0	1	I_I
Resistance:	1	$-R$	0
Conductance:	$-G$	1	0

For an independent voltage source in the adjoint circuit,

$$\varphi_V = \eta_V + \Phi_V \tag{9.9.23}$$

is an arbitrary response, and

$$\psi_V = \Psi_V \tag{9.9.24}$$

is nonzero only if i_V is explicit in f.

For an independent current source in the adjoint circuit,

$$\varphi_I = \Phi_I \tag{9.9.25}$$

is nonzero only if v_I is explicit in f, and

$$\psi_I = -\eta_I + \Psi_I \tag{9.9.26}$$

is an arbitrary response.

For a resistance,

$$\begin{aligned} \varphi_R &= \eta_R + \Phi_R \\ \psi_R &= R\eta_R + \Psi_R \end{aligned} \tag{9.9.27}$$

and, if the resistance branch voltage and current do not appear explicitly in the performance function f, we have

$$\psi_R = R\varphi_R \tag{9.9.28}$$

For a conductance,

$$\begin{aligned} \varphi_G &= -G\eta_G + \Phi_G \\ \psi_G &= -\eta_G + \Psi_G \end{aligned} \tag{9.9.29}$$

and, again, if the conductance branch voltage and current do not appear explicitly in the performance function f, we have

$$\varphi_G = G\psi_G \tag{9.9.30}$$

Note in general that the inclusion of resistance or conductance branch voltages or currents in the performance function leads merely to composite (Thevenin or Norton equivalent) branches in the adjoint circuit.

The interpretation is equally straightforward for multi-terminal elements. To illustrate this statement, we consider as an example the current-controlled voltage source shown in Figure 9.23.

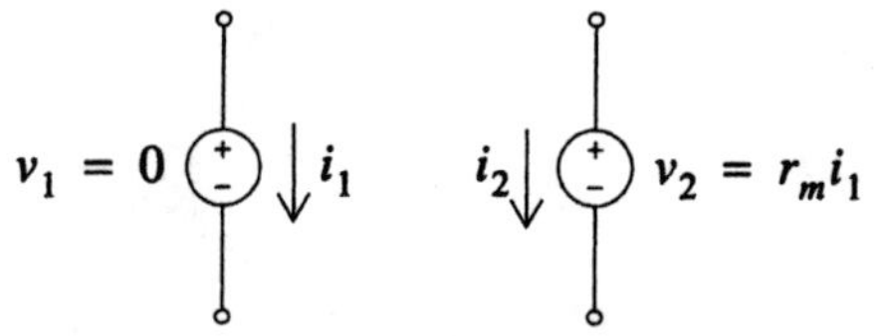

Figure 9.23 Current-controlled voltage source.

The generalized BCRs for this two-branch composite element

$$\alpha v + \beta i = \gamma \tag{9.9.31}$$

are

$$\begin{bmatrix} 1 & 0 \\ 0 & 1 \end{bmatrix} \begin{bmatrix} v_1 \\ v_2 \end{bmatrix} + \begin{bmatrix} 0 & 0 \\ -r_m & 0 \end{bmatrix} \begin{bmatrix} i_1 \\ i_2 \end{bmatrix} = \begin{bmatrix} 0 \\ 0 \end{bmatrix} \tag{9.9.32}$$

And the related adjoint circuit BCRs are

$$\begin{bmatrix} \varphi_1 \\ \varphi_2 \end{bmatrix} = \begin{bmatrix} 1 & 0 \\ 0 & 1 \end{bmatrix} \begin{bmatrix} \eta_1 \\ \eta_2 \end{bmatrix} + \begin{bmatrix} \Phi_1 \\ \Phi_2 \end{bmatrix} \tag{9.9.33}$$

and

$$\begin{bmatrix} \psi_1 \\ \psi_2 \end{bmatrix} = \begin{bmatrix} 0 & r_m \\ 0 & 0 \end{bmatrix} \begin{bmatrix} \eta_1 \\ \eta_2 \end{bmatrix} + \begin{bmatrix} \Psi_1 \\ \Psi_2 \end{bmatrix} \quad (9.9.34)$$

Assuming that v_1, v_2, i_1, and i_2 do not appear explicitly in the performance function, we can ignore Φ_1, Φ_2, Ψ_1, and Ψ_2 to obtain

$$\psi_1 = r_m \varphi_2 \qquad \text{and} \qquad \psi_2 = 0 \quad (9.9.35)$$

The sensitivity circuit is shown in Figure 9.24, which is the same adjoint circuit element that we obtained earlier via Tellegen's Theorem.

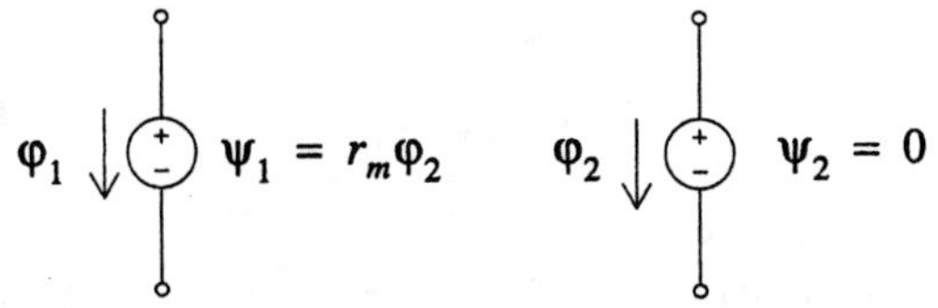

φ_1 + − $\psi_1 = r_m \varphi_2$ φ_2 + − $\psi_2 = 0$

Figure 9.24 Sensitivity circuit for a current-controlled voltage source.

Thus the adjoint circuit is the same whether it is obtained by Tellegen's Theorem or from the transpose of the Sparse Tableau equations. The Sparse Tableau equations have the advantage that they permit more general performance functions, and the introduction of zero-valued sources in the original circuit to define responses for such functions is not necessary. It is therefore easier to include branch voltages, branch currents, and node voltages in the performance function with the Sparse Tableau formulation.

Although the algebra in this section may seem a little daunting at first, the actual procedure for sensitivity computation is quite straightforward. The original analysis (of the form $\boldsymbol{Mx} = \boldsymbol{b}$) is first completed in terms of LU factors. Then $\boldsymbol{M}^T\xi = \left[\frac{\partial f}{\partial x}\right]$ is solved in terms of the $\boldsymbol{U}^T$ and $\boldsymbol{L}^T$ factors of the original solution. Finally, sensitivities are computed from (9.9.8). Most element values occur only once in each Sparse Tableau BCR, so picking off the sensitivity results is relatively easy. The transposed system computations are especially straightforward, with all the circuit variables explicitly represented. Thus, many parameters of interest may enter the Sparse Tableau equations linearly and in only one place, rendering the actual sensitivity computation trivial once ξ has been determined.

9.10 Some Possible Applications of Adjoint Sensitivity

In "sensitivity refined analysis," we may initiate a circuit simulation using the simplest possible device models. Then we can use sensitivities to adjust the response(s) of interest to bring it (them) into closer correspondence with the results that better models might produce.

For example, for a forward-biased diode, initially we might use

$$v = V_D(\text{on}) \tag{9.10.1}$$

as shown in Figure 9.25. Then we can analyze the circuit and obtain a value for the current i flowing through the ideal voltage source model. We recognize that for this current i the actual diode voltage is

$$v = \frac{nkT}{q} ln\left(\frac{i}{I_s} + 1\right) \tag{9.10.2}$$

So, we multiply the sensitivity of our performance function with respect to v by the voltage change

$$\delta v = \frac{nkT}{q} ln\left(\frac{i}{I_s} + 1\right) - V_D(\text{on}) \tag{9.10.3}$$

to approximate the change in a response of interest. We can adjust for all such (small) discrepancies simultaneously and efficiently because we can find all pertinent sensitivities of a single response from only one Forward and Back Substitution with the original LU factors.

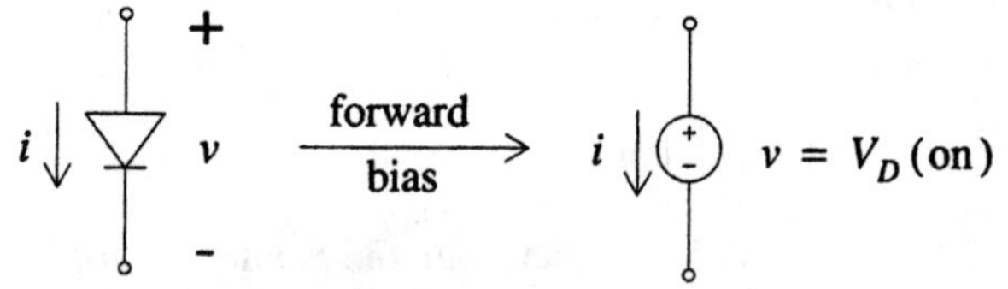

Figure 9.25 Forward-biased diode.

Similarly, for a back-biased diode we can solve for v and then adjust from $I = 0$ to

$$\delta i = I_s\left(e^{\frac{qv}{nkT}} - 1\right) \tag{9.10.4}$$

as shown in Figure 9.26. In general we need not use these idealized equations. We could use tables of measured data, if available, to find the deviations more accurately.

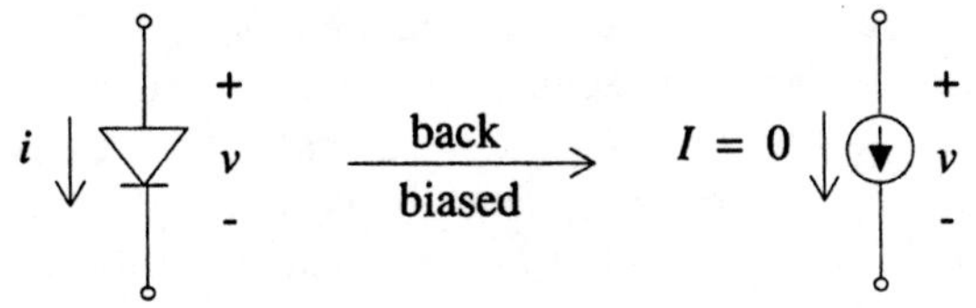

Figure 9.26 Reverse-biased diode.

If in such an approach some sensitivities turn out to be large, we could choose to substitute more sophisticated models for these instances, and then to re-analyze. Often the cost of two or more such approximate analyses is far less than that of a single accurate analysis, which is overburdened with unnecessarily detailed models.

This approach is the automated equivalent of the analysis that designers often carry out for dc bipolar circuit analysis: first use a crude switching model for the BJTs and then adjust for voltage changes by 60 mV per decade of current change at room temperature. In the case of MOS devices, the simplified $i - v$ model shown in Figure 9.27 may be adequate for a first-cut analysis.

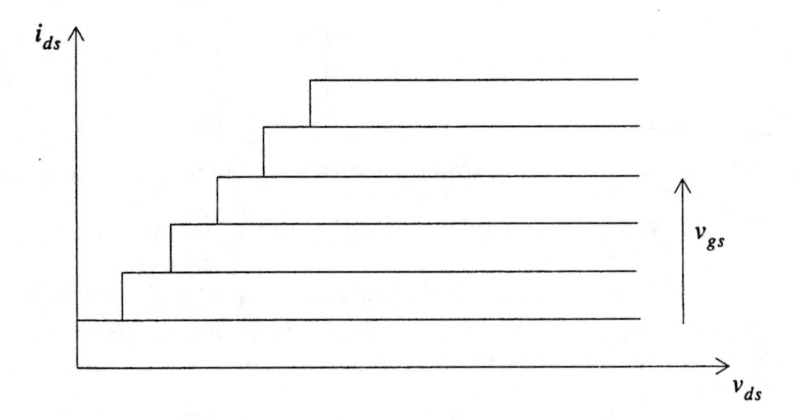

Figure 9.27 I-v characteristics of a MOSFET.

This approach has been applied to operational amplifier circuit analysis [Hage82]. First the circuit is analyzed with simplified ideal op-amp models consisting of nullators and norators. Then the sensitivities with respect to manufacturer specified model parameters are computed. The tolerances of these parameters are indicated on the specification sheets of the manufacturer. This approach works well because operational amplifiers are designed and manufactured to be nearly ideal, so the ideal model used in the first-cut analysis yields an answer quite close the final answer.

If all circuit element parameter value deviations were small or Gaussian in distribution, we could even consider the approximation of pseudo-statistical analysis via sensitivities. Unfortunately, these assumptions are seldom true in practice. We can, however, attempt to exploit sensitivity analysis in worst case tolerance analysis.

Assume that we have a circuit with a number of toleranced elements (resistors, for example). We are interested in finding the worst-case and best-case performance of the circuit. For each toleranced variable p_j in the circuit suppose that it has a nominal value

$$p_j = \bar{p}_j \tag{9.10.5}$$

a lower bound

$$p_j = \bar{p}_j - \Delta p_j^- \tag{9.10.6}$$

and an upper bound

$$p_j = \bar{p}_j + \Delta p_j^+ \tag{9.10.7}$$

In other words

$$\bar{p}_j - \Delta p_j^- \le p_j \le \bar{p}_j + \Delta p_j^+ \tag{9.10.8}$$

with $\bar{p}_j$ being the nominal value of the parameter p_j.

First, find the sensitivity of the desired response variable (or scalar function of response variables) with respect to all toleranced parameters p_j, i.e., $\left.\frac{\partial x_i}{\partial p_j}\right|_{\bar{p}_j}$. We can compute these sensitivities efficiently in terms of the nominal solution for all $p_j = \bar{p}_j$ and the LU factors of the original circuit equations. Next, simultaneously change all toleranced parameters in those directions that the signs of the sensitivities indicate will be mutually reinforcing. Upper tolerance bound:

$$\begin{aligned} \left.\frac{\partial x_i}{\partial p_k}\right|_{\bar{p}_k} &> 0 \Rightarrow p_k \to \bar{p}_k + \Delta p_k^+ \\ \left.\frac{\partial x_i}{\partial p_l}\right|_{\bar{p}_l} &< 0 \Rightarrow p_l \to \bar{p}_l - \Delta p_l^- \end{aligned} \tag{9.10.9}$$

Lower tolerance bound:

$$\left.\frac{\partial x_i}{\partial p_k}\right|_{\bar{p}_k} > 0 \Rightarrow p_k \rightarrow \bar{p}_k - \Delta p_k^-$$
$$\left.\frac{\partial x_i}{\partial p_l}\right|_{\bar{p}_l} < 0 \Rightarrow p_l \rightarrow \bar{p}_l + \Delta p_l^+ \tag{9.10.10}$$

Next, we must reanalyze at both the upper and lower tolerance bounds to ascertain the worst case responses. Moreover, we must check the sensitivities at those worst case boundaries to ensure that they haven't changed in sign from what they were at the nominal point. Figures 9.28 and 9.29 show situations where the response curve is monotonic in the region of interest. In these two cases we may assume that we are at the worst case boundaries. Figure 9.30 shows the situation with an extremity in the middle of the response curve. In this case, we know that we are not at the worst case because of the change in sign of sensitivity from the nominal point to one of the tolerance bounds. We can fit a fifth order polynomial through the computed data (three points and three derivatives), and then find the minimum or maximum of that polynomial within the tolerance bounds. Or, if we lack confidence in the accuracy of this procedure, we may conduct a search in the space of the offending parameter. The latter may be necessary if there are multiple extremities in the performance function within a parameter's tolerance bounds.

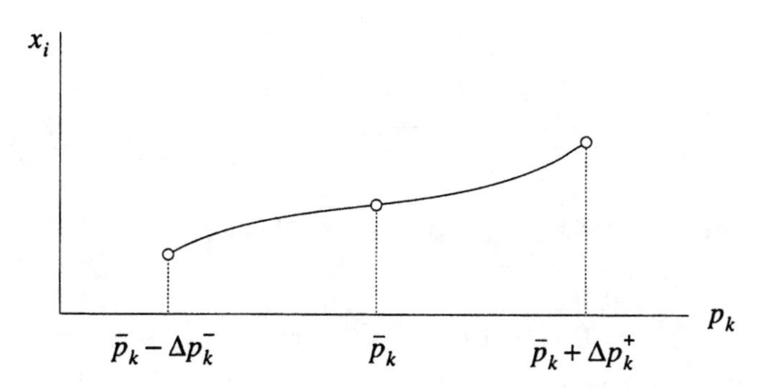

Figure 9.28 Monotonically increasing response curve.

9.11 *Time and Frequency Domain Sensitivity Analysis*

The adjoint circuit concept generalizes easily to the frequency domain. Returning to the basic sensitivity relation (9.5.10), we can re-derive it in terms of phasor quantities to obtain

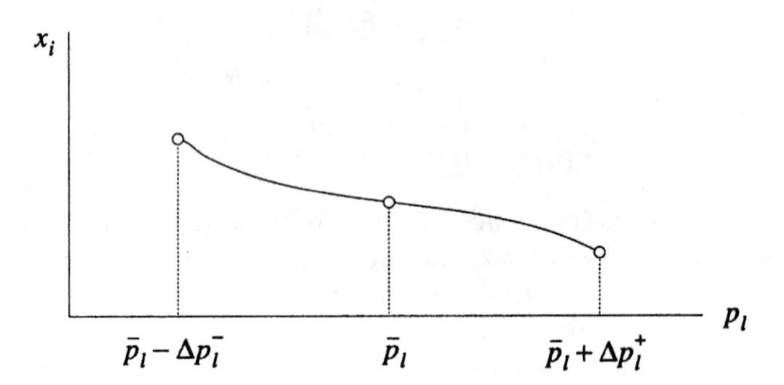

Figure 9.29 Monotonically decreasing response curve.

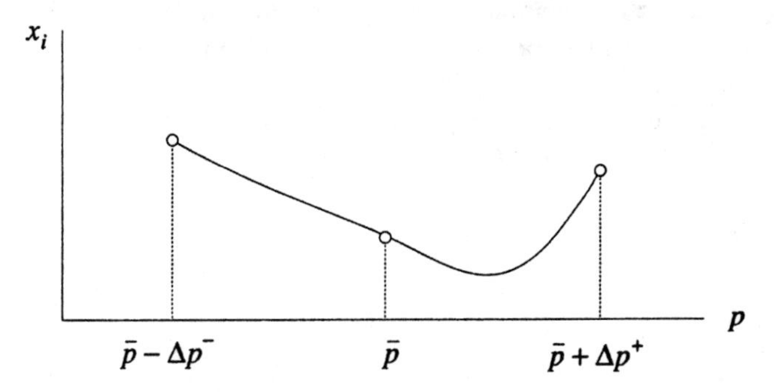

Figure 9.30 Response curve with an extremity in the middle.

$$\sum_{\text{all branches}} [\Phi(j\omega)\,\delta V(j\omega) - \Psi(j\omega)\,\delta I(j\omega)] = 0 \tag{9.11.1}$$

Then, upon considering capacitances to be represented by complex admittances of value $j\omega C$ and inductances to be complex impedances of value $j\omega L$, it is straightforward to reproduce the earlier arguments we made for dc circuits, as shown below.

For a capacitor, we have

$$\begin{aligned} I_C(j\omega) &= j\omega C V_C(j\omega) \\ \delta I_C(j\omega) &= j\omega C \delta V_C(j\omega) + j\omega V_C(j\omega)\,\delta C \end{aligned} \tag{9.11.2}$$

so the contribution to the sensitivity relation (9.11.1) is

$$\Phi_C(j\omega)\,\delta V_C(j\omega) - \Psi_C(j\omega)\,[j\omega C\delta V_C(j\omega) + j\omega V_C(j\omega)\,\delta C] \tag{9.11.3}$$

By choosing the BCR for the adjoint element as

$$\Phi_C(j\omega) = j\omega C\Psi_C(j\omega) \tag{9.11.4}$$

(or in other words retaining the element as a capacitor in the adjoint circuit), the sensitivity of a complex valued function $F(j\omega)$ with respect to C can be computed by

$$\frac{\partial F(j\omega)}{\partial C} = -j\omega\Psi_C(j\omega) V_C(j\omega) \tag{9.11.5}$$

Similarly, for an inductor we have,

$$\begin{aligned} V_L(j\omega) &= j\omega L I_L(j\omega) \\ \delta V_L(j\omega) &= j\omega L\delta I_L(j\omega) + j\omega I_L(j\omega)\delta L \end{aligned} \tag{9.11.6}$$

so the contribution to the sensitivity relation (9.11.1) is

$$\Phi_L(j\omega)[j\omega L\delta I_L(j\omega) + j\omega I_L(j\omega)\delta L] - \Psi_L(j\omega)\delta I_L(j\omega) \tag{9.11.7}$$

By choosing the BCR for the adjoint element as

$$\Psi_L(j\omega) = j\omega L\Phi_L(j\omega) \tag{9.11.8}$$

(or in other words retaining the element as an inductor in the adjoint circuit), the sensitivity can be computed by

$$\frac{\partial F(j\omega)}{\partial L} = j\omega\Phi_L(j\omega) I_L(j\omega) \tag{9.11.9}$$

Thus we find that capacitances and inductances modeled in the frequency domain as complex admittances and impedances, respectively, are self-adjoint (their adjoint circuit equivalents are the same as them). The composite relation for the sensitivities is

$$\delta F(j\omega) = j\omega\left[\sum_C \{-\Psi_c(j\omega) V_c(j\omega)\delta C\} + \sum_L \Phi_L(j\omega) I_L(j\omega)\delta L\right] \tag{9.11.10}$$

It is even possible to obtain a sensitivity with respect to a small frequency variation:

$$\frac{\partial F(j\omega)}{\partial \omega} = j\left[\sum_C \{-C\Psi_C(j\omega) V_C(j\omega)\} + \sum_L L\Phi_L(j\omega) I_L(j\omega)\right] \tag{9.11.11}$$

It is relatively easy to obtain and apply such sensitivities at a single frequency in terms of the LU factors that arise in the original analysis. These sensitivities can be extended to apply to performance measures defined over a range of frequencies, since circuit performance specifications are often in terms of the frequency response.

In section 6.13, we saw that when global frequency domain considerations are involved it may be more efficient -- and more revealing -- to work with the approximate poles of the circuit in terms of AWE sensitivity. Further, the time domain transient sensitivities of linear circuits are most efficiently obtained in terms of AWE-derived approximate poles.

We will now turn our attention to *time-domain sensitivity* or *transient sensitivity*. First, we extend Tellegen's theorem to a more general form in the time domain. For any two circuits N and η that share the same topology:

$$v_b^T(t)\varphi_b(\tau) = [A^T v_n(t)]^T \varphi_b(\tau) = v_n^T(t) A \varphi_b(\tau) = 0 \tag{9.11.12}$$

and

$$\psi_b^T(\tau) i_b(t) = [A^T \psi_n(\tau)]^T i_b(t) = \psi_n^T(\tau) A i_b(t) = 0 \tag{9.11.13}$$

Thus Tellegen's theorem can be applied to a pair of circuits with the same topology by combining the branch currents and voltages at any two instants of time. There is no restriction whatsoever on t and τ in (9.11.12) and (9.11.13). Hence we can generalize (9.5.10) to

$$\sum_{\text{all branches}} [\varphi(\tau_1)\delta v(t_1) - \psi(\tau_2)\delta i(t_2)] = 0 \tag{9.11.14}$$

Since we want to solve and store the results of only one original circuit and one adjoint circuit, we choose $t_1 = t_2 = t$ and $\tau_1 = \tau_2 = \tau$. Further, integrating over a time period of interest from t_0 to t_f,

$$\sum_{\text{all branches}} \int_{t_0}^{t_f} [\varphi(\tau)\delta v(t) - \psi(\tau)\delta i(t)]\,dt = 0 \tag{9.11.15}$$

Now let us apply (9.11.15) to a capacitance. For a capacitance,

$$\begin{aligned} i_C(t) &= C\dot{v}_C(t) \\ \delta i_C(t) &= C\delta\dot{v}_C(t) + \dot{v}_C(t)\delta C \end{aligned} \tag{9.11.16}$$

so the contribution to the sensitivity term of (9.11.15) is

$$\int_{t_0}^{t_f} [\varphi_C(\tau)\,\delta v_C(t) - \psi_C(\tau)\,\{C\delta\dot{v}_C(t) + \dot{v}_C(t)\,\delta C\}\,]\,dt \tag{9.11.17}$$

To remove signal variations from the overall sensitivity expression we can integrate by parts the $\delta\dot{v}_C$ term to obtain

$$\begin{aligned}&-\psi_C(\tau)\,C\delta v_C(t)\,\big|_{t_0}^{t_f}\\ &+\int_{t_0}^{t_f} [\varphi_C(\tau)\,\delta v_C(t) - \psi_C(\tau)\,\delta C\dot{v}_C(t) + \dot{\psi}_C(\tau)\,C\delta v_C(t)\,]\,dt\end{aligned} \tag{9.11.18}$$

We can drop the signal variation $\delta v_C(t)$ from the integral relation if we define the corresponding adjoint circuit branch as

$$\varphi_C(\tau) = -\dot{\psi}_C(\tau)\,C \tag{9.11.19}$$

We would like to avoid negative energy storage element values because of their inherent instability. But we are free to choose τ, since Tellegen's theorem is valid irrespective of the choice of τ. So we choose τ to be *backward time*. In other words,

$$\tau = t_0 + t_f - t_i \tag{9.11.20}$$

that is

$$\begin{aligned} t = t_0 &\Rightarrow \tau = t_f\\ t = t_f &\Rightarrow \tau = t_0\\ dt &= -d\tau \end{aligned} \tag{9.11.21}$$

Thus the BCR for the capacitor in the adjoint circuit becomes

$$\varphi_C(\tau) = C\frac{d\psi_C(\tau)}{d\tau} \tag{9.11.22}$$

an ordinary capacitor. The actual sensitivity computation is a convolution between the forward-in-time voltage slope of the capacitor in the original circuit and the backward-in-time voltage across the capacitor in the adjoint circuit.

$$\int_{t_0}^{t_f} \frac{\partial f(t)}{\partial C} dt = -\int_{t_0}^{t_f} [\psi_C(t_0 + t_f - t)\, \dot{v}_C(t)]\, dt \quad \textbf{(9.11.23)}$$

where $f(t)$ is the performance function. But we are not done yet; we must deal with the first term that we obtained in the integration by parts:

$$-\psi_C(t_0)\, C\delta v_C(t_f) + \psi_C(t_f)\, C\delta v_C(t_0) = 0 \quad \textbf{(9.11.24)}$$

If the initial condition is specified, then $\delta v_C(t_0) = 0$, and we can choose the adjoint circuit initial condition

$$\psi_C(t_0) = 0 \quad \textbf{(9.11.25)}$$

in order to make the term in (9.11.24) zero, and the sensitivity equation in (9.11.23) valid.

There are circumstances too where we may want to find the periodic steady state response of a circuit via adjoint sensitivity; then for one period

$$v_C(t_f) = v_C(t_0) \quad \textbf{(9.11.26)}$$

and

$$\delta v_C(t_f) = \delta v_C(t_0) \quad \textbf{(9.11.27)}$$

so from (9.11.24)

$$\psi_C(t_f) = \psi_C(t_0) \quad \textbf{(9.11.28)}$$

is the appropriate adjoint boundary condition. Thus capacitors are self-adjoint in the time domain, too. Note that a similar development of sensitivity relations can be undertaken for inductance elements.

Sensitivity functions which can be expressed as integral functions are particularly amenable to transient adjoint sensitivity. An example of such a function is the Elmore delay of a waveform. (For a linear circuit, it would be much more efficient to find the sensitivity of an Elmore delay by using AWE-based sensitivity.) Some functions cannot be expressed in the required form, in which case adjoint sensitivity cannot be used. For example, a transient node or branch voltage waveform is not an allowed function. Such functions can be accommodated using direct sensitivity. However, the value of a voltage or current at a particular instant of time T is an allowed function (the function can be expressed as $f = \int_{t_0}^{t_f} v(t)\, \delta(t-T)\, dt$ where $\delta(t-T)$ is a shifted impulse; the adjoint excitation will be an impulse at time T) and the time at which a voltage crosses a particular value is an allowed function and the sensitivity is computed using the equation

$$\frac{\partial T_{cross}}{\partial p} = \frac{\frac{\partial v(T_{cross})}{\partial p}}{\left.\frac{\partial v}{\partial t}\right|_{t = T_{cross}}} \tag{9.11.29}$$

In the case of linear circuits, the following steps are involved in sensitivity computation. First the original circuit (with appropriate zero-valued sources added) must be analyzed. All pertinent original circuit response waveforms must be stored over the time interval of interest. Then the adjoint circuit is analyzed backwards in time. Note that the adjoint circuit analysis is not as computationally expensive as the original. It is possible to store the LU factors in the original analysis and to resurrect them (with appropriate interpolation) as $U^T L^T$ factors for the adjoint circuit. But such a strategy may entail a prohibitive amount of storage for a transient analysis of a large circuit involving many time points. With fixed time step schemes, there is only one set of LU factors, but most practical simulators use variable time step methods. Finally, a separate convolution for each parameter of interest must be carried out.

With nonlinear circuits, the above approach to time domain sensitivity computation is seldom used in practice because it is prohibitively expensive. The Jacobians could change radically from one time step to another and the LU factors must either be stored or recomputed for the adjoint circuit. Further, the analysis time steps of the forward and backward simulation may not coincide, so interpolation must be employed to match the waveforms up for convolution.

Time-domain sensitivity can be very useful in critical path analysis, reliability analysis, circuit optimization, delay tuning, and so on. In the case of linear circuits, approaching the problem in terms of AWE sensitivities may be the most efficient method. In the case of nonlinear circuits, time-domain adjoint sensitivity computation may be too expensive in full-blown detailed circuit simulators. However, in the context of event-driven methods involving simplified device models and no Jacobians, transient sensitivity is efficient enough to be useful in practical applications (see Chapter 11 for details). Thus these efficient timing simulation methods can render techniques like transient sensitivity feasible, whereas they would be too computationally intensive for detailed simulators.

9.12 References

[Director69] S. W. Director and R. A. Rohrer. The Generalized Adjoint network and Network Sensitivities. *IEEE Transactions on Circuit Theory*, vol. CT-16(3), August 1969.

[Desoer69] C. A. Desoer and E. S. Kuh. *Basic Circuit Theory.* McGraw-Hill, 1969.

[Brayton75] R. K. Brayton and S. W. Director. Computation of Time Delay Sensitivities for Switching Circuit Optimization. *IEEE Transactions on Circuits and Systems*, vol. CAS-22, December 1975.

[Brayton80] R. K. Brayton and R. Spence. *Sensitivity and Optimization*. Elsevier Scientific Publishing Company, vol. 2, 1980.

[Hage82] C. J. Hage and R. A. Rohrer. Efficient Op Amp Circuit Analysis with Manufacturer Specified Macromodel Parameters. *IEEE Transactions on CAD of ICs and Systems*, vol. CAD-1(3), July 1982.

[Hocevar85] D. A. Hocevar, P. Yang, T. N. Trick, and B. D. Epler. Transient Sensitivity Computation for MOSFET Circuits. *IEEE Transactions on CAD of ICs and Systems*, vol. CAD-4(4), October 1985.

Chapter 10 *Simulation of Nonlinear Circuits*

Finally, we now have all of the background that we need to discuss nonlinear circuit analysis. The essence of nonlinear circuit simulation was covered briefly in Chapter 1. In this chapter we will elaborate on that exposition and consider as well some of the subtleties that arise in the course of nonlinear circuit simulation. We start with a brief description of the industry standard SPICE.

10.1 SPICE

As conceived originally, SPICE combined (Modified) Nodal Analysis with trapezoidal integration (in its time domain transient mode) and Newton-Raphson (N-R) iteration [Nagel71, Nagel75]. Some variants of SPICE may substitute more sophisticated integration algorithms instead of using trapezoidal. And almost all versions of SPICE employ some "tricks" in attempts to ensure the convergence of the Newton-Raphson (N-R) iterations, especially for dc analysis. Such convergence usually is not an issue in the course of transient analysis, since a sufficiently small time step can almost always be chosen so that the initial guess projected from the previous time point is sufficiently close to the correct solution at the present time point. During a step of the transient analysis, the time step is progressively reduced until convergence is obtained within a reasonable number of N-R iterations.

On occasion, though, the time step may be forced to be too small because of interactions among the algorithms SPICE employs and the nonlinear device models with which it must contend. In such instances, the SPICE transient analysis may be aborted with a message like, "Time step too small." Convergence most often is a problem, however, in the initial dc portion of a circuit simulation, which must be performed prior to either a large signal transient analysis (to obtain an appropriate set of initial conditions) or a small signal ac analysis (to obtain an appropriate bias point). In an attempt to overcome this convergence problem, SPICE allows a user to specify the initial guesses of node voltage values. Good initial guesses both aid and speed the convergence of the nonlinear dc solution.

With minor modification, the combination of algorithms that forms the core of SPICE has been in use since the mid-1960s. Originally, such circuit simulators used Nodal Analysis, and not Modified Nodal Analysis, to handle floating voltage sources. For example, a small resistance could be inserted in series with each floating voltage source, and the combination replaced with a Norton equivalent parallel combination of current source and conductance. But it soon became apparent that the inclusion of such small resistances could cause wide value spreads in the resulting $\boldsymbol{Y}$ matrix and potentially lead to ill-conditioning of the resulting nodal equations. Before sparse matrix manipulation was brought to bear on the solution of the circuit equations, it was recognized that "current variables" could be introduced along with voltage constraint equations to characterize voltage sources and other "impedance-basis" elements. In [Nagel71] these current variables were eliminated in the final course of the formulation of the nodal equations as described in Chapter 2 of this book. Once sparse matrix techniques become accepted and entrenched some questioned the elimination of the current variables since a good sparse matrix solver could take care of them too. It can be argued that a good sparse matrix solver should eliminate those excess currents at the outset, essentially providing normal nodal equations as an intermediate form on the way to solution. But that doesn't appear to be the case, since SPICE tends to struggle when it encounters a circuit which has a large number of inductance elements, which it treats on an impedance basis. We suspect this behavior to be a consequence of the SPICE preference for diagonal pivot selection and the *a priori* row swapping that entails. A sparse matrix solution strategy that did not favor diagonal pivoting probably would improve the performance of SPICE for circuits with a large number of inductances. But so far there has not been much motivation to provide such an alternative since integrated circuits, the primary province of SPICE, typically do not include (many, if any) inductance elements in their models. But with higher signal frequencies and smaller feature size this situation is bound to change as integrated circuit technology evolves. And, beyond the integrated circuit chip, for IC packages, printed circuit boards, and back planes, inductance element models are already a well established part of the circuit models employed. In spite of its idiosyncracies, most nonlinear transient circuit simulators today use Modified Nodal Analysis.

Stated succinctly, there appears to be no alternative to (the) SPICE (combination of algorithms and detailed circuit models) if the goal is to obtain extreme accuracy from a nominal circuit simulation. But the computational price that must be paid to attain such accuracy may become prohibitive for large circuits.

In the remainder of this chapter, we will discuss nonlinear equation solution and all the entailed complications required to deal with real circuits. Multiple coupled nonlinear equations, multi-terminal nonlinear elements, and nonlinear energy storage elements as functions of one or more circuit variables will be discussed. Reasons for non-convergence and methods for dealing with them will be pointed out along the way.

10.2 Nonlinear dc Analysis

Consider the circuit in Figure 10.1, which has no energy storage elements, but has one branch for which the current is a nonlinear function of the voltage. Namely,

$$i_b = f_1(v_b) \tag{10.2.1}$$

where f_1 is a nonlinear function, and i_b and v_b are the branch current and voltage, respectively.

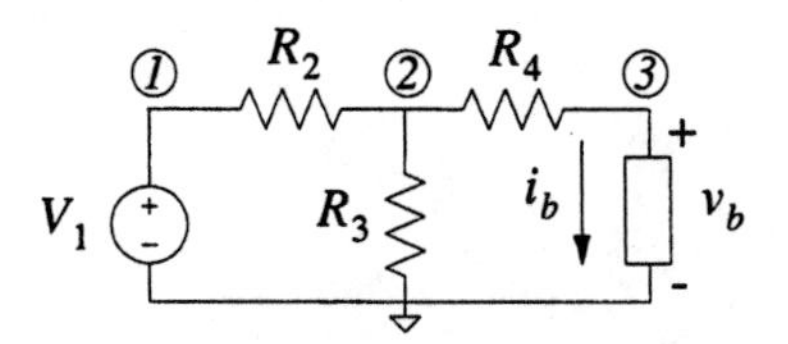

Figure 10.1 Nonlinear dc circuit.

We can write the nodal equations for this circuit the same way that we would if all of the branches were linear. For example, applying KCL at every non-datum node results in the following set of three nonlinear equations in terms of the three node voltages:

$$\text{node 1:} \quad \frac{v_1}{R_2} - \frac{v_2}{R_2} - I_1 = 0$$

$$\text{node 2:} \quad \frac{v_2}{R_2} - \frac{v_1}{R_2} + \frac{v_2}{R_3} + \frac{v_2}{R_4} - \frac{v_3}{R_4} = 0 \tag{10.2.2}$$

$$\text{node 3:} \quad \frac{v_3}{R_4} - \frac{v_2}{R_4} + f_1(v_3) = 0$$

The equation set in (10.2.2) is comprised of two linear equations and one nonlinear equation. Of course we could reduce (10.2.2) to a single nonlinear equation in terms of a single unknown, v_3. This procedure is not possible in general when there are numerous nonlinear models in the circuit. In either case, closed form solutions are not available and we must solve the nonlinear equation, or coupled set of equations, using some form of iteration.

10.3 Newton-Raphson Iteration

Newton-Raphson iteration is at the heart of SPICE (nonlinear dc and transient analysis). The Newton-Raphson algorithm seeks to solve the nonlinear equation

$$f(x) = 0 \tag{10.3.1}$$

iteratively by successive solution of a set of linearized approximations to this equation. Suppose that we have at the kth iteration, x_k, a value of x such that

$$f(x_k) \neq 0 \tag{10.3.2}$$

Then we wish to find

$$x_{k+1} \equiv x_k + \Delta x_k \tag{10.3.3}$$

such that $f(x_{k+1})$ is closer in value to zero than is $f(x_k)$. To do so, we resort to Taylor's Theorem:

$$f(x_k + \Delta x_k) = f(x_k) + f'(x_k)\Delta x_k + \frac{1}{2}f''(\xi)\Delta x_k^2 \tag{10.3.4}$$

where ξ lies between x_k and x_{k+1} in value. For this result to apply, it is necessary that $f'''(\xi)$ exist over that entire interval as well. (This necessary existence of the third derivative of the characteristic becomes a device modeling requirement in SPICE.) If $\frac{1}{2}f''(\xi)\Delta x_k^2$ is small, and this is critical to the convergence of the iteration, then we can ignore it and write

$$f(x_k + \Delta x_k) \approx f(x_k) + f'(x_k)\Delta x_k \tag{10.3.5}$$

In search of solution we consider that

$$f(x_k) + f'(x_k)\Delta x_k = 0 \tag{10.3.6}$$

therefore,

$$\Delta x_k = \frac{-f(x_k)}{f'(x_k)} \tag{10.3.7}$$

Inserting (10.3.7) into (10.3.3) we have

$$x_{k+1} = x_k - \frac{f(x_k)}{f'(x_k)} \tag{10.3.8}$$

This iteration is repeated until Δx_k, the change between iteration is "acceptably small," or lower in absolute value than a pre-set tolerance. A graphical depiction of the iterations involved in equation (10.3.8) is shown in Figure 10.2. From an initial guess, x_0, we linearize the curve using the first two terms of the Taylor series expansion, equation (10.3.5). We solve for the single root of this linear approximation, x_1, using equation (10.3.8). Then, we linearize about this new value, x_1, by projecting to the point on the curve, $f(x_1)$, and linearizing as before. Eventually, we will stop iterating when the linear approximation is accurate enough, such as for the projection which determines x_3.

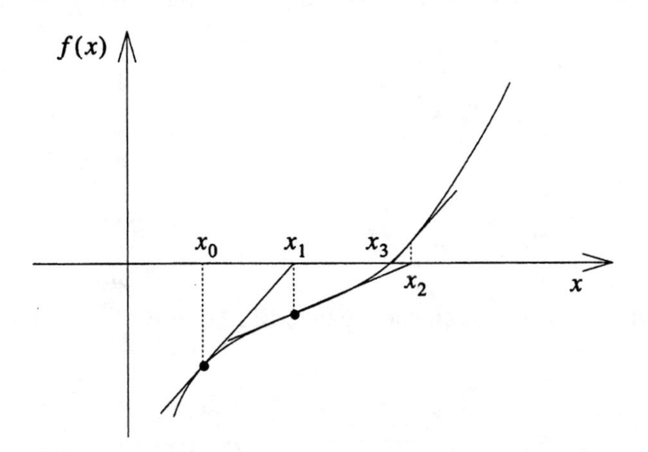

Figure 10.2 Graphical depiction of N-R iterations.

As an example, consider once again the simple diode circuit shown in Chapter 1, repeated here in Figure 10.3. The diode equation is

$$i_d = I_{SAT}\left[exp\left(\frac{qv_d}{\eta kT}\right) - 1\right] \tag{10.3.9}$$

We can characterize the circuit in Figure 10.3 in terms of the following nonlinear equation:

$$f(v_d) = \frac{V_s}{R} - \frac{v_d}{R} - I_{SAT}\left[exp\left(\frac{qv_d}{\eta kT}\right) - 1\right] = 0 \tag{10.3.10}$$

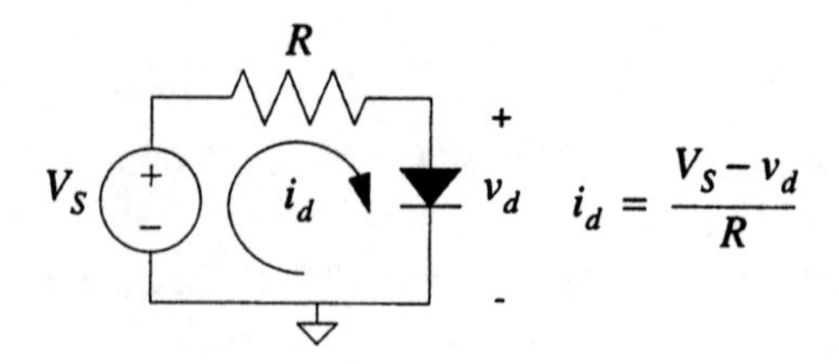

Figure 10.3 Simple diode circuit.

Our goal is to solve (10.3.10) by finding the value of v_d that makes $f(v_d) = 0$. We iterate on (10.3.10) using the N-R equation, (10.3.8). Setting $x_k = v_d$, we have,

$$f(x_k) = \frac{V_s}{R} - \frac{x_k}{R} - I_{SAT}\left[exp\left(\frac{qx_k}{\eta kT}\right) - 1\right] \tag{10.3.11}$$

and

$$f'(x_k) = -\frac{1}{R} - \frac{qI_{SAT}}{\eta kT} exp\left(\frac{qx_k}{\eta kT}\right) \tag{10.3.12}$$

Finally, equation (10.3.8) for the circuit in Figure 10.3 becomes:

$$x_{k+1} = x_k - \frac{\left\{\frac{V_s}{R} - \frac{x_k}{R} - I_{SAT}\left[exp\frac{qx_k}{\eta kT} - 1\right]\right\}}{-\frac{1}{R} - \frac{qI_{SAT}}{\eta kT} exp\left(\frac{qx_k}{\eta kT}\right)} \tag{10.3.13}$$

We demonstrated in Chapter 1 that we could carry out this N-R iteration by linearizing the elements individually (as opposed to formulating the nonlinear equations and linearizing them) and iteratively solving the resulting linearized circuit. For example, for the diode in Figure 10.3 we showed in Chapter 1 that each linearization of this nonlinear model can be represented by the Norton equivalent in Figure 10.4. Here the diode is linearized about the kth iteration value, v_d^k, using the first two terms of the nonlinear diode equation:

$$i_d = i_d^k + \left.\frac{di_d}{dv_d}\right|_k (v_d - v_d^k) \tag{10.3.14}$$

where i_d is given by (10.3.9). The solution of the circuit using the linearization in

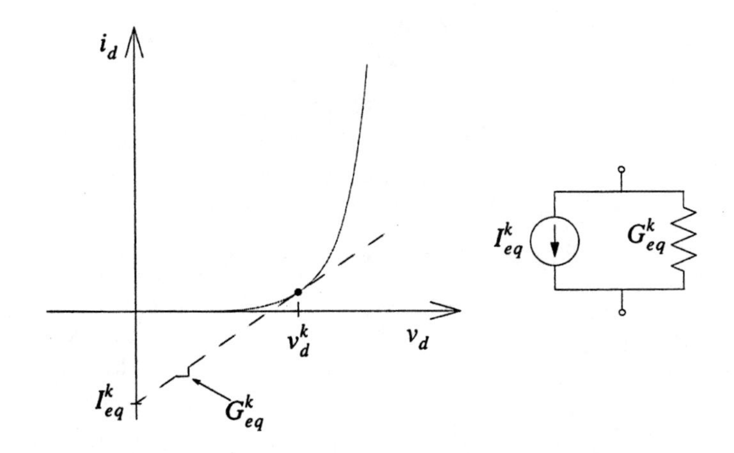

Figure 10.4 Norton equivalent model for diode N-R linearization.

(10.3.14) would give us v_d^{k+1} and i_d^{k+1}, the next iteration values. From (10.3.14) and Figure 10.4 we can see that

$$G_{eq}^k = \left.\frac{di_d}{dv_d}\right|_k \tag{10.3.15}$$

and

$$I_{eq}^k = i_d^k + \left.\frac{di_d}{dv_d}\right|_k (-v_d^k) \tag{10.3.16}$$

The linearized N-R equivalent for the circuit in Figure 10.3 is the circuit shown in Figure 10.5. Note that the dc solution for this circuit at the kth iteration can be shown to be equivalent to equation (10.3.13). The graphical interpretation of these iterations is shown in Figure 10.6. The iterations are carried out until convergence.

10.4 Damped Newton-Raphson Iteration

Newton-Raphson iteration may not always converge, and various modifications on the pure form must be employed to assist in convergence. Most of these tricks amount to limitations on the magnitude of Δx_k so that the remainder term is "sufficiently small." In

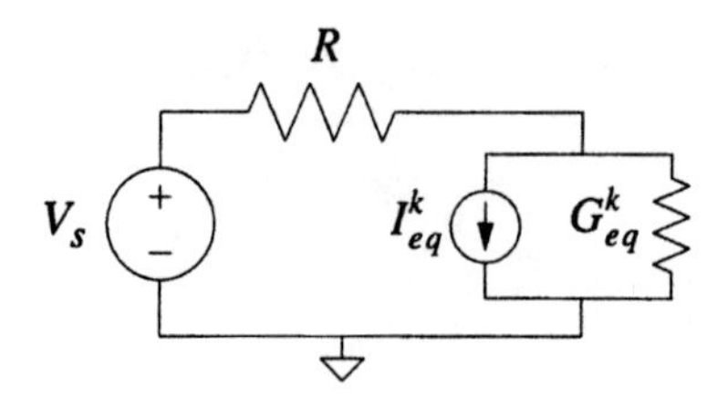

Figure 10.5 Linearized N-R circuit.

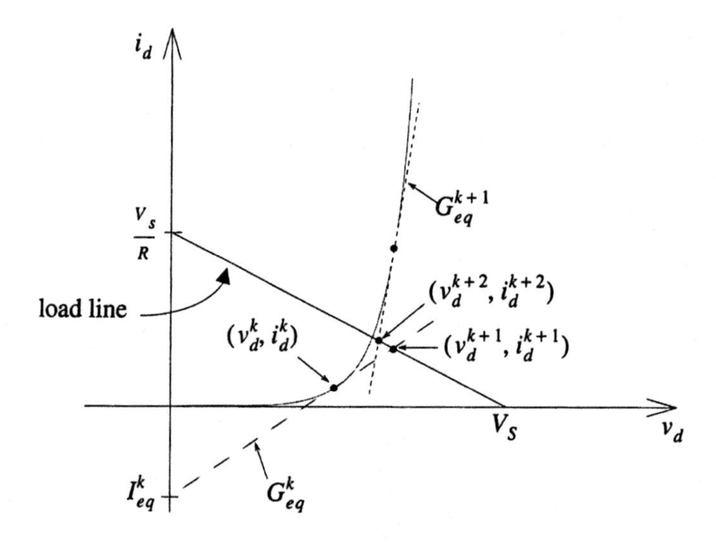

Figure 10.6 N-R iterations.

other words, x is not allowed to change by more than a certain maximum value between successive iterations. Such *limiting* is warranted even for the simple diode example shown above. For instance, given the initial guess and the linear projection shown in Figure 10.7, the voltage at the $k+1$ iteration is calculated and used as the next operating point for Newton-Raphson. However, in the case shown in this figure, due to the exponential nature of the diode curve, the current at this value of voltage may result in a floating point overflow. To address such problems we would instead follow the direction of the first linearized projection, but limit the amount of change that is allowed in one N-R step.

Limiting the magnitude of Δx_k to below its Newton-Raphson value is called *damping*. Almost always when damping is used, the sign of Δx_k is retained, although sophisticated variants that are employed in multiple dimensions may not even do that for all of the vari-

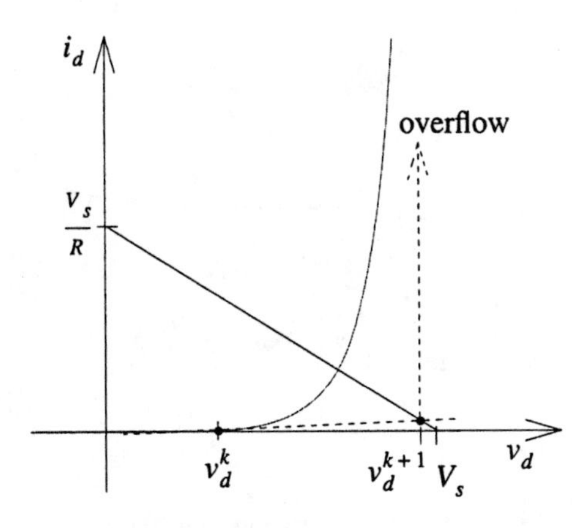

Figure 10.7 Potential for overflow using N-R without damping.

ables of interest. In [Nagel71], a damping factor of $2V_T$ was found to be sufficient for bipolar circuits, where V_T is the thermal voltage (0.0259 V at room temperature). Damping aids convergence, but some times at the cost of extra N-R iterations. During transient analysis, SPICE-like simulators typically give up after a certain number of iterations and try to reduce the time step before starting up the iterations again. There are a number of user-defined settings that can be used to aid in convergence (more on that topic in section 10.9).

Bipolar devices, with their exponential characteristics, are more difficult for N-R than MOSFETs, which in the worst case follow a square law behavior in their $i - v$ characteristics. Thus, the damping factor is usually dictated by the device models with the most abrupt nonlinearities. Another difficulty with bipolar circuits is the diode characteristic in the third quadrant (when the diode is off). The current is nearly a constant, I_{SAT}, and the N-R projections are nearly horizontal lines which makes G_{eq} nearly zero. In order to overcome this, successive secant iteration is often used in practice. (Note that N-R iteration can be thought of as a successive tangent iteration.) The secant is a linearization of the diode which passes through the origin as shown in Figure 10.8. Note that the convergence may be slightly slower, but the iterations will be more robust. Since the secant model passes through the origin, the equivalent circuit model is simply an equivalent conductance

$$G_{eq} = \frac{i_d^k}{v_d^k} \quad \textbf{(10.4.1)}$$

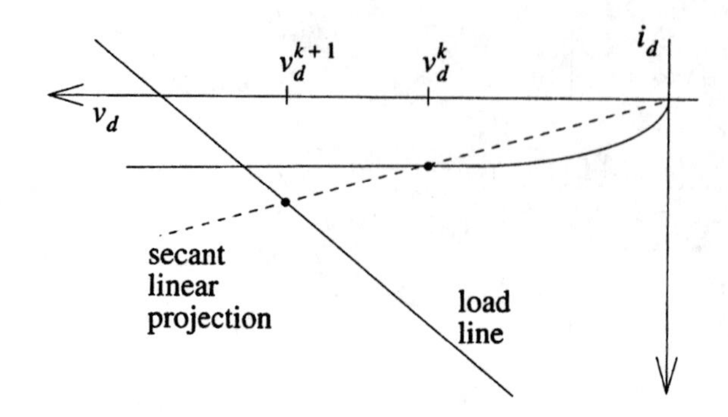

Figure 10.8 Successive secant iteration.

It should be apparent from this diode example that nonlinear iteration algorithms are "model dependent" to some extent. That is, we must adjust the N-R implementation to account for peculiarities in the nonlinear device equations. It is possible that the nonlinear function has more than one solution, as shown in Figure 10.9. In this case, we must rely on an appropriate initial guess to ensure that the "right one" prevails. Convergence problems become even more involved when we consider multiple nonlinear elements, which is the subject of the next section.

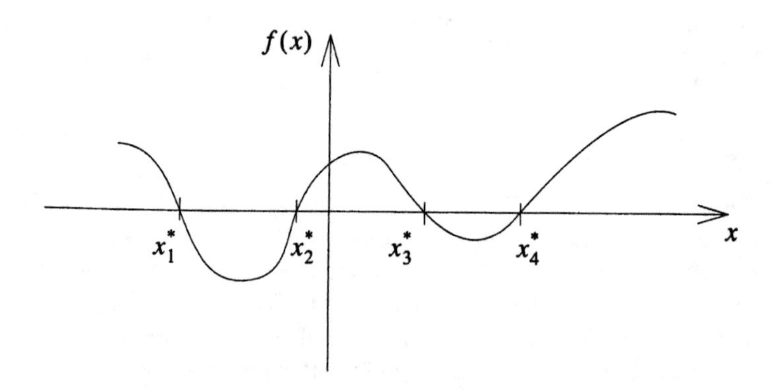

Figure 10.9 A nonlinear function which has multiple roots.

10.5 Multi-Dimensional Newton-Raphson Iteration

Suppose we would like to solve n nonlinear equations in n unknowns:

$$\begin{aligned} f_1(x_1, x_2, \ldots, x_n) &= 0 \\ f_2(x_1, x_2, \ldots, x_n) &= 0 \\ \vdots \quad & \quad \vdots \\ f_n(x_1, x_2, \ldots, x_n) &= 0 \end{aligned} \tag{10.5.1}$$

or

$$\boldsymbol{f}(\boldsymbol{x}) = \mathbf{0} \tag{10.5.2}$$

Then

$$\Delta\boldsymbol{x} = -\left(\frac{\partial \boldsymbol{f}}{\partial \boldsymbol{x}}\right)^{-1} \boldsymbol{f}(\boldsymbol{x}) \tag{10.5.3}$$

is (undamped) Newton-Raphson iteration [Carnahan69, Ortega70]. As in the one-dimensional case, $\|\Delta\boldsymbol{x}\|$ must be "sufficiently small" and appropriate third derivatives must exist. The $n \times n$ matrix $\frac{\partial \boldsymbol{f}}{\partial \boldsymbol{x}}$ is the *Jacobian* of $\boldsymbol{f}$ with respect to $\boldsymbol{x}$ and is defined as:

$$\boldsymbol{J} \equiv \frac{\partial \boldsymbol{f}}{\partial \boldsymbol{x}} \equiv \begin{bmatrix} \frac{\partial f_1}{\partial x_1} & \frac{\partial f_1}{\partial x_2} & \cdots & \frac{\partial f_1}{\partial x_n} \\ \frac{\partial f_2}{\partial x_1} & \frac{\partial f_2}{\partial x_2} & \cdots & \frac{\partial f_2}{\partial x_n} \\ \vdots & \vdots & & \vdots \\ \frac{\partial f_n}{\partial x_1} & \frac{\partial f_n}{\partial x_2} & \cdots & \frac{\partial f_n}{\partial x_n} \end{bmatrix} \tag{10.5.4}$$

To perform nonlinear circuit simulation, we do not actually formulate a set of nonlinear equations and then take their partial derivatives with respect to the solution variables of interest. Rather we build up the Jacobian in terms of linearized stamps that represent the individual circuit elements. We showed how this process happens for a two-terminal nonlinear diode in section 10.3. Next, consider the circuit with two diodes in series shown in Figure 10.10. Using the diode equation from (10.3.9), we can write the two nonlinear nodal equations for this circuit as follows:

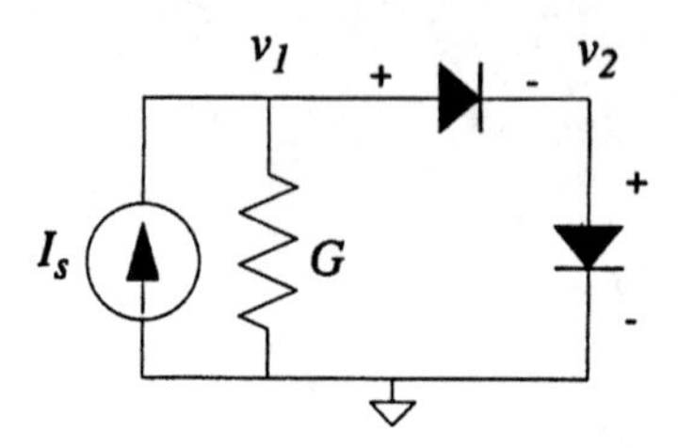

Figure 10.10 Two diode circuit.

$$f_1(v_1, v_2) = -I_s + G_1 v_1 + I_{SAT}\left[exp\left(\frac{q(v_1 - v_2)}{\eta kT}\right) - 1\right] = 0$$

$$f_2(v_1, v_2) = -I_{SAT}\left[exp\left(\frac{q(v_1 - v_2)}{\eta kT}\right) - 1\right] + I_{SAT}\left[exp\left(\frac{qv_2}{\eta kT}\right) - 1\right] = 0 \tag{10.5.5}$$

From (10.5.3) the N-R equations would be

$$\begin{bmatrix} v_1^{k+1} \\ v_2^{k+1} \end{bmatrix} = \begin{bmatrix} v_1^{k} \\ v_2^{k} \end{bmatrix} - \left(\left. \begin{bmatrix} \frac{\partial f_1}{\partial v_1} & \frac{\partial f_1}{\partial v_2} \\ \frac{\partial f_2}{\partial v_1} & \frac{\partial f_2}{\partial v_2} \end{bmatrix}^{-1} \right|_{v_1^k, v_2^k} \begin{bmatrix} f_1(v_1^k, v_2^k) \\ f_2(v_1^k, v_2^k) \end{bmatrix} \right) \tag{10.5.6}$$

where $\boldsymbol{v}$ is a vector of dimension 2, and $\boldsymbol{J}$ is the square Jacobian matrix of dimension 2×2. The entries of the Jacobian matrix are:

$$\left.\frac{\partial f_1}{\partial v_1}\right|_{v_1^k, v_2^k} = G_1 + \frac{qI_{SAT}}{\eta kT} exp\left(\frac{q(v_1^k - v_2^k)}{\eta kT}\right) \tag{10.5.7}$$

$$\left.\frac{\partial f_1}{\partial v_2}\right|_{v_1^k, v_2^k} = -\frac{qI_{SAT}}{\eta kT} exp\left(\frac{q(v_1^k - v_2^k)}{\eta kT}\right) \tag{10.5.8}$$

$$\left.\frac{\partial f_2}{\partial v_1}\right|_{v_1^k,\, v_2^k} = -\frac{qI_{SAT}}{\eta kT} exp\left(\frac{q\,(v_1^k - v_2^k)}{\eta kT}\right) \quad \textbf{(10.5.9)}$$

$$\left.\frac{\partial f_2}{\partial v_2}\right|_{v_1^k,\, v_2^k} = \frac{qI_{SAT}}{\eta kT} exp\left(\frac{q\,(v_1^k - v_2^k)}{\eta kT}\right) + \frac{qI_{SAT}}{\eta kT} exp\left(\frac{qv_2^k}{\eta kT}\right) \quad \textbf{(10.5.10)}$$

But, just as for the case of linear circuits, we do not want to actually find the inverse of $\boldsymbol{J}$. Instead of generating (10.5.6), we would solve

$$\left.\begin{bmatrix} \frac{\partial f_1}{\partial v_1} & \frac{\partial f_1}{\partial v_2} \\ \frac{\partial f_2}{\partial v_1} & \frac{\partial f_2}{\partial v_2} \end{bmatrix}\right|_{v_1^k,\, v_2^k} \begin{bmatrix} v_1^{k+1} \\ v_2^{k+1} \end{bmatrix} = \left(\left.\begin{bmatrix} \frac{\partial f_1}{\partial v_1} & \frac{\partial f_1}{\partial v_2} \\ \frac{\partial f_2}{\partial v_1} & \frac{\partial f_2}{\partial v_2} \end{bmatrix}\right|_{v_1^k,\, v_2^k} \begin{bmatrix} v_1^k \\ v_2^k \end{bmatrix}\right) - \begin{bmatrix} f_1\,(v_1^k, v_2^k) \\ f_2\,(v_1^k, v_2^k) \end{bmatrix} \quad \textbf{(10.5.11)}$$

or more generally

$$\boldsymbol{J}^k \boldsymbol{v}^{k+1} = \boldsymbol{J}^k \boldsymbol{v}^k - \boldsymbol{F}\,(\boldsymbol{v}^k) \quad \textbf{(10.5.12)}$$

or

$$\boldsymbol{J}^k \Delta \boldsymbol{v}^k = -\boldsymbol{F}\,(\boldsymbol{v}^k) \quad \textbf{(10.5.13)}$$

We then solve (10.5.13) by LU factorization. Note that we can arrive at this same set of equations by linearizing the diodes individually as shown in Figure 10.4, and then stamping in the Norton equivalents and solving the resulting linearized circuit. There are, however, some subtleties involved in obtaining the stamps for multi-terminal elements, e.g., transistors, which we will cover in the next section.

10.6 Multi-Terminal Elements

Given a three-terminal element, as shown in Figure 10.11, we can treat the bottom terminal as "common" and characterize the element -- if possible -- on an admittance basis as follows:

$$\begin{aligned} i_1 &= g_1\,(v_1, v_2) \\ i_2 &= g_2\,(v_1, v_2) \end{aligned} \quad \textbf{(10.6.1)}$$

Then, linearizing these relations we obtain

$$i_1 + \Delta i_1 = g_1(v_1, v_2) + \frac{\partial g_1}{\partial v_1}\Delta v_1 + \frac{\partial g_1}{\partial v_2}\Delta v_2 + \text{error}$$
$$i_2 + \Delta i_2 = g_2(v_1, v_2) + \frac{\partial g_2}{\partial v_1}\Delta v_1 + \frac{\partial g_2}{\partial v_2}\Delta v_2 + \text{error} \tag{10.6.2}$$

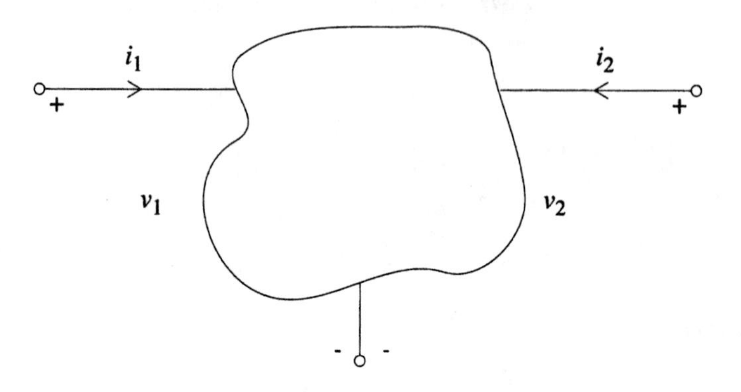

Figure 10.11 Port description of a three-terminal element.

To be general, we must treat all terminals on an equal basis, so we recast the port description of the element into the terminal description in Figure 10.12. Then

$$\begin{aligned}
v_1 &= v_a - v_c & \text{and} \quad \Delta v_1 &= \Delta v_a - \Delta v_c \\
v_2 &= v_b - v_c & \text{and} \quad \Delta v_2 &= \Delta v_b - \Delta v_c \\
i_c &= -(i_1 + i_2) & \text{and} \quad \Delta i_c &= -(\Delta i_1 + \Delta i_2) \\
i_a &= i_1 & \text{and} \quad \Delta i_a &= \Delta i_1 \\
i_b &= i_2 & \text{and} \quad \Delta i_b &= \Delta i_2
\end{aligned} \tag{10.6.3}$$

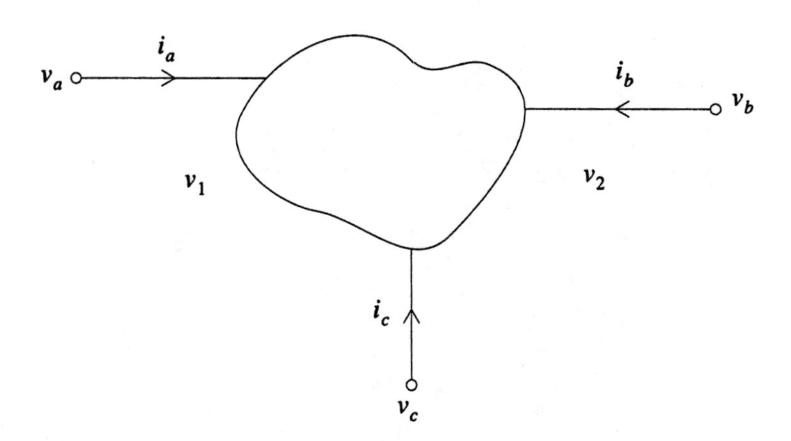

Figure 10.12 Terminal description of a three-terminal element.

So,

$$\begin{aligned}
i_a + \Delta i_a &\approx g_1(v_1, v_2) + \frac{\partial g_1}{\partial v_1}\Delta v_a + \frac{\partial g_1}{\partial v_2}\Delta v_b + \left(-\frac{\partial g_1}{\partial v_1} - \frac{\partial g_1}{\partial v_2}\right)\Delta v_c \\
i_b + \Delta i_b &\approx g_2(v_1, v_2) + \frac{\partial g_2}{\partial v_1}\Delta v_a + \frac{\partial g_2}{\partial v_2}\Delta v_b + \left(-\frac{\partial g_2}{\partial v_1} - \frac{\partial g_2}{\partial v_2}\right)\Delta v_c \\
i_c + \Delta i_c &\approx -g_1(v_1, v_2) - g_2(v_1, v_2) + \left(-\frac{\partial g_1}{\partial v_1} - \frac{\partial g_2}{\partial v_1}\right)\Delta v_a \\
&\quad + \left(-\frac{\partial g_1}{\partial v_2} - \frac{\partial g_2}{\partial v_2}\right)\Delta v_b + \left(\frac{\partial g_1}{\partial v_1} + \frac{\partial g_1}{\partial v_2} + \frac{\partial g_2}{\partial v_1} + \frac{\partial g_2}{\partial v_2}\right)\Delta v_c
\end{aligned} \tag{10.6.4}$$

If we consider normal nodal equations

$$\boldsymbol{Y}\Delta \boldsymbol{v} = \boldsymbol{I} \tag{10.6.5}$$

we can take from the above relations the elements to be stamped into the $\boldsymbol{Y}$ matrix:

$$\begin{array}{r} a\text{-}row \\ \\ b\text{-}row \\ \\ c\text{-}row \end{array}
\begin{bmatrix}
\dfrac{\partial g_1}{\partial v_1} & \dfrac{\partial g_1}{\partial v_2} & \left(-\dfrac{\partial g_1}{\partial v_1}-\dfrac{\partial g_1}{\partial v_2}\right) \\
\dfrac{\partial g_2}{\partial v_1} & \dfrac{\partial g_2}{\partial v_2} & \left(-\dfrac{\partial g_2}{\partial v_1}-\dfrac{\partial g_2}{\partial v_2}\right) \\
\left(-\dfrac{\partial g_1}{\partial v_1}-\dfrac{\partial g_2}{\partial v_1}\right) & \left(-\dfrac{\partial g_1}{\partial v_2}-\dfrac{\partial g_2}{\partial v_2}\right) & \left(\dfrac{\partial g_1}{\partial v_1}+\dfrac{\partial g_1}{\partial v_2}+\dfrac{\partial g_2}{\partial v_1}+\dfrac{\partial g_2}{\partial v_2}\right)
\end{bmatrix} \tag{10.6.6}$$

$$\begin{array}{ccc} a\text{-}column & b\text{-}column & c\text{-}column \end{array}$$

and the $\boldsymbol{I}$ vector:

$$\begin{array}{r} a\text{-}row \\ b\text{-}row \\ c\text{-}row \end{array}
\begin{bmatrix}
-g_1(v_1, v_2) \\
-g_2(v_1, v_2) \\
g_1(v_1, v_2) + g_2(v_1, v_2)
\end{bmatrix} \tag{10.6.7}$$

We must then LU factor the resulting overall $\boldsymbol{Y}$ matrix to obtain $\Delta \boldsymbol{v}$. (Whether we solve for $\Delta \boldsymbol{v}$ or $\boldsymbol{v} + \Delta \boldsymbol{v}$ depends on how we choose to handle independent source contributions to these equations.)

Note that the derivatives like $\partial g_1 / \partial v_1$ are usually provided as part of the device model, so they are in terms of the port voltages. However, in (Modified) Nodal Analysis, we solve for the node voltages, so all the other terms must be expressed with terminal voltages. At each N-R iteration, the node voltage solution is used to compute the port voltages and then the appropriate derivatives.

We see in (10.6.6) and (10.6.7) that the stamps are singular. Each column of both the $\boldsymbol{Y}$ and $\boldsymbol{I}$ stamps adds to zero, and each row of the $\boldsymbol{Y}$ stamp adds to zero. This result should not be surprising; a two-terminal element provides a 2×2 singular stamp to the $\boldsymbol{Y}$ matrix, and an n-terminal element an $n \times n$ singular stamp. And, of course, all elements of the current source vector must add to zero so as not to violate Kirchhoff's Current Law. The $\boldsymbol{Y}$ stamp provides the "floating" (or indefinite) admittance characterization of the linearized element. We can choose a common terminal and cross out its corresponding row and column to obtain the definite admittance characterization. But it is the indefinite characterization that we want to stamp into the $\boldsymbol{Y}$ matrix and $\boldsymbol{I}$ vector, unless one of the terminals just happens to be ground.

We will now consider two common nonlinear multi-terminal elements. Our discussion will be in terms of simple models, but the concepts can easily be extended to more complex device models.

Bipolar Junction Transistors (BJTs)

Consider, for example, the common-base Ebers-Moll model for a bipolar junction transistor (BJT) [Ebers54] shown in Figure 10.13. For this NPN transistor shown we have

$$i_e = -I_{es}\left(e^{\frac{v_{be}}{V_{Te}}} - 1\right) + \alpha_R I_{cs}\left(e^{\frac{v_{bc}}{V_{Tc}}} - 1\right) \quad \textbf{(10.6.8)}$$

and

$$i_c = \alpha_F I_{es}\left(e^{\frac{v_{be}}{V_{Te}}} - 1\right) - I_{cs}\left(e^{\frac{v_{bc}}{V_{Tc}}} - 1\right) \quad \textbf{(10.6.9)}$$

where I_{es} and I_{cs} are the emitter and collector junction saturation currents respectively, V_{Te} and V_{Tc} are the thermal voltages, kT/q, scaled by the appropriate nonideality factor for each junction, and α_F and α_R are the forward and reverse current gains.

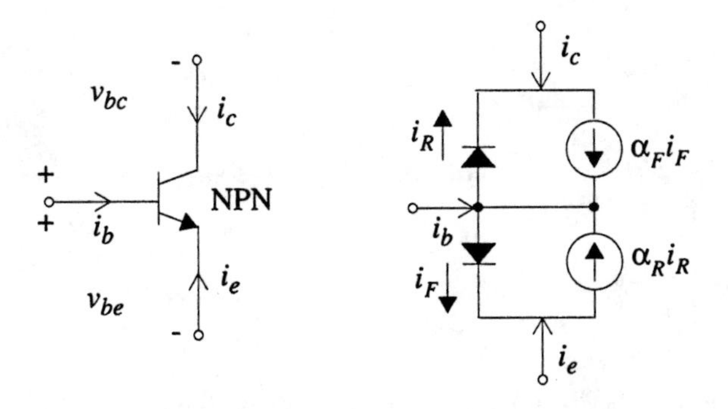

Figure 10.13 NPN type Bipolar Junction Transistor and its model.

We begin by evaluating the partial derivatives for this three port model:

$$\frac{\partial i_e}{\partial v_{be}} = -\frac{I_{es}}{v_{Te}} e^{\frac{v_{be}}{v_{Te}}} \equiv g_{ee}$$

$$\frac{\partial i_e}{\partial v_{bc}} = \alpha_R \frac{I_{cs}}{v_{Tc}} e^{\frac{v_{bc}}{v_{Tc}}} \equiv g_{ec}$$

$$\frac{\partial i_c}{\partial v_{be}} = \alpha_F \frac{I_{es}}{v_{Te}} e^{\frac{v_{be}}{v_{Te}}} \equiv g_{ce} \tag{10.6.10}$$

$$\frac{\partial i_c}{\partial v_{bc}} = -\frac{I_{cs}}{v_{Tc}} e^{\frac{v_{bc}}{v_{Tc}}} \equiv g_{cc}$$

Hence

$$\begin{aligned} \Delta i_e &= g_{ee}\Delta v_{be} + g_{ec}\Delta v_{bc} \\ \Delta i_c &= g_{ce}\Delta v_{be} + g_{cc}\Delta v_{bc} \end{aligned} \tag{10.6.11}$$

or

$$\begin{aligned} \Delta i_e &= -g_{ee}\Delta v_e - g_{ec}\Delta v_c + (g_{ee} + g_{ec})\,\Delta v_b \\ \Delta i_c &= -g_{ce}\Delta v_e - g_{cc}\Delta v_c + (g_{ce} + g_{cc})\,\Delta v_b \\ \Delta i_b &= -(\Delta i_e + \Delta i_c) \end{aligned} \tag{10.6.12}$$

So, we have the ***Y*** matrix stamp:

$$\begin{array}{c} \\ e\text{-row} \\ c\text{-row} \\ b\text{-row} \end{array} \begin{array}{c} \begin{array}{ccc} e\text{-column} & c\text{-column} & b\text{-column} \end{array} \\ \begin{bmatrix} -g_{ee} & -g_{ec} & (g_{ee}+g_{ec}) \\ -g_{ce} & -g_{cc} & (g_{ce}+g_{cc}) \\ (g_{ee}+g_{ce}) & (g_{ec}+g_{cc}) & (-g_{ee}-g_{ec}-g_{ce}-g_{cc}) \end{bmatrix} \end{array} \tag{10.6.13}$$

Note the negative signs in (10.6.13) because the original definition uses v_{be} and v_{bc}, not v_{eb} and v_{cb}. And for the ***I*** stamp:

$$\begin{array}{c} e\text{-row} \\ c\text{-row} \\ b\text{-row} \end{array} \begin{bmatrix} -i_e \\ -i_c \\ i_e + i_c \end{bmatrix} \tag{10.6.14}$$

In summary, we wrote the nodal equations at the three nodes which were connected to the element in Figure 10.13 using the linearized expressions in (10.6.4). We then stamped these linearized equations directly into $\boldsymbol{Y}$. From (10.6.4) we can also generate an equivalent circuit model which represents the linearized three-terminal BJT, and then stamp it into $\boldsymbol{Y}$. While this step is unnecessary, it is helpful in gaining insight and it is left as an exercise to the reader. Note that the linearized circuit model is very similar to the familiar small signal transistor model.

Field Effect Transistors (FETs)

We consider the generation of a three-terminal linearized N-R model for the case of a MOSFET, such as that shown in Figure 10.14. A simple dc MOSFET model can be described by the following set of equations [Muller77]:

$$i_{ds} = \beta\left[(v_{gs} - v_T)\, v_{ds} - \frac{v_{ds}^2}{2}\right] \quad \text{for} \quad v_{ds} < v_{gs} - v_T \quad \text{(linear)} \tag{10.6.15}$$

$$i_{ds} = \frac{\beta}{2}(v_{gs} - v_T)^2 \quad \text{for} \quad (v_{ds} > v_{gs} - v_T) \quad \text{(saturation)} \tag{10.6.16}$$

and

$$i_{ds} = 0 \quad \text{for} \quad v_{gs} < v_T \quad \text{(cutoff)} \tag{10.6.17}$$

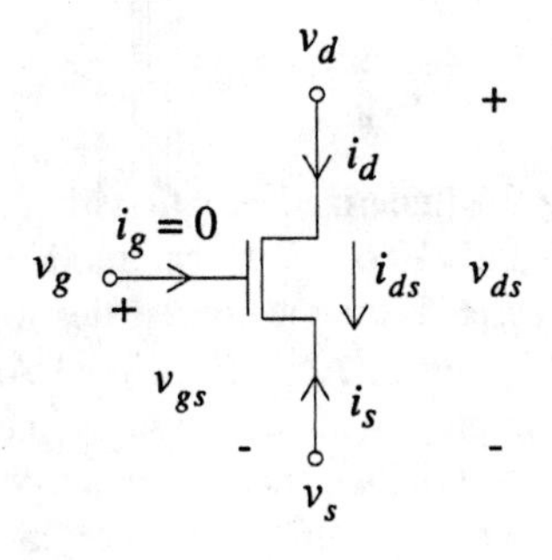

Figure 10.14 An N-type Metal Oxide Semiconductor Field Effect Transistor (NMOSFET).

In terms of the port description of the NMOSFET in Figure 10.14,

$$i_g = f_1(v_{gs}, v_{ds}) = 0$$

$$i_d = f_2(v_{gs}, v_{ds}) = i_{ds} \tag{10.6.18}$$

$$i_s = -i_g - i_d = -i_{ds}$$

We linearize these port currents by characterizing them with the first two terms of their Taylor series expansion:

$$i_g^k + \Delta i_g^k = 0$$

$$i_d^k + \Delta i_d^k = f_2(v_{gs}^k, v_{ds}^k) + \left.\frac{\partial f_2}{\partial v_{gs}}\right|_k (v_{gs}^{k+1} - v_{gs}^k) + \left.\frac{\partial f_2}{\partial v_{ds}}\right|_k (v_{ds}^{k+1} - v_{ds}^k) \tag{10.6.19}$$

$$i_s^k + \Delta i_s^k = -(i_d^k + \Delta i_d^k)$$

where the partial derivative terms are well known as the small-signal transconductance,

$$\frac{\partial f_2}{\partial v_{gs}} = \frac{\partial i_{ds}}{\partial v_{gs}} = g_m \tag{10.6.20}$$

and the small-signal drain to source conductance,

$$\frac{\partial f_2}{\partial v_{ds}} = \frac{\partial i_{ds}}{\partial v_{ds}} = G_{ds} \tag{10.6.21}$$

as shown in Figure 10.15. Using the linearized equations in (10.6.19) along with the partial derivative terms in (10.6.20) and (10.6.21) we can model the linearized MOSFET by the equivalent circuit in Figure 10.16. Note the similarities to the small-signal MOSFET model shown in Figure 2.2 of Chapter 2. Applying Nodal Analysis with this linearized model is equivalent to directly stamping in the expressions in (10.6.6).

Before leaving the topic of MOSFETs, we should further point out that all of the above discussions and equations have been for NMOSFETs. PMOSFETs are similarly handled, except, of course, the voltages are of opposite sign. Moreover, for both N- and P-MOSFETs, the drain, source, and gate voltages must be monitored continuously to recognize the region of operation, and the direction of drain-source current flow. For example, when the source is at a higher voltage than the drain for the NMOSFET in Figure 10.14, the current sources in the linearized model in Figure 10.16 must be negative to note that the currents are directed oppositely. Simply stated, the source node becomes the drain and the drain node becomes the source. In other words, irrespective of the labeling of an NMOSFET, the lower potential node is the (effective) source and the higher potential node the

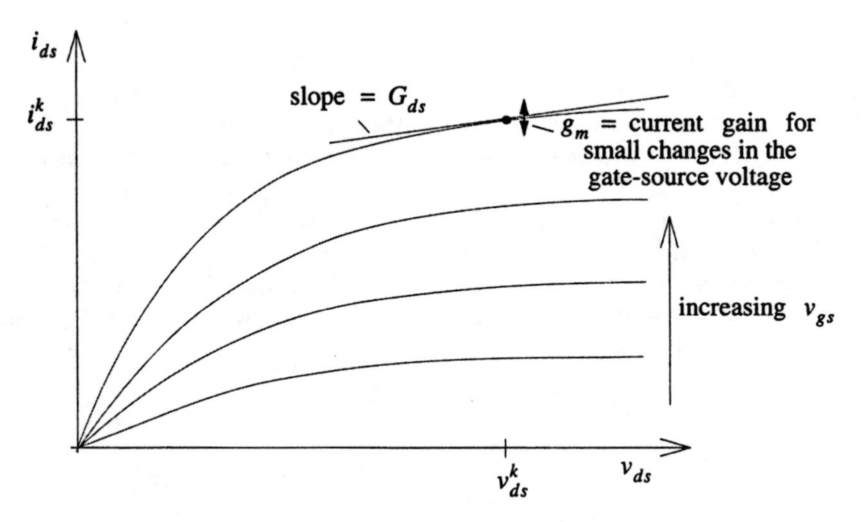

Figure 10.15 Linearization of NMOSFET curve about an operating point.

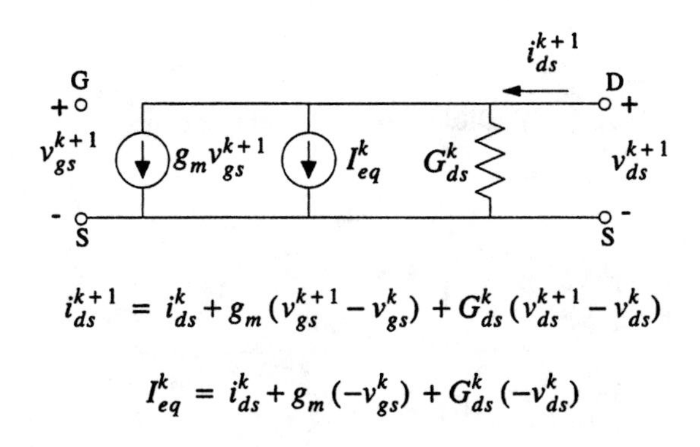

$$i_{ds}^{k+1} = i_{ds}^k + g_m (v_{gs}^{k+1} - v_{gs}^k) + G_{ds}^k (v_{ds}^{k+1} - v_{ds}^k)$$

$$I_{eq}^k = i_{ds}^k + g_m (-v_{gs}^k) + G_{ds}^k (-v_{ds}^k)$$

Figure 10.16 N-R linearized MOSFET model.

(effective) drain, and vice versa for a PMOSFET.

As a final note on Newton-Raphson iteration, we should point out that it is not always easy to obtain the correct partial derivatives that constitute the stamps. Consequently, many circuit simulators resort to "numerical differentiation." A typical approach is to fit a cubic (spline) through four appropriate points on the characteristic curve and then to employ its symbolic derivative. The advantage of a cubic over lower order polynomial fits is that $f'''(\xi)$ is guaranteed to exist.

10.7 Nonlinear Transient Analysis

Up to now we have considered only nonlinear dc analysis. Now we will consider nonlinear transient analysis. As was briefly outlined in Chapter 1, nonlinear transient analysis begins by linearizing the energy storage elements with difference approximations as described in Chapters 4 and 5 and then performing N-R iterations on the resulting nonlinear dc equivalent circuit.

To consider this procedure in more detail, we start with the diode circuit in Figure 10.17. Notice that there are two capacitors in this circuit along with an independent current source which is a function of time. We know that the capacitors and the diode will be represented by linear Norton equivalents for the TR integration and the N-R iteration respectively, as shown in Figure 10.18.

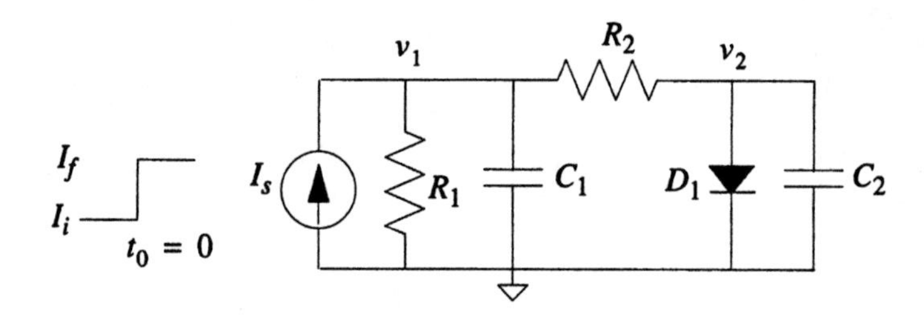

Figure 10.17 Diode circuit with linear capacitors.

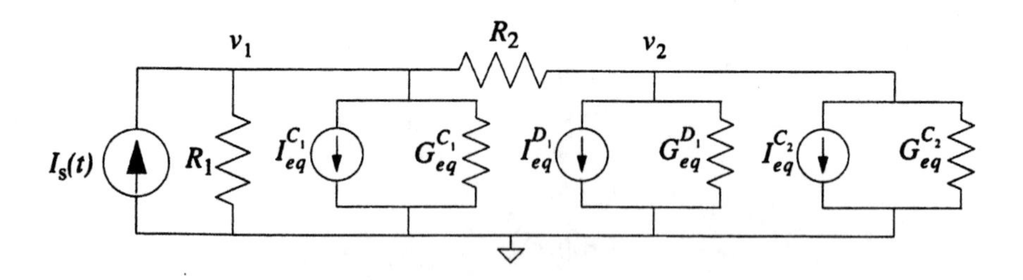

Figure 10.18 Linear dc equivalent circuit for the circuit in Figure 10.17.

To begin, we would perform a dc analysis for $t < 0$. The input current source would be set to its initial value of I_i, and the capacitors would be opened by setting their I_{eq}'s and G_{eq}'s equal to zero, as shown in Figure 10.18. The diode is first linearized about some initial voltage guess, probably zero, and then the iterations proceed as described previously for a dc analysis. Once convergence is reached, we know $v_1(0^+)$ and $v_2(0^+)$ since capacitance voltages cannot change instantaneously. However, one idiosyncracy of Trapezoidal integration, and Forward Euler for that matter, is the necessity to compute $i_C(0^+)$ and $v_L(0^+)$ for the companion models of capacitance and inductance elements, respectively, since a step change in excitation occurs at time t_0. Calculating these currents would require us to replace the Norton equivalents in Figure 10.18 with independent voltage sources as described in Chapter 4. Changing the circuit topology to handle steps is not always warranted. For this reason, some versions of SPICE handle only piecewise linear (continuous) input waveforms. In addition, the Backward Euler integration approximation poses no such problem with steps, so some versions of SPICE provide that option to accommodate step functions.

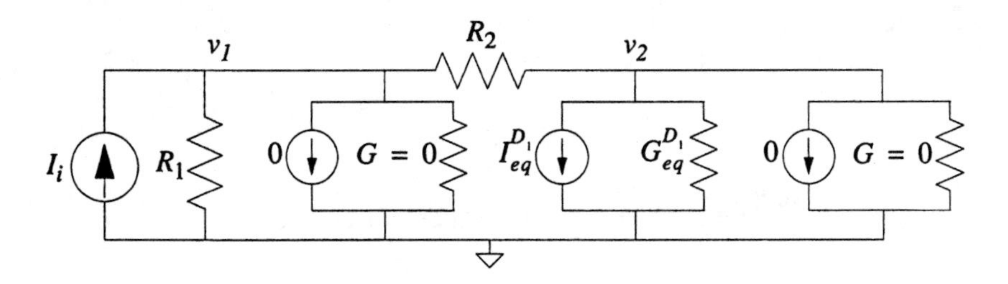

Figure 10.19 Linear dc equivalent circuit for finding solution at $t = 0$.

Once the discontinuity is handled for this step example, we begin the transient analysis by taking a step in time to $t = \Delta t$. First, the input current source is set equal to I_f, its value for $t > 0$. Second, using TR integration, the two capacitors are linearized. Referring to Figure 10.18, the companion model values for the capacitors are:

$$I_{eq}^{C_1} = i_{C1}(0^+) + \frac{2C_1 v_1(0^+)}{\Delta t} \qquad G_{eq}^{C_1} = \frac{2C_1}{\Delta t}$$

$$I_{eq}^{C_2} = i_{C2}(0^+) + \frac{2C_2 v_2(0^+)}{\Delta t} \qquad G_{eq}^{C_2} = \frac{2C_2}{\Delta t} \tag{10.7.1}$$

Then, the diode model is repeatedly linearized and the circuit is iteratively solved until convergence is reached at time $t = \Delta t$. Notice that the initial solution values appear in the TR companion models, and these voltages are also used as the starting guess for the first N-R iteration. Once convergence is reached, time is advanced and the whole process is repeated.

10.8 Nonlinear Energy Storage Elements

Although we have covered nonlinear transient analysis which considers circuits with nonlinear devices and linear energy storage elements, we must also consider such analyses when there are nonlinear energy storage elements. Diodes, for example, are characterized by a nonlinear i-v characteristic, and a nonlinear charge storage (capacitance) characteristic. As shown in Figure 10.20, there are two nonlinear capacitance terms associated with diodes. Both the depletion capacitance and the diffusion capacitance are nonlinear functions of the diode voltage, v_d.

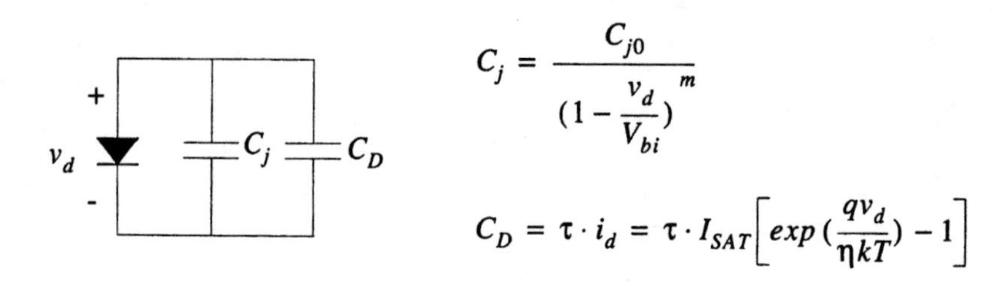

+
v_d
-
C_j
C_D
$C_j = \frac{C_{j0}}{\left(1 - \frac{v_d}{V_{bi}}\right)^m}$
$C_D = \tau \cdot i_d = \tau \cdot I_{SAT}\left[exp\left(\frac{qv_d}{\eta kT}\right) - 1\right]$

Figure 10.20 Nonlinear diode capacitances.

For these nonlinear capacitors, C is a nonlinear function of the voltage which appears across it, therefore,

$$i_C = C(v_d)\frac{dv_d}{dt} \tag{10.8.1}$$

There are subtle problems associated with the use of integration algorithms for the expression in (10.8.1). To begin to appreciate them, consider the general case which can be described by the following set of nonlinear state equations

$$\dot{x}(t) = f(x(t), u(t)) \tag{10.8.2}$$

where f is a nonlinear operator. Of course, we would never consider the actual formulation of such state equations; the linear case is complicated enough. This form serves our purpose, however, which is to point out the problems we can encounter if we are not careful.

If we apply Trapezoidal integration to the above, we obtain

$$x(t+\Delta t) = x(t) + \frac{\Delta t}{2}[f(x(t), u(t)) + f(x(t+\Delta t), u(t+\Delta t))] \tag{10.8.3}$$

or

$$\boldsymbol{x}(t+\Delta t) - \frac{\Delta t}{2}[\boldsymbol{f}(\boldsymbol{x}(t+\Delta t), \boldsymbol{u}(t+\Delta t))] = \boldsymbol{x}(t) + \frac{\Delta t}{2}[\boldsymbol{f}(\boldsymbol{x}(t), \boldsymbol{u}(t))] \quad (10.8.4)$$

Of course, $\boldsymbol{x}(t)$, $\boldsymbol{f}(\boldsymbol{x}(t), \boldsymbol{u}(t))$, and therefore the right hand side of (10.8.4) are known. But we must perform Newton-Raphson iteration to find $\boldsymbol{x}(t+\Delta t)$, and the appropriate Jacobian is

$$\boldsymbol{J} = \boldsymbol{1} - \frac{\Delta t}{2}\left(\frac{\partial f}{\partial x}\right)\bigg|_{\boldsymbol{x}(t+\Delta t), \boldsymbol{u}(t+\Delta t)} \quad (10.8.5)$$

So we must be careful to update the Jacobian to reflect the presumed value of $\boldsymbol{x}(t+\Delta t)$ at each Newton-Raphson iteration. If we don't, we may converge to a close but incorrect solution. This point is easy to appreciate in terms of the state equations, but it may not be so easy to implement in Modified Nodal equations -- or any other practicable form, for that matter.

To illustrate the point, consider a nonlinear capacitor that is characterized by the charge-voltage relation

$$q = f(v) \quad (10.8.6)$$

Then we have the $i - v$ relation

$$i = \dot{q} = \frac{df}{dv}\dot{v} \quad (10.8.7)$$

and we might write

$$\dot{v} = \left(\frac{df}{dv}\right)^{-1} i \quad (10.8.8)$$

Trapezoidal integration of (10.8.8) would yield

$$v(t+\Delta t) \approx v(t) + \frac{\Delta t}{2}\left(\left[\frac{df}{dv}\bigg|_t\right]^{-1} i(t) + \left[\frac{df}{dv}\bigg|_{t+\Delta t}\right]^{-1} i(t+\Delta t)\right) \quad (10.8.9)$$

Some simulators have erroneously attempted to use

$$v(t+\Delta t) \approx v(t) + \left[\frac{df}{dv}\bigg|_t\right]^{-1} \frac{\Delta t}{2}\{i(t) + i(t+\Delta t)\} \quad (10.8.10)$$

which does not appropriately update the capacitance companion model, leading to a violation of the conservation of charge. The problem, of course, is that the companion model depends on df/dv evaluated at $v_{t+\Delta t}$, which in itself in an unknown and for which we are tying to iteratively solve.

Referring back to the capacitors in Figure 10.20, the TR integration expressions are changed slightly to reflect the fact that the capacitances change with voltage, and hence time:

$$v_c(t_{n+1}) = v_c(t_n) + \frac{\Delta t_n}{2}\left[\frac{i_c(t_{n+1})}{C(t_{n+1})} + \frac{i_c(t_n)}{C(t_n)}\right] \tag{10.8.11}$$

We can rearrange (10.8.11) as follows

$$i_c(t_{n+1}) = \frac{2C(t_{n+1})}{\Delta t_n}[v_c(t_{n+1}) - v_c(t_n)] - \frac{C(t_{n+1})}{C(t_n)} i_c(t_n) \tag{10.8.12}$$

and from (10.8.12) derive the companion model for this nonlinear capacitor, as shown in Figure 10.21.

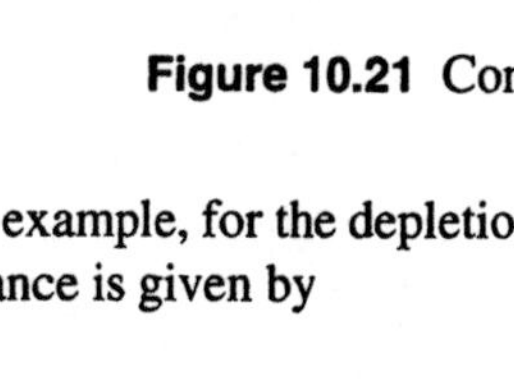

$i_c(t_{n+1})$

I_{eq} G_{eq}

$$I_{eq} = \frac{C(t_{n+1})}{C(t_n)} i(t_n) + \frac{2C(t_{n+1})v_c(t_n)}{\Delta t_n}$$

$$G_{eq} = \frac{2C(t_{n+1})}{\Delta t_n}$$

Figure 10.21 Companion model for a nonlinear capacitor.

As an example, for the depletion capacitance in Figure 10.20, the equivalent (nonlinear) conductance is given by

$$G_{eq} = \frac{2}{\Delta t_n} \frac{C_{j0}}{\left(1 - \frac{v_c(t_{n+1})}{V_{bi}}\right)^m} \tag{10.8.13}$$

and the equivalent (nonlinear) current source is

$$I_{eq} = \left[\frac{i(t_n)}{C(t_n)} + \frac{2v_c(t_n)}{\Delta t_n}\right] \frac{C_{j0}}{\left(1 - \frac{v_c(t_{n+1})}{V_{bi}}\right)^m} \tag{10.8.14}$$

We could model the nonlinear diffusion and deplction capacitances on an individual basis using the companion model in Figure 10.21. We then treat this Norton equivalent just as we would treat any other nonlinear device model and we evaluate it via Newton-Raphson.

It is a bit more difficult in practice to deal with a multi-terminal capacitance element, such as those which might arise in the detailed modeling of an MOS transistor. For MOSFETs, for example, the nonlinear capacitances may be functions of several voltages, not just the voltage across them. For instance, referring to the MOSFET and its associated capacitances in Figure 10.22, C_{gs} can be shown to be a nonlinear function of v_{gs}, v_{ds}, and v_{sb}. Therefore, C_{gs} is a nonlinear capacitance which is a function of three voltages.

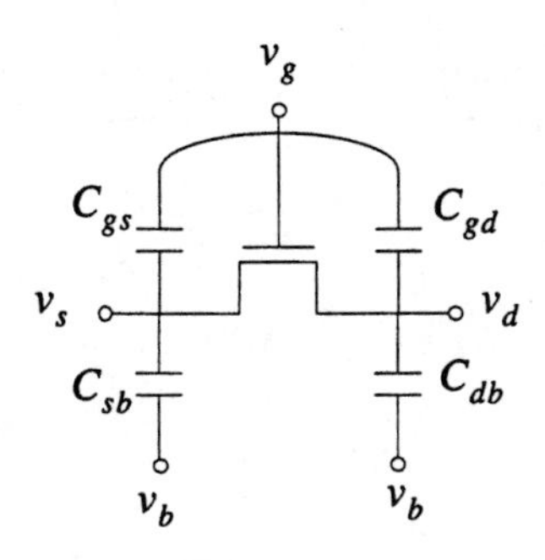

Figure 10.22 A MOSFET and its associated capacitances.

As a simple example, if we assume that $v_{sb} = 0$ and choose to ignore the drain to substrate capacitance, there are two remaining MOSFET capacitances which can be expressed as nonlinear functions of two voltages.

$$C_1 = f_1(v_1, v_2)$$
$$C_2 = f_2(v_1, v_2) \tag{10.8.15}$$

We can no longer write

$$i_C = C(v)\frac{dv}{dt} \tag{10.8.16}$$

but must now write

$$i_{C1} = \frac{dq_1}{dt} = \frac{\partial q_1}{\partial v_1}\dot{v}_1 + \frac{\partial q_1}{\partial v_2}\dot{v}_2$$
$$i_{C2} = \frac{dq_2}{dt} = \frac{\partial q_2}{\partial v_1}\dot{v}_1 + \frac{\partial q_2}{\partial v_2}\dot{v}_2 \tag{10.8.17}$$

We may also express these relations in terms of the charge as a nonlinear function of the voltage

$$q_1 = f_1(v_1, v_2) \tag{10.8.18}$$

and

$$q_2 = f_2(v_1, v_2) \tag{10.8.19}$$

since the most straightforward way to deal with this situation is to take

$$i_{C1} = \dot{q}_1 \approx \frac{f_1(v_1(t+\Delta t), v_2(t+\Delta t)) - f_1(v_1(t), v_2(t))}{\Delta t} \tag{10.8.20}$$

and

$$i_{C2} = \dot{q}_2 \approx \frac{f_2(v_1(t+\Delta t), v_2(t+\Delta t)) - f_2(v_1(t), v_2(t))}{\Delta t} \tag{10.8.21}$$

Then, generically, we can take

$$\begin{aligned} i_C &= i_C(t) \quad \text{for Forward Euler} \\ i_C &= i_C(t+\Delta t) \quad \text{for Backward Euler} \end{aligned} \tag{10.8.22}$$

or

$$i_C = \frac{1}{2}[i_C(t) + i_C(t+\Delta t)] \quad \text{for Trapezoidal} \tag{10.8.23}$$

Stated another way, the TR equation is

$$q_C(t_{n+1}) = q_C(t_n) + \frac{\Delta t_n}{2}(\dot{q}_C(t_{n+1}) + \dot{q}_C(t_n)) \tag{10.8.24}$$

and therefore,

$$C(t_{n+1})v(t_{n+1}) = C(t_n)v(t_n) + \frac{\Delta t_n}{2}(i_C(t_{n+1}) + i_C(t_n)) \tag{10.8.25}$$

or

$$v(t_{n+1}) = \frac{C(t_n)v(t_n)}{C(t_{n+1})} + \frac{\Delta t_n}{2C(t_{n+1})}(i_C(t_{n+1}) + i_C(t_n)) \tag{10.8.26}$$

Note, finally, that these integration algorithms merely transform the nonlinear capacitor to an equivalent nonlinear resistance, which still must be linearized to produce the appropriate stamp.

10.9 The Bottom Line

Newton-Raphson iteration is the cause of most of SPICE-like simulators' non-convergence problems. N-R assumes that the third derivative of all constituent model equations with respect to all circuit variables exists over the range of solution values. This assumption serves as a guide to a device model developer, who has to make sure not only that there are no discontinuities in $i-v$ and $q-v$ characteristics, but that the third derivative of the characteristic functions exist at all operating points. This restriction can be difficult to satisfy, especially in the regions where a device moves from one mode of operation to another.

The evaluation of device models and stamping of the Jacobian matrix is done once per device for each N-R iteration, and the procedure is repeated for each time step. Hence the speed of the model evaluation code often determines the speed of the overall simulator. There is a costly price to be paid for model complexity. In the interests of efficiency, transcendental function evaluations should be avoided in the model evaluation code wherever possible.

Some simulators use a *device model bypass* algorithm to reduce the number of model evaluations required. Hence, if the terminal voltages of a device have not changed by more than a pre-set tolerance since the last time the device was evaluated, the new evaluation is skipped and the previous model is used. Note that the previous evaluation could have been during a previous N-R iteration at the present time point, or the final N-R iteration at the previous time point.

SPICE-like simulators often have convergence problems while attempting a dc solution. One way to tackle this problem is to supply intelligent guesses to be used for the first N-R iteration (in SPICE3, .NODESET cards can be used for this purpose). SPICE fails to converge during a dc solution if the number of N-R iterations exceeds a pre-set limit. The user can increase the limit to aid convergence (using the ITL1 parameter in SPICE3). Another way of dealing with the situation is to tell SPICE not to carry out a dc solution prior to transient analysis, and to supply initial conditions for the transient analysis (using .IC cards and the UIC command in SPICE3).

During transient analysis, non-convergence can occur for a number of reasons. When a pre-set number of N-R iterations is exceeded, the time step is reduced and another attempt is made. The time step is not reduced below a certain limit. There is also a limit on the total number of N-R iterations (ITL5 in SPICE3) which can be changed by the user. In addition, SPICE allows control of the default time step (TSTEP), the tolerance used in determining whether N-R has converged (ABSTOL and RELTOL), and the truncation error tolerance (TRTOL). There are also controls provided that can influence pivoting strategies (PIVTOL and PIVREL). Knowing what these parameters do and how they influence the convergence and accuracy of the simulation are key to making good use of a simulation tool.

In summary, when properly implemented, the combination of Modified Nodal equations, Newton-Raphson iteration, and Trapezoidal integration can yield nonlinear transient circuit simulation with high accuracy -- well beyond the precision with which integrated circuits can be built. But the price exacted in program maintenance and support, device model development, and CPU resource utilization is very high. The latter has tended to preclude the use of circuit simulators for the transient analysis of very large circuits. If one is willing to sacrifice accuracy, some compromises may be made. Logic and switch level simulation are very efficient because they employ extremely simple models, and produce crude or no timing information. For many high performance digital designs, a more detailed appreciation of timing is critical to the successful operation of the circuit. Consequently, there is a great deal of research that attempts to address the gap between highly accurate but highly inefficient circuit simulation, and highly efficient but highly inaccurate logic/switch level simulation. Some of these alternatives are explored in the following chapter.

10.10 References

[Ebers54] J. J. Ebers and J. L. Moll. Large Signal Behavior of Junction Transistors. *IRE*, vol. 42-12, December 1954.

[Carnahan69] B. Carnahan, H. A. Luther, and J. O. Wilkes. *Applied Numerical Methods.* Wiley, 1969.

[Ortega70] J. M. Ortega and W. C. Rheinboldt. *Iterative Solution of Nonlinear Equations in Several Variables.* Academic Press, 1970.

[Nagel71] L. W. Nagel and R. A. Rohrer. Computer Analysis of Nonlinear Circuits, Excluding Radiation (CANCER). *IEEE Journal of Solid State Circuits*, vol. SC-6, pp. 162-182, August 1971.

[Nagel75] L. W. Nagel. *SPICE2, A Computer Program to Simulate Semiconductor Circuits.* Technical Report ERL-M520, UC-Berkeley, May 1975.

[Muller77] R. S. Muller and T. I. Kamins. *Device Electronics for Integrated Circuits.* Wiley, 1977.

Chapter 11 *Timing Simulation*

Traditional circuit simulation is impractical for the functional and timing verification of large integrated circuits and systems. This chapter explores some alternatives to "full-blown" detailed circuit simulation. Timing simulation is an active and evolving topic of research and this chapter does not cover all the approaches suggested in the literature. Rather, this chapter seeks to demonstrate how the fundamentals that we studied in the previous ten chapters form the foundation for these alternative approaches. In the process, we will review some representative timing simulation techniques.

11.1 The Quest for Other Methods of Simulation

There are two main factors that make circuit simulation impractical as a means of verifying a full chip or system. The first is the complexity of such chips and systems. With advances in integrated circuit fabrication, it is not uncommon for processor chips to contain in excess of a million transistors and memory chips many times that. Further, device models tend to increase in their complexity from each generation of fabrication technology to the next. The asymptotic order of complexity of circuit simulation is superlinear in the number of elements in the circuit. Hence, as circuit size grows, the CPU time requirement for simulation grows superlinearly and quickly becomes prohibitive for circuits consisting of tens of thousands of transistors. Since each element contributes a fixed "stamp" to the system matrix, the memory requirements of these programs typically grow linearly with the size of the underlying circuit.

The second factor that makes circuit simulation impractical is the rapid increase in the speed of chips and systems. With each new generation of hardware, the simulation required to solve a circuit for a *fixed period of time*, say $1\,\mu s$, keeps increasing. Luckily, the growth of CPU time with the time period of simulation is linear. Even in the case of simulators that store all waveforms in memory, the growth of memory requirements with the period of simulation is linear, too.

The quest for high-performance circuits has led to an increased emphasis on custom design. Custom-crafted circuits are often tricky and require careful analysis and simulation to guarantee correct operation. It has often been noted that one constantly faces the challenge of designing and verifying tomorrow's ever more complex and ever faster hardware with the aid of today's computers. In the push to design more complex circuits in shortening design cycles, simulation is often the bottleneck. Circuit simulation, while essential in its own right, is not enough to face this challenge successfully. The combined pressures of needing more simulation for greater time periods of larger circuits with tighter schedules has led to a significant number of alternative approaches. This chapter will quickly review some of these methods, and then concentrate on one class of techniques called *timing simulation.*

11.2 Static vs. Dynamic Simulation

Dynamic simulation implies the simulation of a circuit from a given start time to a given end time. The inputs to the circuit during this period of time are fully specified. (In the case of digital or logic circuits, these input signals are called "input patterns" or "input vectors.") The initial state of the circuit is usually at least partially specified and/or computed. The goal is to solve the circuit for the required time interval. All the methods described so far in this book deal with dynamic simulation methods.

Static simulation, on the other hand, attempts to characterize a circuit for all time, independent of its input signals. While this book will not cover static simulation in any detail, this section will provide a quick overview. Static simulation, or *static timing analysis* (as it is more commonly known) [Hitchcock82], is often used to characterize the delay of an interconnected set of combinational logic blocks between the flip-flops of a digital circuit. Figure 11.1 shows a simple circuit consisting of two banks of flip-flops (FF_1 and FF_2) and four combinational blocks (B_1 through B_4). Each combinational block's delay is precharacterized. The delay from each input to each output is described either as an equation or stored as a look-up table. The delays are functions of such variables as input slope, fanout, and output load capacitance. The precharacterization phase consists of many circuit simulation runs at different temperatures, power levels, loading conditions, and so on. Delay data from these runs are abstracted into a *timing model* for each block.

The actual analysis is carried out in two phases. In the first phase, the delay of each signal is propagated through the combinational blocks, using the precharacterized delay figures for each block. Thus each signal is labeled with an *arrival time* at which its correct digital signal can be guaranteed. In the second phase, the *required arrival time* is propagated backwards from the target bank of flip-flops, FF_2. The required arrival time on the signal is the latest time by which that signal must have its correct value in order for the required worst-case delay between the banks of flip-flops to be met. The difference between the required arrival time and the actual arrival time of each signal is termed the *slack* of the signal. All the signals are listed in increasing order of their slack. This analysis yields a wealth of information about the timing of the circuit.

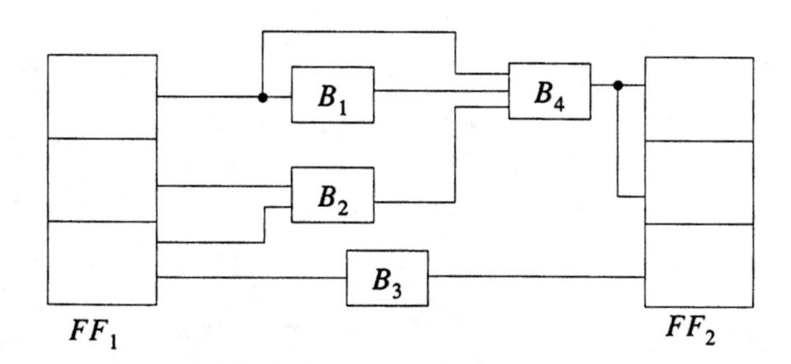

Figure 11.1 Static timing analysis of circuits containing combinational blocks.

Clearly, if there is negative slack on any of the signals, the circuit will not meet the performance requirements. The path with the least (perhaps most negative) slack on all of its signals is the *critical path*. The signals on this path will all have the same slack. The slacks also contain clues needed to redesign the circuit to cause it to function correctly. The above analysis can also be carried out with a minimum and maximum delay for each block. In that case, a set of *early* and *late arrival times* can be computed. The early mode is computed using the best possible case for the arrival of all input signals to a block and the late mode considers the most pessimistic scenario. Then two sets of slacks are computed for each signal. These slacks yield valuable information about the timing properties of the circuit including possible violations of flip-flop setup or hold times. For more details on the calculation and interpretation of slacks, see [Hitchcock82].

Static timing analysis is a highly efficient method to characterize the timing of circuits. It can be used to determine the critical path of a circuit and obtain valuable information on other timing characteristics. However, it assumes that all paths in the circuit are *active* or *sensitizable*. In reality, however, there are certain paths in logic circuits that are not sensitizable because of the nature of the logic or the manner in which the circuit is exercised. These paths are called *false paths*. Because it ignores the false path problem, static timing analysis often predicts pessimistic (or overly conservative) worst-case path delays.

11.3 Dynamic Simulation

Depending on the level of modeling abstraction, dynamic simulation of circuits can be carried out at many levels. The different levels of simulation are qualitatively depicted in Figure 11.2

Circuit simulation:

The device models in circuit simulation [Sze85, Muller86, Roulston90] consist of equiva-

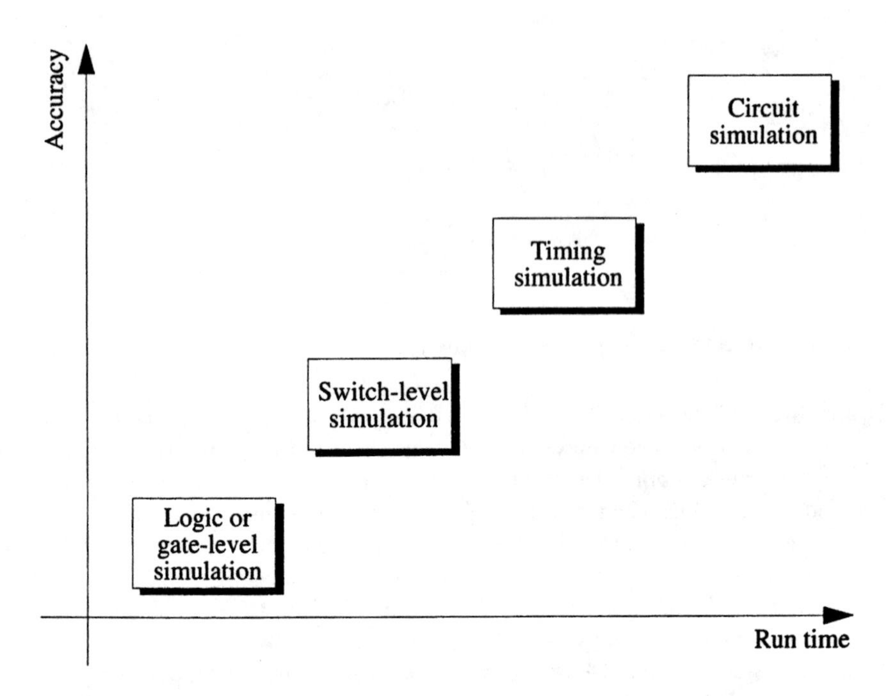

Figure 11.2 Trade-offs between the different levels of simulation.

lent circuits for complex devices and analytical equations describing the characteristics of these constituent elements. Algorithms used for solution consist typically of integration (as described in Chapter 4 and Chapter 5), linearization (as described in Chapter 10), and sparse solution of the resulting equations (see Chapter 7 and Chapter 3). The unknowns are typically element and node voltages and element currents as a function of time. These *waveforms* are usually represented as floating point numbers. Circuit simulation is by far the most detailed, accurate, and general of the methods described in this chapter. It is also the slowest and is limited to relatively small circuits for relatively short intervals of time compared to the other methods.

Timing simulation

Timing simulators [Chawla75, de Geus82, de Geus84, Chen84, Tsao85, Sakallah85, Vidigal86, Ruan88, Visweswariah89, Ruan90, Lin93, Devgan94] use simplified device models in order to obtain a speedup over traditional circuit simulators. The simplifications in the device models, while allowing rapid turnaround, often limit the applicability of timing simulators. Many timing simulators, for example, are limited to MOS digital circuits. Like circuit simulators, timing simulators attempt to compute all the voltages and currents of a circuit. The algorithms involved usually include event-driven methods, by which only *changes* in the circuit quantities are computed. These methods allow the exploitation of *latency* to render the simulation more efficient. The run time of this class of simulators usually increases linearly with the size of the circuit being simulated, thus permitting the simulation of relatively large circuits for relatively long intervals of time. Timing simulation is only as accurate as its modeling assumptions permit and there is a trade-off between model complexity and run time. Timing simulation is described in detail in Sections 11.4 to 11.7.

Switch-level simulation

The basic model for a transistor in switch-level simulation is an ideal switch [Bryant80, Bryant81]. This level of simulation is confined to digital MOS circuits. If the signal at the gate of an MOS transistor causes it to turn ON, it is conducting and there is a path between the drain and the source of the transistor. If not, there is no connection between the drain and source. The simplest signal representation consists of three discrete states: '0' (logic low), '1' (logic high) and 'X' (unknown or uninitialized). An N (P) type MOS transistor with a signal of '1' ('0') on its gate is conducting whereas a gate signal of '0' ('1') causes it to be open. A transistor with an 'X' on the gate can either be conducting or nonconducting and is said to be *potentially conducting*.

A switch-level simulator operates as follows. To begin, the circuit is partitioned into *strongly connected components* (SCCs) which are subcircuits consisting of transistors that are channel-connected, as shown in Figure 11.3. The communication at the boundary of SCCs is through gates of transistors only. Each SCC is represented by an undirected graph. Each vertex of the graph corresponds to a signal in the circuit and each edge corresponds to the drain-source channel of a transistor.

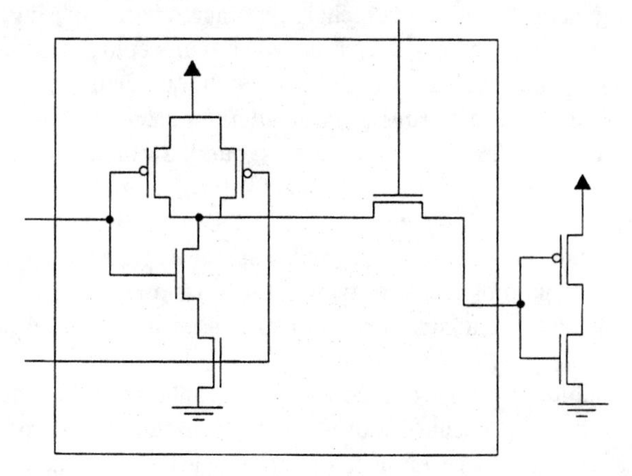

Figure 11.3 An example of a strongly connected component (SCC).

The actual simulation consists of a simple procedure. All nodes of the circuit except the primary inputs are initialized to 'X'. Input signals are then applied. For each vertex of each SCC, all conducting paths to VDD (the power supply) and GROUND are traced. Paths consisting of transistors that are turned ON are called *definite paths*. Paths consisting of transistors that are potentially conducting are called *potential paths*. If all the conducting paths from a node are to VDD (GROUND), the state of that signal is '1' ('0'). If there are no paths to either VDD or GROUND, the signal is assigned its previous state. But in the case where paths exist to both VDD and GROUND, there is a *resolution* process to determine the state of the signal.

Resolution of signals is based on the concept of *strength*. Each transistor is assigned a strength or current-driving capability. The strength is usually chosen from a small set of integers to reflect the width and type of the transistor (N or P type). The strength can be thought of as being indicative of the conductance of the transistor. The strength of a series path consisting of ON transistors is the strength of the weakest ON transistor. The strength of multiple parallel paths consisting of ON transistors is the strength of the strongest path. If the strength of the paths to VDD (GROUND) is "much more" than the strength of the paths to GROUND (VDD), the signal is assigned a '1' ('0'). If the two strengths are "comparable," the signal is assigned an 'X'. Of course, empirical heuristics are necessary to determine what "much more" and "comparable" mean. Once the states of all of the signals of all the SCCs have been determined, they are propagated to their fanout SCCs and used as gate inputs for the next cycle of simulation. To proceed to the next cycle, the new primary inputs for that cycle are assigned, the fanouts across the borders of the SCCs are propagated, and then the simulation procedure is repeated. Switch-level simulators are called *unit delay simulators* since they have no concept of time except for moving from one set of inputs to the next (presumably at the next "clock tick"). The analysis assumes that sufficient time elapses between such cycles for the circuit to completely settle down.

The procedure outlined above is adequate to handle static CMOS ratioed logic. However, it does not in general accommodate charge storage, charge sharing, or bidirectional signal flow that is common in dynamic circuits or in circuits containing pass transistors. A simple enhancement can be made to extend the algorithm to handle these situations. Each node in the circuit is assigned a *storage node strength* or *storage node size*, which represents the value of the capacitance of that node to ground. Storage node strengths are chosen from among a small set of integers. VDD, GROUND, and the primary inputs are assigned the highest strength. Whenever paths to VDD and GROUND are traced from a node of an SCC, the node is not only assigned a signal state, but also a *signal strength*, being the strength of the path by virtue of which it was assigned its state. Then the storage node strengths, signal states, and signal strengths are used in a rule-based resolution process.

Switch-level simulation possesses some unique advantages. It is efficient because its MOS model is so simple and because it uses an event-driven algorithm. Hence large circuits can be simulated for a large number of time cycles. The switch-level behavior of some commonly used circuit blocks can be precompiled for additional efficiency [Bryant87]. Parallel fault simulation is also possible at the switch level. However, switch-level simulators have some fundamental limitations linked to their simplistic modeling of transistors and time. First, they provide limited or no timing information. Second, their handling of analog situations like charge sharing, glitches, and bidirectional signal flow is inaccurate and some times even wrong. Finally, they depend on partitioning the circuit into SCCs, which may result in some very large circuit blocks. For example, a barrel shifter could be one large SCC, and the repeated tracing of paths in that graph can be computationally expensive.

Logic or gate-level simulation

Logic simulation is at a higher level of abstraction than switch-level simulation. In its simplest form, the circuit is modeled as a connection of simple logic primitives like NAND gates, NOR gates, and so on. Each primitive has a *logic behavior* by which the output(s) of the gate can be computed given the input(s). In addition, each primitive has a *delay model* that represents the delay from each of the inputs to each of the outputs. Signal representation, as in the case of switch-level simulators, consists of a '1' (logic high), '0' (logic low), and 'X' (unknown or uninitialized state). Some logic simulators have special signal representations for high impedance states, tristate signals, and so on.

The simulation algorithm consists of a simple event-driven or *selective trace* mechanism. All primary inputs are assigned their values and the fanouts of the primary inputs (those primitives to which they are connected) are placed on an event queue for evaluation. When the primitives are evaluated, their outputs and delays are computed. The change on each output is scheduled to take effect at a future point in time depending on the delay of the primitive. The fanouts of the primitive are scheduled to be evaluated when the fanout signals change. One way to implement such a simulator is to have a *time wheel* with a list of *events* at each time point.

There are two kinds of events, *update events* and *evaluate events*. The update events are first processed, during which signals are given their new values (either because a primary input changed at that time or because a previous evaluation of a primitive caused this signal update to be scheduled). Then the evaluate events are processed, during which primitives are evaluated. The fanout signals and primitives are then scheduled for a future time. Zero-delay primitives can require the update-evaluate combination to be repeated multiple times at a given point in the time wheel. When the list of events to be processed at a given time is empty, time is moved to the next time at which there is at least one event pending, and the simulation continues.

A number of extensions over the simple procedure described above are commonly found in logic simulators. The user is often afforded the capability to write his/her own behaviors for building blocks of the circuitry. A state-strength model for signal representation is used (as in switch-level simulation) to better model bidirectional signals, tristate conditions, and buses. Various high-level primitives, circuit debugging aids, display options, and so on, are integral parts of most logic simulators. Many logic simulation engines have special purpose hardware to accelerate their performance.

Logic simulators are very fast and can handle large chips and/or systems. They are the mainstay of circuit verification of large, digital systems. The primitive models are simple and hence quick to evaluate. The accuracy of the results is, of course, only as good as the accuracy of the behaviors of the primitives and the timing model provided for those primitives.

Now that we have a high-level understanding of circuit, timing, switch-level and gate-level simulation, we will delve into the topic of timing simulation in more detail.

11.4 Motivation for Timing Simulation

Device modeling

The analytical equations that describe electronic device models are complex, and their evaluation is time consuming. In fact, over half the computer time of circuit simulation programs can be spent evaluating device models. In such a situation, improvements in the algorithms that formulate and solve equations yield very little in the way of improved turnaround time. In order to gain significant speed over traditional simulation methods, this "modeling bottleneck" has to be overcome by using simplified device models. Simplified models lead to more efficient simulation, but with a concomitant loss of simulation accuracy. This trade-off is shown in Figure 11.2.

Device models typically consist of $i-v$ characteristics and parasitic capacitances. In the case of MOS transistors, there is only one current to be modeled since the gate current is negligible. The channel current computation is often simplified using table look-up, piecewise constant, or piecewise linear models. In addition to the $i-v$ characteristics, the equivalent circuit for the MOS transistor consists of nonlinear (voltage-dependent) capacitances to model the gate oxide, junction, and overlap capacitances. Figure 11.4 shows a simplified equivalent circuit where the five nonlinear, floating capacitances associated with the MOS transistor have been replaced by three linear grounded capacitances. Each of these equivalent capacitances has a width, length, and area coefficient. All the coefficients are determined by a fitting and optimization process. Miller effects (e.g., gate to drain capacitance coupling) are not simulated with this simplified model. However, for most digital circuits, this model has been shown to produce sufficient accuracy. A number of timing simulators use this linear grounded capacitance model to simplify their computations. The advantage of this model is that node and branch variables can be solved in a decoupled fashion in the absence of floating capacitors. Of course, there is an accuracy penalty to be paid for this modeling simplification.

When nonlinear capacitors are modeled by timing simulators, piecewise linear functions in the $q-v$ plane are most often used. These approximations correspond to piecewise constant functions in the $C-v$ plane. Such a model can be constructed in a charge and energy conserving manner as shown in Figure 11.5. The curved lines in the figure

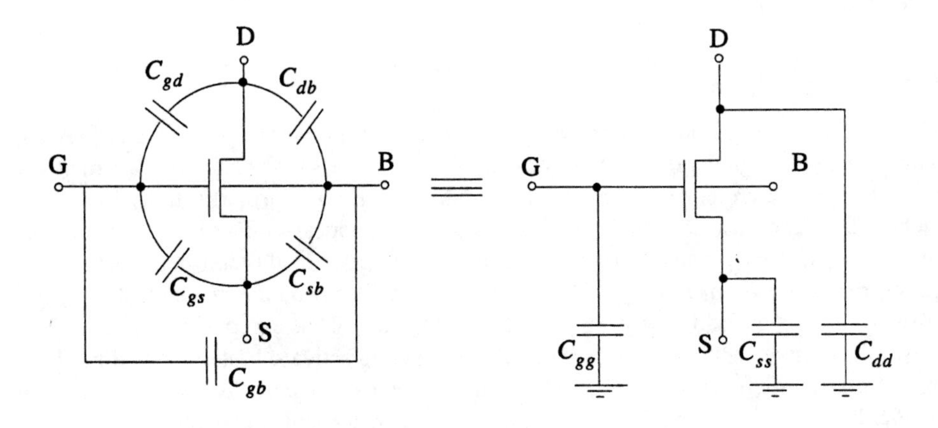

Figure 11.4 Simplified equivalent circuit for the capacitances associated with MOSFETs.

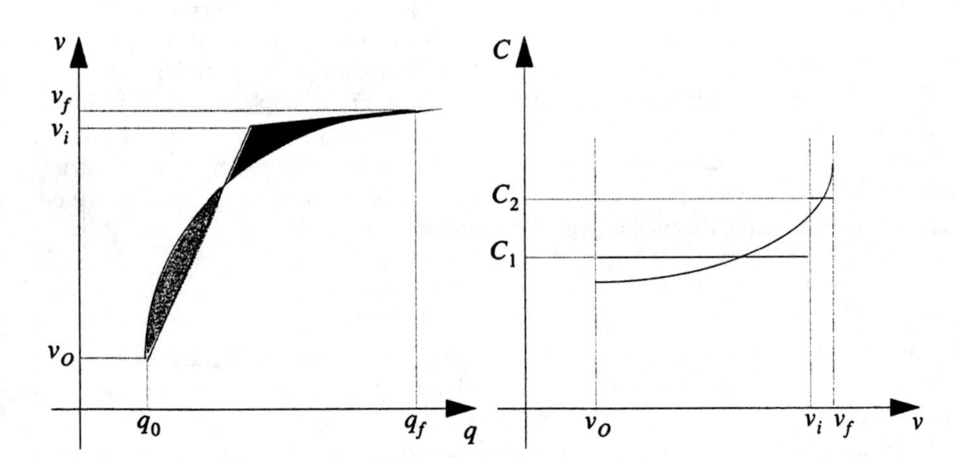

Figure 11.5 Modeling of nonlinear capacitances.

show the actual $q-v$ and $C-v$ characteristics of the capacitor. Since the end points of the analytical curve and piecewise linear approximation in the $q-v$ plane are the same (q_0 and q_f), charge is conserved. If the two shaded areas in the figure on the left are equal, the area under the analytical curve and the piecewise linear approximation are equal, implying conservation of energy. The piecewise linear table is easily converted to a piecewise constant table in the $C - v$ plane.

Exploitation of latency

Most circuits display both *spatial* and *temporal latency*. In fact, larger circuits often display a relatively high degree of latency. *Spatial latency* refers to the situation where at any given time, there are portions of the circuit with no activity. Further, activity in a small section of a circuit usually affects only the neighboring portions rather than the entire circuit. Likewise, for a given portion of a circuit there are periods of time when it is inactive, giving rise to *temporal latency*. Timing simulators take advantage of inherent temporal latency by using event-driven or activity-driven methods. Thus computation time is spent only on the active portions of the circuit, and only during active subintervals of time. Timing simulators often take advantage of spatial latency, too, by *partitioning* the circuit into smaller blocks called *subcircuits* that can be solved in a decoupled fashion.

Variable accuracy

In the accuracy-speed trade-off, it would be beneficial to be able to trade accuracy for more speed and vice versa. In other words, the ability to accommodate variable accuracy device models is desirable. Traditional circuit simulators are accurate, but do not allow the reduction of accuracy requirements in return for faster simulation (adjustment of tolerances does affect run time, but not by much). Variable accuracy can be useful both across portions of the circuit and across runs of the simulator. For example, a critical path or a sensitive portion of a circuit could be assigned higher accuracy without paying the price of slow run time for the entire circuit. Or a new design can be quickly simulated with crude models as a sanity check, and the device models can be replaced by progressively more accurate models during the final tuning of the circuit.

High capacity

Using traditional circuit simulation methods, designs are often manually broken into smaller chunks and simulated separately. This design partitioning is necessary since circuit simulators typically cannot accommodate a full chip or system. The process of splitting circuits into smaller ones, verifying them separately and thus guaranteeing correct operation of the overall design is tedious and error-prone. Hence any increase in capacity is a welcome aid in the design of integrated circuits.

We will first understand how these motivating principles are applied in the MOS Timing Simulator (MOTIS), and then discuss some examples of more recent research activity in the domain of timing simulation.

11.5 The MOS Timing Simulator (MOTIS)

MOTIS [Chawla75], developed at AT&T Bell Laboratories, was one of the first timing simulators. It pioneered such techniques as event-driven simulation, table models, and simplified device models. It was developed for use on circuits containing MOS transistors only. It was designed to simulate digital logic circuits including transmission gates (or pass gates) to accurately predict the timing of the circuits, including subtle characteristics such as glitches.

Device models

The model for an MOS transistor in MOTIS is simple. The MOSFET $i - v$ characteristics consist of a two-dimensional table representing the drain to source current,

$$i_{ds} = \frac{W_{eff}}{L_{eff}} f(v_{gs} - V_{Th}, v_{ds}) \quad \textbf{(11.5.1)}$$

with 64 equally spaced voltage values on each axis. An auxiliary one-dimensional table for threshold voltage,

$$V_{Th} = f(v_{sb}) \quad \textbf{(11.5.2)}$$

uses 64 values to represent the *body effect* or the variation of threshold voltage due to *back-gate bias* or v_{sb}. The threshold voltage is used as a shift in the main two-dimensional table. Different tables are built for different types of transistors (N type, P type, saturation loads, etc.) as well as for different temperatures and process bias conditions. The evaluation of the channel current of a transistor from analytical equations in the innermost loop of a simulator can be quite time consuming. MOTIS replaces that evaluation by the two simple table look-ups discussed above and scaling of the result by the W_{eff}/L_{eff} ratio of the device.

The circuit to be simulated is assumed to consist of simple building blocks like NAND gates, registers, flip-flops, and so on. Each gate is precharacterized to have a certain equivalent load capacitance C_L which is linear (voltage-independent) and grounded. Hence Miller effects (gate to source and gate to drain capacitance coupling) are ignored.

Gate current evaluation

Using the look-up table to evaluate the current of each MOS transistor of a logic gate, currents of transistors in parallel are added, and currents of transistors in series are combined in the following fashion:

$$\frac{1}{I_{total}} = \frac{1}{I_1} + \frac{1}{I_2} \tag{11.5.3}$$

I_1 and I_2 are computed by table look-up as though the *total series voltage* of the two transistors were applied across each of the transistors in turn. Referring to Figure 11.6, I_1 and I_2 are computed as though all of v_{out} were applied in turn to the channels of transistors M_1 and M_2. From (11.5.3),

$$\frac{v_{out}}{I_{total}} = \frac{v_{out}}{I_1} + \frac{v_{out}}{I_2} \tag{11.5.4}$$

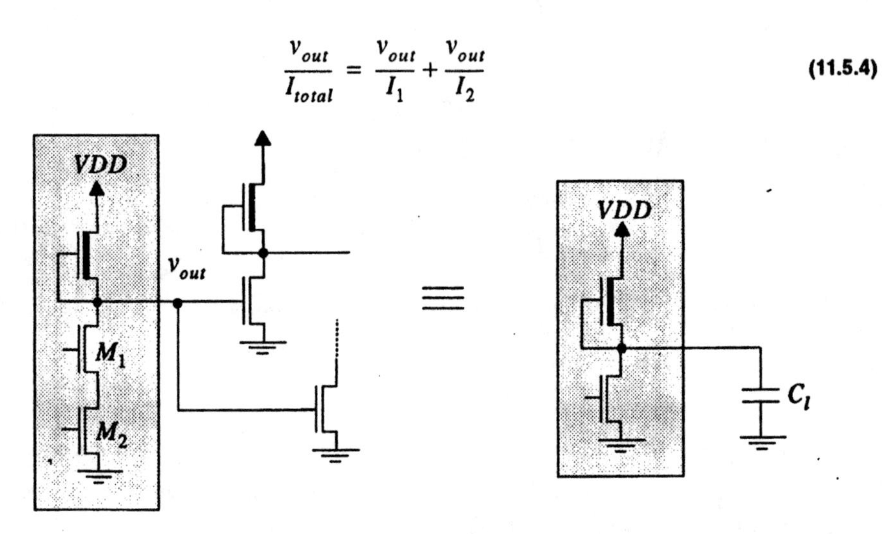

Figure 11.6 Equivalent model for an NMOS NAND gate in MOTIS.

Or

$$\frac{1}{G_{total}} = \frac{1}{G_1} + \frac{1}{G_2} \tag{11.5.5}$$

which is exactly like combining conductances in series. The total current flowing out of a logic block I_{net} is computed as the difference between the total current of the pull-up chain ($I_{pull-up}$) and the pull-down chain ($I_{pull-down}$), as illustrated in Figure 11.7.

Event-driven simulation

The model for a gate is shown in Figures 11.6 and 11.7. In the first step, each gate is replaced by a single equivalent pull-up and pull-down device, and the output load is modeled by a single grounded capacitance. Then each of the pull-up and pull-down transistor

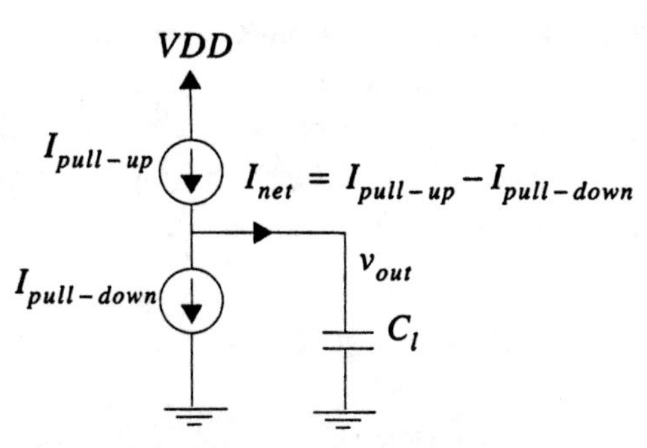

Figure 11.7 MOTIS electrical model for a logic gate.

chains is replaced by a single current source using the gate current evaluation procedure described above (see Figure 11.7).

Any time the inputs to the gate change, the gate is re-evaluated and then its fanouts are scheduled to be re-evaluated, too. The equations for evaluating logic gates are as follows:

$$\Delta v_{out} = \frac{1}{C_l} \int_{t_1}^{t+\Delta t} I_{net} dt \tag{11.5.6}$$

Using the Backward Euler formula to integrate I_{net}:

$$\Delta v_{out} = \frac{1}{C_l} I_{net}(t+\Delta t)\,\Delta t \tag{11.5.7}$$

During the period t to $t+\Delta t$, the variations of the gate input voltages are ignored. However, I_{net} is a function of v_{out}, therefore

$$I_{net}(t+\Delta t) = I_{net}(t) + G_{eq}\Delta v_{out} \tag{11.5.8}$$

where G_{eq} is the equivalent conductance at the output of the gate. Combining (11.5.7) and (11.5.8) yields

$$\Delta v_{out} = \frac{I_{net}(t)}{\frac{C_l}{\Delta t} - G_{eq}} \tag{11.5.9}$$

where C_l is known and $I_{net}(t)$ is computed using table look-up and series/parallel combinations. Hence we need only G_{eq} to compute Δv_{out}.

G_{eq} can be computed using expensive Newton-Raphson iterations, but in the interest of speed, MOTIS uses a secant approximation as shown in Figure 11.8. A secant is drawn from the present operating point to the origin of the $i-v$ plane and the slope of that line is used as G_{eq}. Once Δv_{out} is computed, the change is propagated to the fanouts of the gate. Simulation then continues in an event-driven fashion.

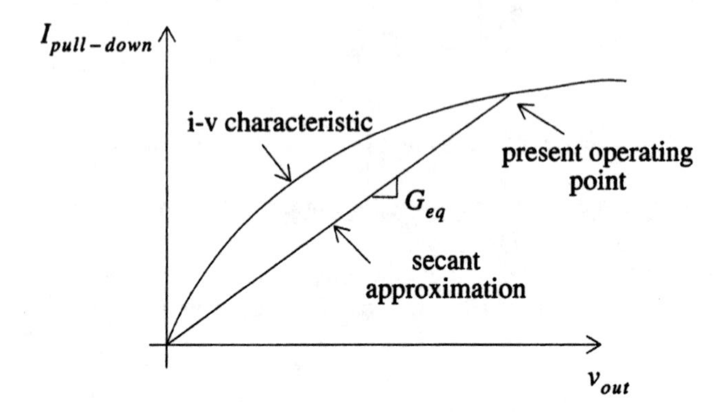

Figure 11.8 The secant approximation in MOTIS.

MOTIS was the first timing simulator and pioneered such techniques as event-driven timing simulation, simplified device models and table models. It only accommodates circuits consisting of carefully precharacterized logic blocks and some limited pass transistor configurations. It ignores gate-to-drain capacitance coupling because of its grounded capacitor assumption. However, it is very efficient and sufficiently accurate for the types of circuits and technologies that were present at the time it was first developed. Many enhancements on the basic MOTIS algorithms have been carried out after the original development [Chen84, Tsao85].

11.6 ELogic and SAMSON

This section and the next will briefly review some other interesting timing simulation strategies that were reported in the literature since the time that MOTIS was originally published. The approaches presented here were motivated by the need for a general, fast, large-capacity and reasonably accurate timing simulator for digital circuits. The review here is not exhaustive; some representative timing simulation efforts are briefly described.

SAMSON

SAMSON [Sakallah85] is a mixed circuit/logic simulator. It uses an event-driven algorithm to maximally exploit temporal latency. First, SAMSON partitions the circuit into loosely connected subcircuits. Each subcircuit is allowed to have its own time step, dictated by its activity pattern and any changes in the inputs to that subcircuit. Each subcircuit is evaluated at discrete time points, and a *circuit event* is said to occur at each of these time points.

Each subcircuit, at any given time, is either *alert* or *dormant*. Alert circuits are modeled by a set of nonlinear algebraic differential equations, as in detailed SPICE-like simulators. A subcircuit is always alert during its circuit events, and dormant otherwise. Dormant subcircuits do not require linearization or discretization in their solution process. Instead, the previous behavior of the subcircuit is extrapolated using a predictor/corrector integration scheme. Hence evaluation of dormant subcircuits is relatively inexpensive. The higher the proportion of dormant subcircuits and the longer the subintervals of time for which they are dormant, the more efficient the simulation. If any input to a dormant subcircuit changes, the subcircuit is replaced by an alert model, with a reduced time step from the one previously computed using the dormant model. The outputs of dormant models could drive other dormant subcircuits and cause those subcircuits to change to an alert state.

To discretize time, a stiffly stable variable order integration scheme is used in SAMSON [van Bokhoven75]. These integration formulas provide efficient extrapolation for dormant circuits. In addition, good truncation error formulas and estimates for the decoupling error between subcircuits are available. Two types of truncation error are used. The first is an *a priori* truncation error estimate that can be applied before the solution at the next time step has been computed. This estimate is useful in time step selection. The second is an *a posteriori* estimate that can be used to calculate the truncation error in a time step after the solution has been computed. It is useful in rejecting time steps that were too large. If a time step is rejected, theoretically, the simulation time has to be *rolled back* to the time of the rejected time step since other subcircuits could have marched ahead with erroneous input waveforms. But rolling back time is computationally expensive and difficult to implement. Instead, SAMSON deals with the problem by being conservative in replacing dormant models with alert ones.

The actual solution of linear(ized) equations uses either a Block LU Factorization or a Block Gauss-Seidel iteration (see Section 11.8 for a discussion of Gauss-Jacobi and Gauss-Seidel iterations) since the global matrix is composed of several loosely connected diagonal blocks.

Thus SAMSON uses partitioning and event-driven algorithms to exploit latency, without significant loss in simulation accuracy. The event-driven nature helps the simulator incorporate logic blocks characterized by behavioral models as subcircuits, in addition to transistor-level subcircuits.

Electrical Logic simulation (ELogic)

Electrical Logic simulation (ELogic) [Kim89] uses an event-driven method to analyze MOS digital circuits. Nodal Analysis is used as the basis of equation formulation. Each node voltage is allowed to take one of a set of discrete values or states. Nodes can move from one state only to the next adjoining state in either direction. The transition from one state to another is called an *event* and the time when that occurs is the *event time*. Simulation proceeds by computing the event time for each node. Simulation time is moved to the time of the most imminent event, which is then processed. At that time, the node voltage is updated to its next state and the expected time that will elapse before its *next* event is computed, and an event scheduled into the future. The number of voltage states can be varied both across portions of the circuit and across runs of the simulator, allowing the user to influence the speed-accuracy trade-off.

Figure 11.9 shows a node voltage v_n that moves from state V_1 through V_4 with transition times t_1, t_2, and t_3. These transition times are the unknowns in ELogic. Event times are determined using the equivalent circuits shown in Figure 11.10, where each Norton equivalent in Figure 11.10(a) represents a linearized MOSFET model. Assume that node A has just had an event and its new voltage is V_n. We must now compute the time it will take for node A to reach its next discrete voltage state, V_{n+1}. ELogic assumes that there is a capacitor to ground from every node. The linearized MOSFET model is computed assuming that the voltage of node A is V_n. The voltages of all other nodes except the one being updated are assumed to be constant, and hence an independent voltage source is used to represent the capacitor to ground from those nodes.

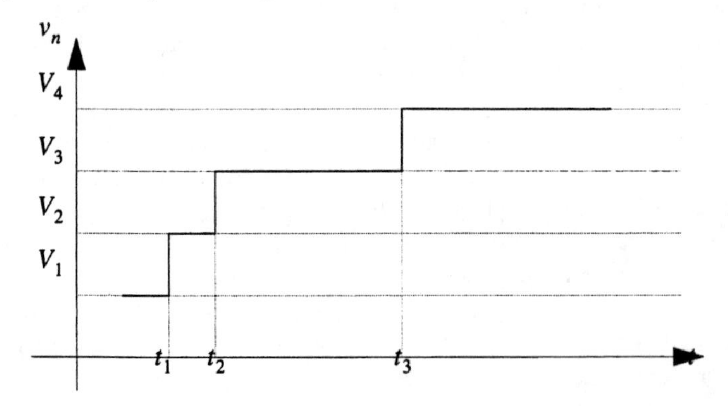

Figure 11.9 In ELogic, node voltages make discrete transitions from one state to another; the transition times are the unknowns.

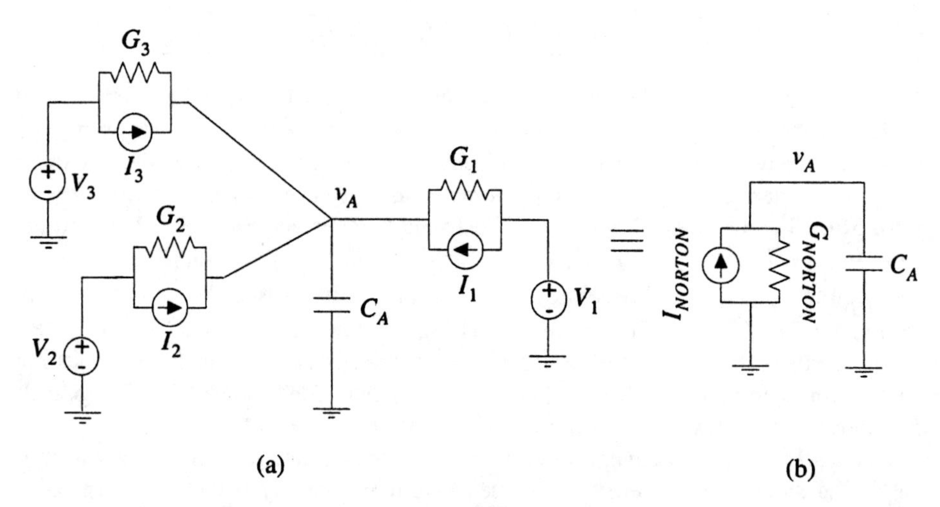

Figure 11.10 Equivalent circuit in ELogic for updating of node A. The equivalent circuit in (a) is combined to compute the event time using (b).

All the elements connected to node A are combined into a single Norton equivalent consisting of a current source I_{NORTON} in parallel with a conductance G_{NORTON}, as shown in Figure 11.10(b). Then a Forward Euler model is used to compute the time to the next event Δt:

$$\Delta t = \frac{(V_{n+1} - V_n)\, C_A}{I_{NORTON} - V_n G_{NORTON}} \tag{11.6.1}$$

At that time into the future, the voltage of node A will be updated to V_{n+1} and the time for it to make its *next* transition will be computed using the voltages at the surrounding nodes and linearized device models in effect at that time. An event queue of transitions is maintained and simulation consists of repeatedly processing the most imminent event. There are also variants of the ELogic algorithm that use Trapezoidal integration for only the capacitor at the node that is being updated, and that permit floating capacitors [Kim89].

The advantages of the ELogic algorithm are that it is fast due to its event-driven nature and it allows for a continuous accuracy-speed trade-off. One of the problems, however, is that discrete voltage states can lead to an oscillation problem whereby two nodes repeatedly cause each other to jump between discrete states, in turn causing an unnecessarily large number of events. ELogic addresses this problem by implementing a cycle detector to detect and suppress such spurious oscillation.

11.7 Piecewise Approximate Timing Simulation

Piecewise approximate simulation refers to a class of timing simulators that model circuit quantities (voltages, currents, $i-v$ characteristics) by piecewise functions. For example, a node voltage could be approximated by a piecewise linear function of time or the current through a nonlinear device could be approximated as a piecewise linear function of voltage. In SPECS (Simulation Program for Electronic Circuits and Systems) [Visweswariah91a, Visweswariah93], $i-v$ characteristics are assumed to be piecewise constant, branch currents piecewise constant in time and branch voltages to be piecewise linear in time. In ACES (Adaptively Controlled Explicit Simulation) [Devgan94], $i-v$ characteristics are modeled by piecewise linear functions and all branch and node quantities are piecewise linear in time. In SWEC (StepWise Equivalent Conductance) [Lin93], branch conductances are approximated by piecewise constant functions of time.

The advantage of piecewise approximate methods is that they lend themselves well to event-driven simulation. Whenever a branch or circuit quantity reaches a corner of a piecewise segment, an event is processed. By varying the number of piecewise segments used in the approximation, variable accuracy is a natural benefit of piecewise approximate models. Thus there is a direct trade-off between accuracy and the number of events processed, or the speed of simulation. This section will describe some piecewise approximate timing simulation methods that were recently reported in the literature. We first begin with a description of SPECS.

Device models for 2-terminal non-energy-storage devices

In SPECS, all non-energy-storage devices have their $i-v$ characteristics modeled by piecewise constant functions. In the case of two terminal elements, these are one-dimensional functions as shown in Figure 11.11. The models can be built at different levels of accuracy as shown in the figure. A large number of stepwise constant segments, as shown on the right, leads to a more accurate model, and vice versa. The implication of this approximation is that as long as the voltage across the element is between v_i and v_{i+1}, it behaves like an independent current source of value $\bar{I}_i$. The v_i's and $\bar{I}_i$'s are computed and stored *a priori* as table models, using the following simple formula:

$$\bar{I}_i = \frac{1}{(v_{i+1} - v_i)} \int_{v_i}^{v_{i+1}} f(v)\, dv \tag{11.7.1}$$

where $i = f(v)$ represents the actual $i-v$ characteristics of the element.

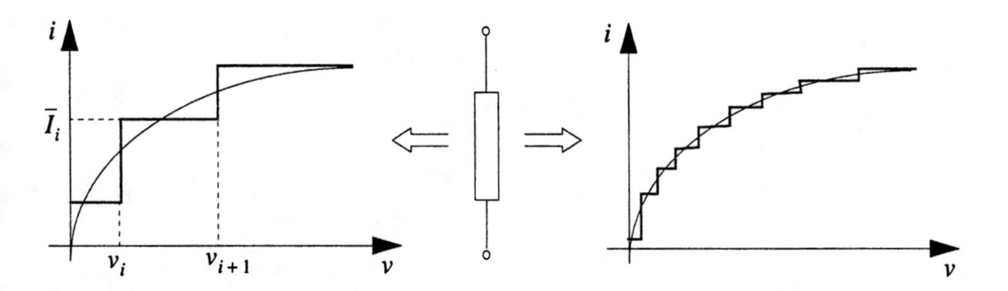

Figure 11.11 $I - v$ characteristics are approximated by piecewise constant functions.

Device models for linear capacitances

Linear (non-voltage-dependent or constant) capacitances are modeled using their constituent equation

$$i = C\frac{dv}{dt} \tag{11.7.2}$$

Nonlinear capacitors will be discussed later in this section.

Simulation algorithm overview

Consider a simple section of a circuit as shown in Figure 11.12. This section will explain the simulation of this circuit qualitatively. See [Visweswariah93] for a more mathematical description. Figure 11.11 assumes that we have a circuit consisting only of two-terminal resistive elements and grounded constant capacitors. Later in this section, we will see that resistive elements are treated as links in a tree/link formulation (see Appendix A for details), which explains the notation L_5, L_6, and L_7 for these elements (not to be confused with the notation of L for inductors in previous chapters). Each of the these link elements is represented by a table model of $i - v$ characteristics, as shown in Figure 11.12.

An *event* is said to occur (or *fire*) whenever a device reaches a corner in its table of $i - v$ characteristics. At that time, the branch current of the element that caused the event makes a step jump to the new current looked up from the table model. Hence the branch current of all nonstorage elements is piecewise constant in time. These currents distribute to the capacitances in accordance with KCL, causing the branch currents of *all elements* to be piecewise constant in time. Let us say branch L_6 has an event causing a change of current Δi_6. Then, by applying KCL,

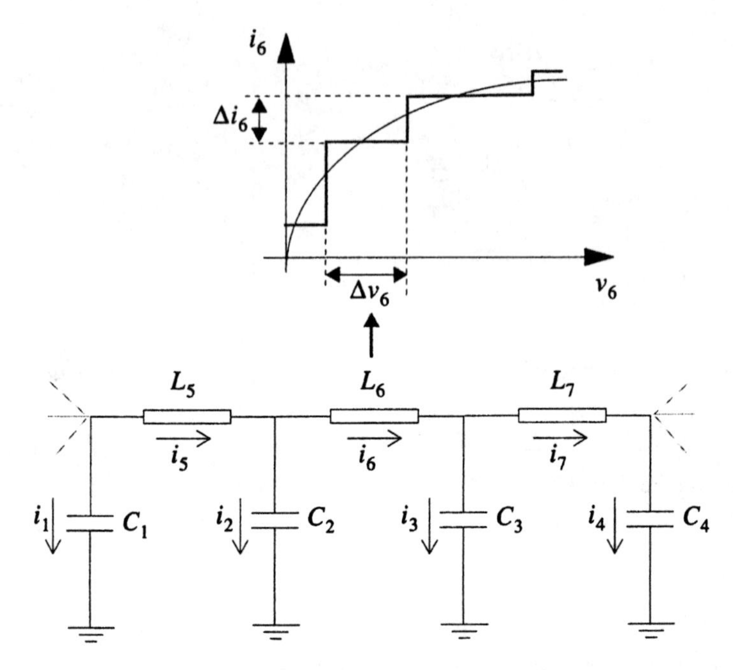

Figure 11.12 Segment of circuit to demonstrate SPECS algorithm.

$$i_2 = i_5 - i_6 \Rightarrow \Delta i_2 = -\Delta i_6 \tag{11.7.3}$$

and

$$i_3 = i_6 - i_7 \Rightarrow \Delta i_3 = \Delta i_6 \tag{11.7.4}$$

No other currents anywhere else in the circuit change, assuming this event is the only one occurring at the present time.

Piecewise constant currents in capacitors cause piecewise linear branch voltages. These voltages distribute via KVL causing *all branch voltages* to be piecewise linear in time. Hence KVL can be applied in terms of voltage slopes, which in turn are piecewise constant in time.

$$\dot{v}_2 = \frac{i_2}{C_2} \Rightarrow \Delta \dot{v}_2 = \frac{-\Delta i_6}{C_2} \tag{11.7.5}$$

$$\dot{v}_3 = \frac{i_3}{C_3} \Rightarrow \Delta \dot{v}_3 = \frac{\Delta i_6}{C_3} \tag{11.7.6}$$

No other capacitor voltage slopes in the circuit change. Applying KVL,

$$\dot{v}_6 = \dot{v}_2 - \dot{v}_3 \Rightarrow \Delta\dot{v}_6 = -\frac{\Delta i_6}{C_2} - \frac{\Delta i_6}{C_3} = -\frac{\Delta i_6}{C_{eq}} \tag{11.7.7}$$

$$\dot{v}_5 = \dot{v}_1 - \dot{v}_2 \Rightarrow \Delta\dot{v}_5 = \frac{\Delta i_6}{C_2} \tag{11.7.8}$$

$$\dot{v}_7 = \dot{v}_3 - \dot{v}_4 \Rightarrow \Delta\dot{v}_7 = \frac{\Delta i_6}{C_3} \tag{11.7.9}$$

where C_{eq} is the series combination of C_2 and C_3. Other than L_5, L_6, and L_7, there is no change in the voltage slope of any other noncapacitor element in the circuit. These slopes allow us to predict the next *event time* of each of L_5, L_6, and L_7 using the simple formula

$$\text{Event time} = \text{Present time} + \frac{\text{Target voltage} - \text{Present voltage}}{\text{Voltage slope}} \tag{11.7.10}$$

The target voltage is determined from the table model. Thus, for L_6,

$$\Delta t_6 = \frac{\Delta v_6}{\dot{v}_6} \tag{11.7.11}$$

is the time to the next event. Hence the event corresponding to L_6 is *scheduled* for a time Δt_6 later as shown in Figure 11.13. L_5 and L_7 are *rescheduled* to reach their target voltages at new event times due to their change in slope, as shown in Figure 11.13. The figure shows L_5's event being rescheduled. Similarly, L_7's event is rescheduled, too.

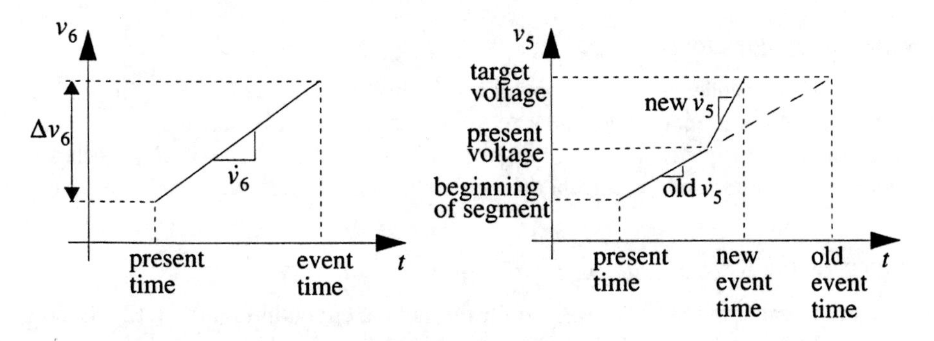

Figure 11.13 Scheduling and rescheduling of events for L_6 and L_5.

An *event queue* is maintained. Time is moved to that of the most imminent event in the queue and that event is processed. As a result, that event is scheduled for a future time and perhaps some other events are rescheduled. The event queue is appropriately juggled to restore time ordering and then the next most imminent event is processed. Event processing is the heart of the simulation process. The equations above show how event processing can be reduced to a small number of arithmetic operations. Thus simplified models (tables) and event-driven algorithms contribute to the efficiency of the timing simulator. Further, although L_5 and L_6 in Figure 11.12 may both be identical devices, they can have table models built with a different number of segments in their table models; hence variable accuracy is possible across portions of the circuit and across runs of the simulator.

Device models for MOS transistors

All piecewise approximate simulators use piecewise functions to approximate device characteristics. In SPECS, the $i-v$ characteristics of MOS transistors are approximated by stepwise constant functions in two dimensions as shown in Figure 11.14. The average current is computed as follows:

$$\bar{I}_{ij} = \left[\frac{1}{v_{gs(i+1)} - v_{gs(i)}}\right]\left[\frac{1}{v_{ds(j+1)} - v_{ds(j)}}\right] \int_{v_{gs(i)}}^{v_{gs(i+1)}} \int_{v_{ds(j)}}^{v_{ds(j+1)}} f(v_{gs}, v_{ds})\, dv_{gs} dv_{ds} \qquad \textbf{(11.7.12)}$$

The channel current is also a function of v_{sb} ("back-gate bias") due to the variation of threshold voltage (also called the "body effect" or "back-gate bias effect"). However, building a three dimensional table would make the table very large. Instead, a two dimensional table is built as shown in (11.7.12) and the effect of the back-gate bias is taken into account dynamically during the simulation. In the simplest case, the MOS parasitic capacitances are modeled as shown in Figure 11.4.

Modeling of independent sources

The voltage of independent sources is approximated in SPECS by a piecewise linear function of time and the current of independent current sources is approximated by a piecewise constant function of time as shown in Figure 11.15. Hence the voltage of *all* branches is piecewise linear in time and the current through *all* branches is piecewise constant in time.

Event processing without loops of Cs

The event processing procedure described for the sample circuit in Figure 11.12 will now be formalized and generalized. In SPECS, the voltage-step of each device is fixed and the unknown is the time necessary for the traversal to the next voltage boundary. Hence event processing is similar to the ELogic concept of finding the next time that a node voltage

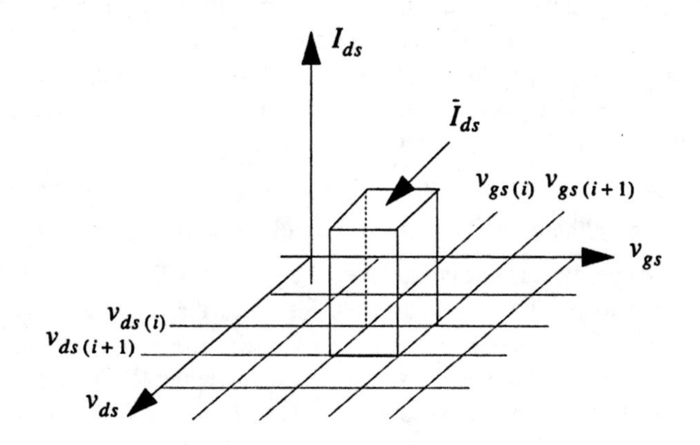

Figure 11.14 Each range of v_{gs} and v_{ds} has a constant current $\bar{I}_{ds}$. The current shown here is only for one subrange of v_{gs} and v_{ds}.

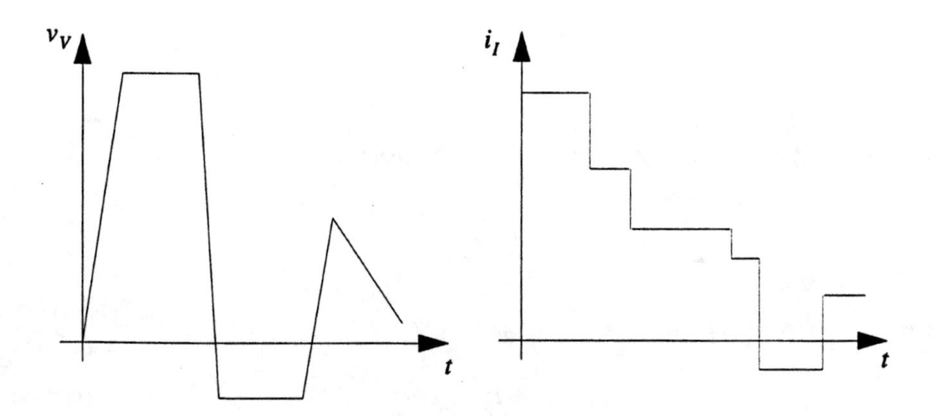

Figure 11.15 Modeling of independent sources.

crosses a threshold, only here we are dealing with branch voltages. This section describes event processing in SPECS with the assumption that all the linear capacitors and independent voltage sources of the circuit form a tree. One way for this assumption to be satisfied is to have a linear capacitor or independent voltage source from each node to ground, and no floating capacitors or voltage sources. Let the circuit contain p independent voltage sources, q linear capacitors, and m cotree links.

Events are usually caused by a (resistive) link moving from one segment of its table model to the next. Events can also be caused by an independent voltage or current source changing its voltage slope or current value, respectively. The list of independent voltage source events is known ahead of time. In fact, the simulation begins by ramping up independent voltage sources to their initial value and stepping up independent current sources to their value at the beginning of the simulation. This method of finding the initial state of a circuit is called a "pseudo-transient analysis" and replaces the dc solution that traditional circuit simulators carry out before starting the transient analysis.

Let us consider an event caused by the jth link having a change of current Δi_{lj}. We will use the subscript t to indicate tree branch values and the subscript l to denote link values. KCL for every fundamental cutset that includes the jth link will dictate how the change of current is distributed to other circuit elements. Since all the links in each cutset have a fixed current dictated by their respective table models, the change of current will be absorbed by the unique tree branch in each of these fundamental cutsets such that KCL continues to be satisfied. We write KCL and KVL in terms of fundamental cutsets and loops, using the $\boldsymbol{F}$ matrix (see section A.5 for details). Each row of the $\boldsymbol{F}$ matrix corresponds to a fundamental cutset and each column a fundamental loop. Writing KCL for the ith tree branch,

$$i_{ti}^{old} = -\sum_{k=1}^{m} \boldsymbol{F}_{ik} i_{lk}^{old} \qquad \textbf{(11.7.13)}$$

where the superscripts old and new are used to denote circuit quantities before and after the processing of the event. Writing KCL after the event,

$$i_{ti}^{new} = -\sum_{k=1}^{m} \boldsymbol{F}_{ik} i_{lk}^{new} \qquad \textbf{(11.7.14)}$$

Subtracting (11.7.14) from (11.7.13), we get

$$\Delta i_{ti} = -F_{ij} \Delta i_{lj} \qquad \textbf{(11.7.15)}$$

(11.7.15) can be applied to compute the new current of every tree branch that includes the jth link in its fundamental cutset. But that set of tree branches constitutes the fundamental loop of the jth link (see Appendix A for details). Hence all the tree branches in the jth link's fundamental loop are assigned a new current and KCL is now satisfied for all cutsets in the circuit.

The next step is to write KVL for the rth link before and after the event:

$$\dot{v}_{lr}^{old} = \sum_{k=1}^{p+q} \boldsymbol{F}_{kr} \dot{v}_{tk} = \sum_{k=1}^{p} \boldsymbol{F}_{kr} \dot{V}_{tk} + \sum_{k=p+1}^{p+q} \boldsymbol{F}_{kr} \frac{i_{tk}^{old}}{C_k} \qquad \textbf{(11.7.16)}$$

$$\dot{v}_{lr}^{new} = \sum_{k=1}^{p} F_{kr}\dot{V}_{tk} + \sum_{k=p+1}^{p+q} F_{kr}\frac{i_{tk}^{new}}{C_k} \tag{11.7.17}$$

Subtracting and substituting from (11.7.15),

$$\Delta\dot{v}_{lr} = \sum_{k=p+1}^{p+q} F_{kr}\frac{\Delta i_{tk}}{C_k} = \sum_{k=p+1}^{p+q} -F_{kr}\frac{F_{kj}\Delta i_{lj}}{C_k} = -\frac{\Delta i_{lj}}{C_{eq\,(r,j)}} \tag{11.7.18}$$

where $C_{eq\,(r,j)}$ is the series equivalent capacitance of the shared capacitances in the fundamental loops of the jth and r^{th} links, taking orientation into account. In particular, for the jth link itself,

$$\frac{1}{C_{eq\,(j,j)}} = \sum_{k=p+1}^{p+q} -\frac{F_{kj}^2}{C_k} = -\sum_{\text{Cs in } j\text{loop}} \frac{1}{C} \tag{11.7.19}$$

which is the series equivalent capacitance of all the capacitors in j's fundamental loop, since the square of every significant element of the $\boldsymbol{F}$ matrix is unity.

The *event time* of every affected link is adjusted in the event queue. The set of affected links is called the *sphere of influence* of the link that had an event. Two links are in each others' sphere of influence if they share at least one capacitance in their fundamental loops. The jth link is scheduled to have its next event when its voltage slope dictates that it will reach the *next* corner in its table model. The links in its sphere of influence will be *rescheduled* to reach their target voltages either earlier or later, as shown in Figure 11.13. In particular, it is possible for a new voltage slope to be zero (in which case the new event time is ∞) and the corresponding event leaves the active event queue. It is also possible that the new voltage slope of a rescheduled link is opposite in sign to the old voltage slope. In that case, the link is rescheduled to return to the start of the segment that it had begun to traverse prior to the event. Note that a change of voltage slope does not affect the next event time of an independent current source, just as the change of current in an independent voltage source does not change its voltage slope.

An event caused by a current source transition is handled in much the same manner as above. The current of each tree branch in the fundamental loop of the source is first updated. Then the new slope of all affected links is computed and events appropriately rescheduled. The source's voltage slope changes, too, but that does not affect any events on the queue.

An event caused by an independent voltage source is a little different. The change in voltage slope of the source results in the change of voltage slope by the same absolute amount in each of the links in the source's fundamental cutset. Each of these links is appropriately rescheduled to take into account its new slope. No currents change in the circuit until the next time that a link has an event and hence a current change.

Thus event processing consists of a few arithmetic operations and the juggling of an event queue, and can be performed efficiently.

Timing error estimation

This section will estimate the timing error incurred due to the piecewise constant approximation in $i - v$ characteristics. Traditional simulators use a local truncation error formula to estimate integration error. In SPECS, however, the actual integration and simulation are exact, whereas the approximation is in the device models. Further, the unknowns we are dealing with are event times, not voltages or currents. Hence we need to develop an error criterion that relates the timing error to the device modeling error.

Consider a link l. A portion of its table model is shown in Figure 11.16. The dark solid line shows the table with one segment in the table model for the voltage range $v_0 \le v < v_f$. The heavy dotted line shows the table if the same voltage range were represented by two voltage segments $[v_0, v_i\ \}$ and $[v_i, v_f\}$. Let the three table model currents be

$$I_1 = \frac{1}{(v_i - v_0)} \int_{v_0}^{v_i} f(v)\,dv$$
$$I_2 = \frac{1}{(v_f - v_0)} \int_{v_0}^{v_f} f(v)\,dv \qquad (11.7.20)$$
$$I_3 = \frac{1}{(v_f - v_i)} \int_{v_i}^{v_f} f(v)\,dv$$

where $f(v)$ represents the true $i - v$ characteristics of l. Let $\Delta i_1 \equiv I_2 - I_1$ and $\Delta i_2 \equiv I_3 - I_2$. Let the slope of the voltage across the link l be s in its transition from v_0 to v_f with the current I_2. We compute the difference in timing Δt for the traversal from v_0 to v_f between a one-segment model and a two-segment model. As the number of segments approaches infinity, the timing error will monotonically decrease to zero. Assume that the transition is completed with no change of slope during the traversal. Obviously the two-segment model is more accurate.

If the voltage slope across l is s on the current segment I_2, then the voltage slope on the segment with current I_1 is $(s + \Delta i_1 / C_{eq})$ and the slope in the segment corresponding to I_3 is $(s - \Delta i_1 / C_{eq})$, where C_{eq} is the series combination of all the capacitors in the fundamental loop of the link l. These voltage slopes are derived using (11.7.18). Now we can write an expression for the timing error Δt:

$$\Delta t = \left(\frac{v_i - v_0}{s + \Delta i_1 / C_{eq}} + \frac{v_f - v_i}{s - \Delta i_2 / C_{eq}} \right) - \left(\frac{v_f - v_0}{s}\right) \qquad (11.7.21)$$

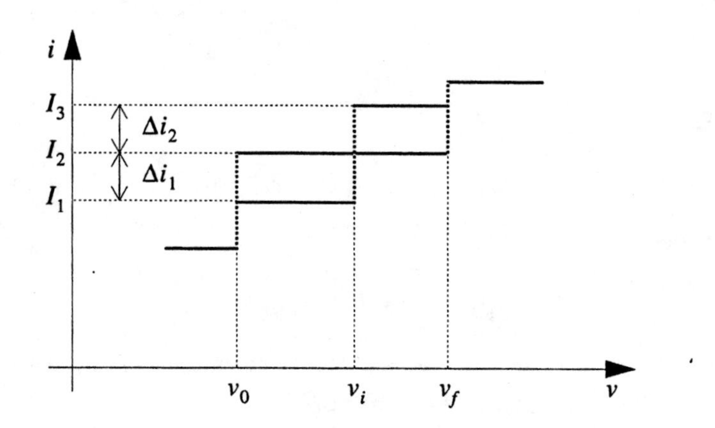

Figure 11.16 Relative timing error between a one- and two-segment model.

or

$$\Delta t = \frac{1}{s}\left(\frac{v_i - v_0}{1 + \frac{\Delta i_1}{sC_{eq}}} + \frac{v_f - v_i}{1 - \frac{\Delta i_2}{sC_{eq}}} - v_f + v_0\right) \tag{11.7.22}$$

Expanding the first two terms into a series valid for small $\Delta i_1 / sC_{eq}$ and $\Delta i_2 / sC_{eq}$, we obtain

$$\Delta t = \frac{1}{s}\Bigg[\{v_i - v_0\}\{1 - \frac{\Delta i_1}{sC_{eq}} + \frac{\Delta i_1^2}{s^2C_{eq}^2} - \ldots\} + \{v_f - v_i\}\{1 + \frac{\Delta i_2}{sC_{eq}} + \frac{\Delta i_2^2}{s^2C_{eq}^2} + \ldots\} - v_f + v_0\Bigg] \tag{11.7.23}$$

Collecting the constant terms, we have

$$\Delta t = \frac{1}{s}(v_i - v_0 + v_f - v_i - v_f + v_0) + O(\Delta i_1^1, \Delta i_2^1) = O(\Delta i_1^1, \Delta i_2^1) \tag{11.7.24}$$

Thus the constant terms cancel out. Collecting the coefficients of Δi_1 and Δi_2 we have

$$\Delta t = \frac{1}{s}\left[-(v_i - v_0)\left(\frac{\Delta i_1}{sC_{eq}}\right) + (v_f - v_i)\left(\frac{\Delta i_2}{sC_{eq}}\right)\right] + O(\Delta i_1^2, \Delta i_2^2) \tag{11.7.25}$$

From (11.7.20), we obtain

$$I_2 (v_f - v_0) = I_1 (v_i - v_0) + I_3 (v_f - v_i) \tag{11.7.26}$$

Substituting for $(v_f - v_0)$ we have

$$I_2 [(v_f - v_i) + (v_i - v_0)] = I_1 (v_i - v_0) + I_3 (v_f - v_i) \tag{11.7.27}$$

Rearranging terms,

$$(v_f - v_i) (\Delta i_2) = (v_i - v_0) (\Delta i_1) \tag{11.7.28}$$

Combining (11.7.25) and (11.7.28) we find that the coefficients of Δi_1 and Δi_2 are zero. The *relative timing error*, $\Delta t / t$ is of the order of

$$\frac{\left(\frac{\Delta v \Delta i^2}{s^3 C_{eq}^2}\right)}{\left(\frac{\Delta v}{s}\right)} = \left(\frac{\Delta i}{s C_{eq}}\right)^2 \tag{11.7.29}$$

Hence, the relative timing error is second order in Δi, and inversely proportional to the square of the transient voltage slope across the device. In digital circuits the timing of transitions from a high to a low voltage and vice versa are very important. Transitions are periods of high slope, so this algorithm produces relatively accurate timing results. On the other hand, when steady state is approached, the voltage slope is relatively low and hence the relative timing error is large. This error manifests itself as a *steady state level error.* Since timing information is more important than level information in most digital circuits, this algorithm is well suited to digital timing verification.

Steady state algorithm

Due to the discretization in device models, there is the possibility of oscillation when steady state is approached because a device can jump back and forth between two current levels in its table model. To deal with this problem, a special steady state algorithm is used. The basic principle of the steady state algorithm is to *never let a device have an event that instantaneously changes the sign of its voltage slope.* Since steady state is defined as the condition of zero voltage slope, no device can cross its steady state. The above principle ensures the stability of the integration (see section 5.4 for a discussion of stability of integration formulas), and is illustrated below by means of an example.

Consider a link l traversing the segment BC in its table model with a positive slope as shown in Figure 11.17. When it reaches the point C, l has an event and its current is updated to the value E (looked up from the table model). If its new voltage slope is posi-

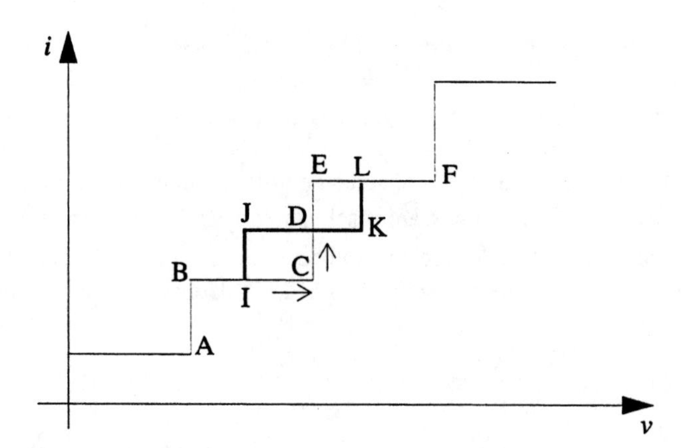

Figure 11.17 Example to demonstrate the steady state algorithm.

tive, l is assigned a new event time to reach F and event processing continues as usual. However, if the new voltage slope is negative, it means that the increase of current corresponding to CE is too much. Then an "in-between" current level D is computed at which the voltage slope of the link l is exactly zero. This current is chosen to be l's new current and l's event leaves the active event queue since its voltage slope is zero. If any link in l's sphere of influence subsequently has an event, l's voltage slope will cease to be identically zero. At that time a *pseudo-segment* JK is created by interpolation and l is scheduled to reach K or J, depending on whether the voltage slope after rescheduling is positive or negative, respectively. If, for example, the new voltage slope is positive, an event occurs when l reaches the voltage K. At that time the current is updated to the value L. Again, if there is a change in sign of voltage slope, an in-between current is computed. If not, l continues along the segment LF. Note that only one pseudo-segment is required at any given time.

In the case of MOS transistors, the above algorithm can easily be extended to two dimensions. Instead of a pseudo-segment, a *pseudo-square* is created in the $v_{gs} - v_{ds}$ plane when necessary.

Event processing in the presence of loops of Cs

The scalar equations that we solved to determine event times in the absence of floating capacitors get replaced by small sets of matrix equations that must be solved simultaneously when the circuit has floating capacitors. The basic idea behind event processing is the same, but the update equations are now coupled. Details of the equation formulation and solution can be found in [Visweswariah93].

Modeling of nonlinear capacitors

The nonlinear capacitance model shown in Figure 11.5 is used to create a piecewise constant $C-v$ model for each nonlinear capacitor. An event is associated with each nonlinear capacitance. Whenever the capacitance reaches a corner in its $C-v$ table model, the capacitance has an event. At that time, the capacitance is updated to its new value from the table. Every link that has the updated capacitance in its fundamental loop has a new voltage slope and a new C_{eq}. Such links are rescheduled based on their new voltage slopes and simulation continues. In addition to nonlinear capacitors, ECL circuits containing bipolar transistors have been incorporated into SPECS using the simple Ebers-Moll BJT model [Visweswariah91b].

Sensitivity in SPECS

The simple device models and event-driven nature of SPECS make the computation of transient sensitivity extremely efficient by both the adjoint and direct methods [Nguyen89, Feldmann91]. For direct sensitivity (see section 9.1), the sensitivity element for a two-terminal link is obtained by directly differentiating the BCR with respect to the parameter of interest p:

$$\begin{aligned} i &= f(v,p) \\ \frac{di}{dp} &= \frac{\partial f}{\partial v}\frac{dv}{dp} + \frac{\partial f}{\partial p} \\ \varphi &= \frac{\partial f}{\partial v}\psi + \frac{\partial f}{\partial p} \end{aligned} \tag{11.7.30}$$

Hence the sensitivity circuit element consists of a current source of value $\frac{\partial f}{\partial p}$ in series with a conductance of value $\frac{\partial f}{\partial v}$. Since the original element is a current source with step changes in current at the voltage segment boundaries, the sensitivity element is a zero-valued conductance (open circuit), except at the voltage boundaries where it is a train of Dirac impulse functions of height equal to the height of the corresponding step in the original $i-v$ characteristics, as shown in Figure 11.18. In the adjoint method, the contribution of a two-terminal link to the sensitivity relation (9.5.10) is

$$\varphi\delta v - \psi\delta i = \varphi\delta v - \psi\left(\frac{\partial f}{\partial v}\delta v + \frac{\partial f}{\partial p}\delta p\right) \tag{11.7.31}$$

Choosing the BCR for the adjoint circuit as $\varphi = \psi\frac{\partial f}{\partial v}$, the sensitivity relation becomes

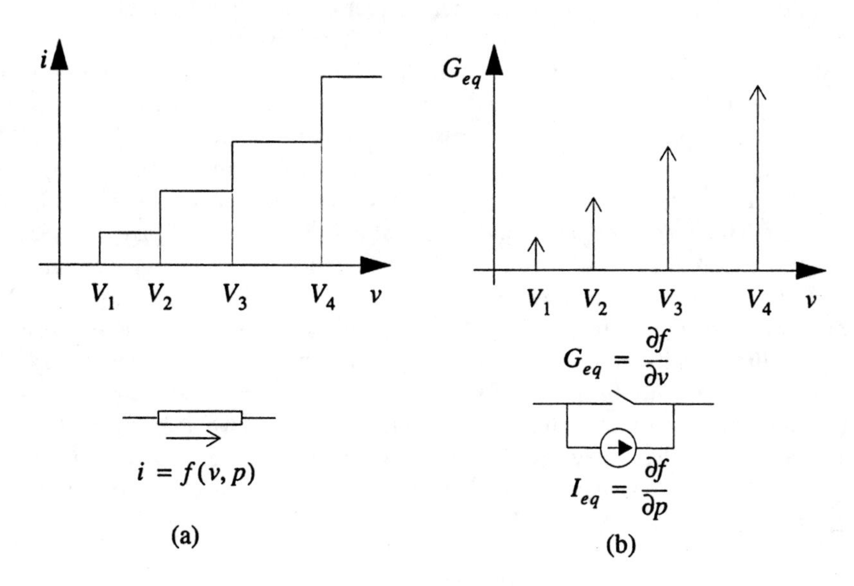

Figure 11.18 (a) Two-terminal element in SPECS and (b) the corresponding sensitivity circuit element.

$$\frac{\partial F}{\partial p} = -\psi \frac{\partial f}{\partial p} \tag{11.7.32}$$

where F is the sensitivity function of interest. Hence the adjoint circuit element is simply a conductance with a series of impulses at voltage segment boundaries.

The sensitivity and adjoint circuit elements for capacitors are the same as those derived in Chapter 9. The simulation of the sensitivity and adjoint circuits is particularly straightforward. The associated circuit in either case consists of disjoint capacitors to ground from each node. At each event time of the original circuit, there is an impulse of current that flows from one capacitor to another, causing an instantaneous change in a pair of capacitor voltages due to charge redistribution. Hence the associated circuit can be solved at a much lower cost than the original circuit. Finally, sensitivity results are obtained by combining the waveforms of the original and associated circuit. In the case of adjoint sensitivity, a convolution between the two sets of waveforms is required to obtain the required sensitivities.

SPECS summary

Simplified device models and event-driven methods give SPECS a speedup of one to two orders of magnitude over classical methods. With the grounded capacitor model shown in Figure 11.4, most digital circuit delays can be predicted to within 5 percent of the correct delays. SPECS can be used to simulate much larger circuits than traditional simulators, and transient sensitivity analysis can be performed efficiently. However, new device models must be derived from the standard available (SPICE-like, analytical) device models; the SPECS algorithm is based on special purpose device models that have to be created and tuned by means of an automated optimization process. Floating capacitors (and thus loops of capacitors) cause a loss of efficiency. In particular, if there are floating capacitors all over the circuit, then every link is coupled to every other link and much of the efficiency is lost. The integration technique used is akin to Forward Euler and hence suffers from requiring extremely small time steps in the presence of *stiff circuits*. Circuits with widely differing time constants are said to be stiff. In particular, circuits derived by extracting transistors and parasitics from layout information can have many tiny resistors in them, resulting in stiff circuits and severe degradation of simulation efficiency. Finally, inductors do not naturally fit into the formulation since piecewise constant currents and piecewise linear voltages are just the opposite of the situation in which inductors would be comfortable. Nevertheless, they can be accommodated by treating inductors as "current-based" and associating events with inductors whenever they cross predetermined current thresholds [Visweswariah90].

Adaptively Controlled Explicit Simulation (ACES)

Explicit integration methods like Forward Euler are efficient, but not stable. We saw in section 5.4 that in order for Forward Euler integration to be stable, the time step is limited by the smallest time constant of the circuit. Thus for stiff circuits (circuits with a large variation in time constants) explicit integration methods can be inefficient. This behavior is also seen in timing simulators that use explicit integration, like SPECS. Adaptively Controlled Explicit Simulation [Devgan94] seeks to render explicit simulation stable by adaptively controlling the *update derivative*, and then takes advantage of the explicit integration method in an efficient timing simulation algorithm.

The essence of ACES is that the Forward Euler formula

$$x(t+\Delta t) = x(t) + \Delta t \dot{x}(t) \tag{11.7.33}$$

is replaced by

$$x(t+\Delta t) = x(t) + \Delta t \dot{x}^{+}(t) \tag{11.7.34}$$

In (11.7.33), Δt is adjusted to keep the integration formula stable. In (11.7.34), the update derivative $\dot{x}^{+}(t)$ is adaptively controlled to render the integration stable. Each state variable of the circuit is considered to be *quiescent* if its time derivative is zero. Once a state variable is quiescent, its update derivative is adjusted to retain the state in quiescence, hence $\dot{x}(t) \neq \dot{x}^{+}(t)$. For a non-quiescent state variable, there is no need to adjust the update derivative and hence $\dot{x}^{+}(t) = \dot{x}(t)$. To begin, all state variables are non-quiescent. The time step required to place each variable into quiescence is computed and time is moved forward to the smallest of these time steps. The state variable with the smallest time step is then placed in quiescence and retained in quiescence from that time on by adaptively adjusting its update derivative. Then the next state variable in placed in quiescence, and so on, until all state variables are quiescent, and steady state is reached. Of course, a primary input to the circuit may change, causing state variables to come out of the quiescent state. This method is illustrated below by means of an example.

Consider the RC circuit shown in Figure 11.19. Assume a step input. The state equations for this circuit can be written as

$$\begin{aligned} \dot{v}_1 &= -2v_1 + v_2 + u \\ \dot{v}_2 &= v_1 - v_2 \end{aligned} \tag{11.7.35}$$

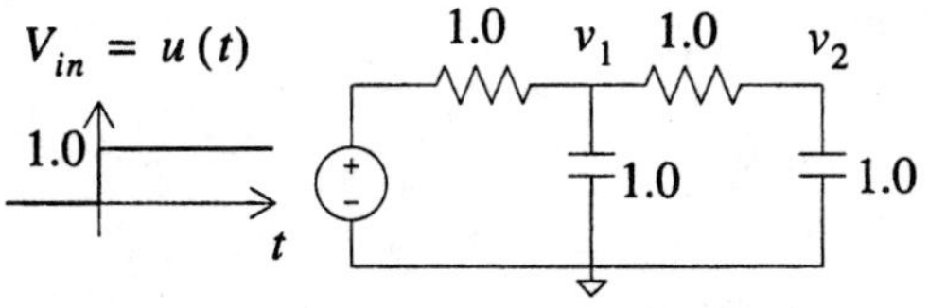

Figure 11.19 Sample circuit to demonstrate ACES integration.

Consider a time step of Δt_1 for v_1 and Δt_2 for v_2. Integrate the two state variables using (11.7.34) with such a time step as to achieve quiescence. Then

$$\dot{v}_1(\Delta t_1) = -2v_1(\Delta t_1) + v_2(\Delta t_1) + 1 \tag{11.7.36}$$

Using the Forward Euler approximation in (11.7.34) and solving for the time at which quiescence is reached,

$$0 = -2v_1(0) - 2\Delta t_1 \dot{v}_1^{+}(0) + v_2(0) + \Delta t_1 \dot{v}_2^{+}(0) + 1 \tag{11.7.37}$$

Initially, $v_1(0) = v_2(0) = 0$ and $\dot{v}_1(0) = \dot{v}_1^{+}(0) = 1$ and $\dot{v}_2(0) = \dot{v}_2^{+}(0) = 0$ since neither v_1 nor v_2 is in quiescence. Hence $\Delta t_1 = 0.5$.

Carrying out a similar analysis for v_2, we have

$$\begin{aligned} \dot{v}(\Delta t_2) &= v_1(\Delta t_2) - v_2(\Delta t_2) \\ 0 &= v_1(0) + \Delta t_2 \dot{v}_1^{+}(0) - v_2(0) - \Delta t_2 \dot{v}_2^{+}(0) \end{aligned} \tag{11.7.38}$$

which cannot be satisfied for any value of Δt_2 indicating that v_2 cannot be placed in quiescence. So now we choose as our time step $\Delta t_1 = 0.5$ (the smallest feasible time step) and proceed to move time to 0.5. Note that at time 0.5, $\dot{v}_1(0.5) = 0$, $v_1(0.5) = 0.5$, $\dot{v}_2^+(0.5) = \dot{v}_2(0.5) = 0.5$, $v_2(0.5) = 0.0$, and $\dot{v}_1^+(0.5)$ is unknown as shown in Figure 11.20.

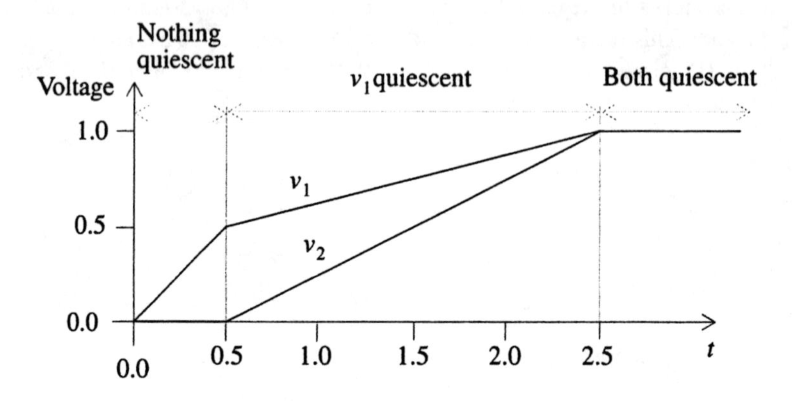

Figure 11.20 Waveforms for sample circuit in Figure 11.19.

Now that v_1 is in quiescence, the next step is to find the time steps $\Delta\hat{t}_1$ and $\Delta\hat{t}_2$ that place v_2 in quiescence, while retaining v_1 in quiescence. We have the following equations for v_1:

$$\begin{aligned} \dot{v}_1(0.5+\Delta\hat{t}_1) &= -2v_1(0.5+\Delta\hat{t}_1) + v_2(0.5+\Delta\hat{t}_1) + 1 \\ 0 &= -2v_1(0.5) - 2\Delta\hat{t}_1\dot{v}_1^+(0.5) + v_2(0.5) + \Delta\hat{t}_1\dot{v}_2^+(0.5) + 1 \end{aligned} \tag{11.7.39}$$

Thus $\Delta\hat{t}_1$ drops out of the equation and we get

$$\dot{v}_1^+(0.5) = 0.25 \tag{11.7.40}$$

For the other state variable

$$\begin{aligned} \dot{v}_2(0.5+\Delta\hat{t}_2) &= v_1(0.5+\Delta\hat{t}_2) - v_2(0.5+\Delta\hat{t}_2) \\ 0 &= v_1(0.5) + \Delta\hat{t}_2\dot{v}_1^+(0.5) - v_2(0.5) - \Delta\hat{t}_2\dot{v}_2^+(0.5) \\ 0 &= 0.5 + \Delta\hat{t}_2(0.25) - 0.0 - \Delta\hat{t}_2(0.5) \end{aligned} \tag{11.7.41}$$

yielding $\Delta\hat{t}_2 = 2.0$ and the final waveforms shown in Figure 11.20. Now both state variables are in quiescence and the simulation is completed, until one of the inputs changes.

The key idea in the above integration method is to place into quiescence the states with the fastest time constants and then adaptively adjust their update derivatives to keep them in quiescence. If there are n states in quiescence and m states not in quiescence, then we would solve $n+m$ equations with n unknown update derivatives to keep the quiescent states in quiescence and m unknown time steps for each of the others to reach quiescence. Of course, the smallest of these time steps is selected. A local truncation error can be derived for this integration method, quite similar to the truncation error for Forward Euler integration.

This integration method has been successfully implemented in a simulator [Devgan94] called ACES. In ACES, the circuit is first partitioned into subcircuits to exploit spatial latency. Devices are modeled with piecewise linear characteristics in the $i-v$ plane, thus permitting variable accuracy by using coarser or finer piecewise linear segments. Adaptively controlled explicit integration is used in each partition. For each partition, the time step is chosen to be the smallest from the following three sets:

- the set of time steps to place each non-quiescent state variable into quiescence
- the set of time steps that would get each nonlinear device to the next boundary in its piecewise linear model
- the set of time steps at which primary inputs of the subcircuit change, including fanout signals from other partitions.

Further, the time step may have to be decreased to satisfy a truncation error check.

Thus ACES uses Forward Euler integration with adaptive control, partitioning, piecewise linear models, and event-driven techniques to implement an accurate, general, and variable accuracy timing simulator.

11.8 Relaxation Methods in Circuit Simulation

This section will describe a class of iterative methods in circuit simulation known as *relaxation methods*. Relaxation can be applied at various levels, including linear algebraic equation level, nonlinear algebraic equation level and nonlinear differential equation level. This section will briefly explain some relaxation-based techniques for circuit simulation.

Waveform Relaxation (WR)

Traditional circuit simulation consists of three nested loops: integration, linearization and sparse solution. The asymptotic order of complexity is superlinear in the size of the circuit. Waveform relaxation [Lelarasmee82, White83, Newton84] attempts to *partition* the circuit into subcircuits. Subcircuits are repeatedly solved and the solutions pieced together by *Waveform Relaxation* (WR). The convergence of the overall solution is linear and the basic idea is that repeated inexpensive solutions of small partitions is more efficient than one solution of a large circuit. A by-product of partitioning is that each subcircuit can have its own time step depending on its activity pattern. Hence latency can be maximally exploited when combined with partitioning.

Consider a circuit represented by the following equations.

$$\boldsymbol{C}(\boldsymbol{v}, \boldsymbol{u})\dot{\boldsymbol{v}} + \boldsymbol{f}(\boldsymbol{v}, \boldsymbol{u}) = \boldsymbol{0} \qquad \text{and} \qquad \boldsymbol{v}(0) = \boldsymbol{V} \tag{11.8.1}$$

where $\boldsymbol{C}$ is the $n \times n$ symmetric diagonally dominant nodal capacitance matrix with

$$C_{ii}(\boldsymbol{v}, \boldsymbol{u}) = \text{sum of capacitances incident on node } i$$
$$C_{ij}(\boldsymbol{v}, \boldsymbol{u}) = -\ \text{total floating capacitance between nodes } i \text{ and } j.$$

$\boldsymbol{v}$ is an $n \times 1$ vector of node voltages and $\dot{\boldsymbol{v}}$ its time derivative. $\boldsymbol{u}$ is an $r \times 1$ vector of primary inputs and $\boldsymbol{V}$ the vector of initial conditions at time $t = 0$. $\boldsymbol{f}$ is a vector function of the net current at each node due to resistive elements (transistors, resistors, diodes, current sources, controlled sources, etc.). Let us assume for simplicity that we partition this system of equations into subcircuits comprising one single node each. Then the steps involved in the waveform relaxation algorithm are as follows.

1. Set the WR iteration count $k = 0$. The iteration count will be indicated by a superscript.
2. Guess a waveform $v^0(t)$ for $t \in [0, T]$ such that $\boldsymbol{v}^0(0) = \boldsymbol{V}$, where T is the period of simulation. It is easiest to choose $\boldsymbol{v}^0(t) = \boldsymbol{V}$ for all time.
3. Increment the iteration count k and solve the following equation for $v_i^k(t)$ for all time for $1 \le i \le n$:

$$\begin{aligned}
&\sum_{j=1}^{i} C_{ij}(v_1^k, \ldots, v_i^k, v_{i+1}^{k-1}, \ldots, v_N^{k-1}, \boldsymbol{u})\,\dot{v}_j^k \\
&+ \sum_{j=i+1}^{N} C_{ij}(v_1^k, \ldots, v_i^k, v_{i+1}^{k-1}, \ldots, v_N^{k-1}, \boldsymbol{u})\,\dot{v}_j^{k-1} \\
&+ f_i(v_1^k, \ldots, v_i^k, v_{i+1}^{k-1}, \ldots, v_N^{k-1}, \boldsymbol{u}) = 0
\end{aligned} \tag{11.8.2}$$

Note that in solving the equations, we are taking the *present iteration* value for the voltages from 1 to i, but the *previous iteration* values for the voltages from $i+1$ to N. Hence when we solve for v_1, we use all previous iteration values and when we solve for v_N, we use all present iteration values. Mathematically, the above procedure is called Gauss-Seidel (GS) relaxation.

4. Repeat step 3 until

$$\text{Max}_{1 \le i \le n} \text{Max}_{\text{for all } t} \left| v_i^k(t) - v_i^{k-1}(t) \right| \le \varepsilon \tag{11.8.3}$$

where ε is a convergence tolerance.

To state the above steps in words: start by guessing a voltage for every node for all time. Solve for the first node for all time using the guess for the rest of the voltages and treating them all as known values. Then solve for the second node's voltage for all time using the just-obtained solution for the first node and using the initial guess for the remaining nodes. Continue in this manner until all the nodes have been solved once. That concludes the first WR iteration. Repeat the procedure for as many iterations as necessary. When *all* waveforms stop changing for *all* time, as defined by a tolerance factor, terminate the process.

Is convergence to a unique answer guaranteed? Can we have partitions that are larger than a single node? How many iterations does it take to converge to an answer? How important is the initial guess? Does the ordering of the nodes matter? These are all legitimate questions and they are addressed below.

It can be shown that under some fairly mild assumptions, the fact of convergence is guaranteed [Lelarasmee82]. The mild assumptions involve requiring nonzero $\boldsymbol{C}$ matrix diagonal entries (i.e., a capacitance to ground from every node) and super-smooth (Lipschitz continuous) $i - v$ and $q - v$ characteristics for all elements. The fact of convergence is not affected by the initial guess or the order of processing of the equations.

There is no reason why each equation in (11.8.2) cannot be replaced by a set of equations corresponding to a subcircuit with more than one node. In practice, subcircuits of several to several hundred transistors are formed.

The number of iterations necessary for convergence, on the other hand, depends on a number of factors. Reduction of the number of waveform relaxation iterations and/or the cost of each iteration is attempted in a number of ways:

- Feedback across subcircuits increases the number of iterations necessary for convergence. Hence strongly coupled circuits and circuits with feedback are lumped into a single subcircuit wherever possible. A straightforward method to accomplish this result is to use Strongly Connected Components (SCCs) as described in our discussion of switch-level simulation (see Figure 11.3).
- Subcircuits are *scheduled* in the direction of information flow to the extent possible. Hence if an input signal of subcircuit 2 is an output of subcircuit 1, it makes sense to process subcircuit 1 before subcircuit 2. If the circuit has global feedback, it is not possible to schedule in this fashion, and the feedback loop has to be "broken" during the scheduling process.
- Long simulation times cause an increase in the number of iterations required for convergence. The concept of *time windows* is introduced to solve this situation. The

total period of simulation is divided into time windows and each time window is solved as an independent simulation problem. The initial conditions for each time window are obtained from the final values of the previous time window. There is a danger with carrying windowing too far. At each window boundary, all the waveforms of all the subcircuits have to be synchronized. So the exploitation of latency or multirate behavior of the circuit is temporarily lost. Time windows, however, generally accelerate convergence, reduce the storage requirements and reduce the number of iterations required.

The simple procedure that we outlined above involves one time window and three nested loops for iteration count, subcircuit number, and time. The actual order in which these tasks are performed can be varied and is called a *relaxation schedule*.

- If one set of subcircuits with global feedback requires a large number of iterations to achieve convergence, the extra iterations may not be necessary for all subcircuits. If the inputs and loading of a subcircuit have not changed from the previous iteration, that subcircuit is said to be *partially converged* and its processing is skipped. The concept of partial convergence can also be applied to subintervals of time. For example, the first 10ns of time may have converged for a given subcircuit, so that part of the simulation can be skipped.
- A good initial guess reduces the number of iterations required. The results of a logic simulation, for example, can be used as an initial guess.
- The first few WR iterations produce results that need further refinement, so there is no point in computing them precisely. Hence an *adaptive error control scheme* can be used, with larger tolerances for the solution of equations during initial iterations. As global convergence is approached, these tolerances can be tightened to meet the required accuracy.
- *Successive over-relaxation* (SOR) is used to accelerate the convergence of the WR iterations. In an SOR scheme, the equation used for applying updates between WR iterations is

$$v^k_{\text{applied}} = v^{k-1} + \alpha(v^k - v^{k-1}) \tag{11.8.4}$$

where α is the overrelaxation factor, and the superscript denotes the iteration count. Hence every time there is an update of any variable v, the change since the last iteration is exaggerated by a factor α. If the movement towards the solution is monotonic, overrelaxation helps reach the actual solution quicker.

A number of variations on WR have been tried and are briefly described here for completeness.

- *Gauss-Jacobi relaxation:* Rewrite (11.8.2) as

$$\sum_{j=1, j \neq i}^{N} C_{ij}(v_1^{k-1}, v_2^{k-1}, \ldots, v_{i-1}^{k-1}, v_i^k, v_{i+1}^{k-1}, \ldots, v_N^{k-1}, u)\, \dot{v}_j^{k-1}$$
$$+ C_{ii}(v_1^{k-1}, v_2^{k-1}, \ldots, v_{i-1}^{k-1}, v_i^k, v_{i+1}^{k-1}, \ldots, v_N^{k-1}, u)\, \dot{v}_i^k \tag{11.8.5}$$
$$+ f_i(v_1^{k-1}, v_2^{k-1}, \ldots, v_{i-1}^{k-1}, v_i^k, v_{i+1}^{k-1}, \ldots, v_N^{k-1}, u) = 0$$

for the solution of $v_i^k(t)$ for all time. The change here is that we are using the *previous iteration* value for all node voltages except for the one being solved. This type of relaxation is called Gauss-Jacobi (GJ) relaxation. GJ relaxation converges under the same conditions as GS relaxation and the convergence is linear, but the number of iterations required for convergence is typically larger. Also, two versions of each waveform need to be stored for GJ relaxation. The advantage of GJ relaxation over GS relaxation is that all the subcircuits can be solved in parallel since they have a dependency only on the previous iteration solutions. Hence GJ relaxation is a good candidate for implementation on parallel processors. Note that the ordering of equations in GJ is immaterial since all solutions are carried out with previous iteration values, even if the present iteration solutions are available.

- *Overlapped waveform relaxation:* One of the reasons that WR needs many iterations to converge is that when a subcircuit is solved, the loading from its fanouts is not accounted for correctly. The number of iterations is sensitive to coupling and feedback between subcircuits. The global WR procedure is used to communicate this coupling information during subsequent iterations. Instead, overlapped waveform relaxation [Mokari-Bolhassan85] includes one stage of fanout in each subcircuit. Thus each subcircuit is solved including its immediate fanouts. Then the waveforms of the fanouts are discarded. The extra time spent solving slightly larger partitions pays dividends in the form of fewer global WR iterations required for convergence. An overlap of "two inverting stages" has been proposed as an optimal amount of overlap.
- *Segmented waveform relaxation:* This form of waveform relaxation is applicable only to synchronous, clocked circuits and addresses the problem of global feedback. Consider a circuit operated with a two phase non-overlapping clock scheme as shown in Figure 11.21. The clock period is divided into four time segments as shown in the figure. Then the standard WR algorithm is used, with the time segments acting as time windows. The difference here is that the partitioning is different for each time segment [Dumlugol86]. If the circuit being simulated has global feedback, the feedback loop is broken based on the clocking scheme of the subcircuits involved in the feedback. For example, with phase ϕ_1 on and ϕ_2 off, the feedback loop may be broken at a certain point because of the way the circuit is clocked. In the other phase of the clocks, the loop could be broken in some other way. So in any given time segment, there is no global feedback.

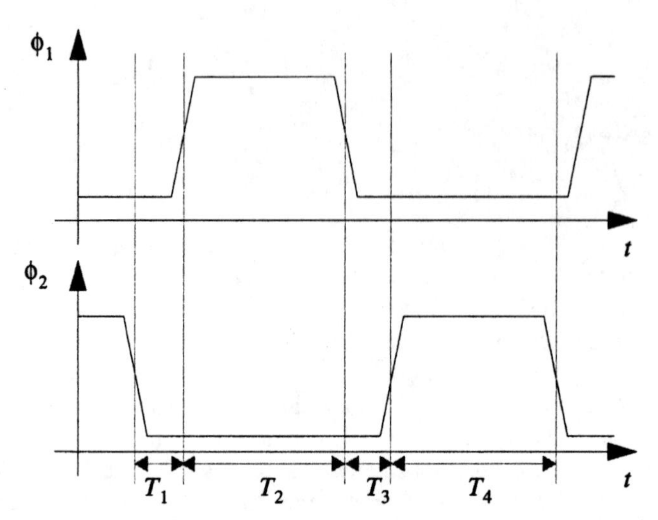

Figure 11.21 Time segments in a two-phase clocking scheme.

Simulation is now simple. Any time segment of a subcircuit S can be scheduled for processing provided (a) all inputs of S have converged in the ith WR iteration of that time segment, (b) all outputs of S have converged in the ith *or* $(i-1)^{st}$ WR iteration of that time segment, and (c) S has converged in the previous time segment. In fact, a parallel processor version can assign subcircuit and time-segment pairs that are ready for simulation to the next available process. Although an elegant method for breaking feedback loops, the applicability of segmented waveform relaxation is limited to digital synchronous circuits.

Waveform relaxation in summary

The main advantage of waveform relaxation is a speedup over conventional simulators with no loss in accuracy. It is true that waveform relaxation results in no loss of accuracy provided the same device models are used. However, in practice, the speedup is sensitive to the partitioning algorithm and the implementation of windowing and partial convergence. Because of the sensitivity to good partitioning, waveform relaxation is only applied to digital MOS circuits. In that realm, there are many prototype implementations of WR and parallel WR reported in the literature that show promising speedups over conventional methods.

One Step Relaxation (OSR)

One of the problems in WR is that circuits with global feedback need many iterations to converge. Time windowing in WR helps reduce the number of relaxation iterations required for convergence. The basic flow in waveform relaxation is:

```
Find an initial guess for all waveforms for all time;
For w = 1, 2, ... until end of simulation time /* time window */
    For i = 1, 2, ... until convergence /* relax iterations */
        For p = 1, 2, ... number of partitions /* partitions */
            While t < T_window /* discretized time */
                Integrate circuit equations
                For NR = 1, 2, ... until convergence /* N-R iterations */
                    Solve linear equations
                End
            End
        End
    End
End
```

Hence windowing "steals" time out of the inner integration loop and brings it out to the outermost loop. *One Step Relaxation (OSR)* [Hennion85] carries the concept of time windowing to the limit. Each time step is made into a window and relaxation is carried to convergence within that window. The basic flow for comparison is as follows:

```
Find an initial guess for all waveforms for all time;
While t < T_simulation /* for all time */
    For i = 1, 2, ... until convergence /* relax iterations */
        For p = 1, 2, ... number of partitions /* partitions */
            Integrate circuit equations
            For NR = 1, 2, ... until convergence /* Newton-Raphson iteration */
                Solve linear equations
            End
        End
        Compute a predictive value
    End
End
```

Note that the predictive value is a prediction for the zeroth relax iteration of the next time step for all subcircuits.

One problem with OSR is that the entire circuit is confined to using the same time step, which is not a good way to exploit latency. Activity in any small part of the circuit will force small time steps for the entire circuit. Further, partial convergence cannot be used to skip re-solving some equations. Global feedback, however, is less of a problem than in waveform relaxation. Also, the memory requirements of OSR are not as prohibitive as WR since waveforms can be written to disk as they are computed.

Iterated Timing Analysis (ITA)

ITA [Newton84] makes three major modifications over OSR. First, in OSR, the Newton-Raphson iteration in the innermost loop is carried forward to convergence. In ITA, however, only one Newton-Raphson iteration is performed. The rationale is that since we carry the relaxation to convergence at each time point, there is no need to solve the nonlinear equations exactly at each relaxation iteration. The second modification is that ITA uses an event-driven approach to recompute only those partitions that have activity during any particular time step. An event queue is maintained. The most imminent event from the event queue is picked and the appropriate partition is solved for its present time step using just one Newton Raphson iteration. If it is converged, it is scheduled for a future time. If not, it is scheduled for the next relaxation iteration at the same time. The process is repeated until the event queue for the present time is empty. The fanouts of any partitions that have changed are also scheduled at the present time. When full convergence is reached at the present time, time is advanced to the next event from the queue and the process is repeated. Finally, ITA usually uses successive over relaxation (SOR) to achieve quick convergence.

In summary, ITA improves its efficiency, both by reducing the total number of Newton-Raphson iterations required and by using an event-driven scheme to exploit latency.

11.9 Conclusions

Many alternatives to detailed circuit simulation are available. Simulation can be performed at different levels like the switch-level, gate-level, and so on. This chapter discussed timing simulators, which use simplified device models to carry out an approximate but quick circuit analysis. Various timing simulation algorithms like MOTIS, piecewise approximate simulation, and adaptively controlled explicit simulation were presented. The area of timing simulation is a topic of current research. Further, the class of relaxation methods can be used to partition a circuit and obtain the solution of the overall circuit in terms of the individual subcircuit analyses.

While this chapter did not cover all of the approximate and relaxation-based methods in the literature, it showed that these methods are based on the fundamentals that were covered in the earlier chapters of this book.

11.10 References

[van Bokhoven75] W. M. G. van Bokhoven. Linear implicit differentiation formulas of variable step and order. *IEEE Transactions on Circuits and Systems*, vol. CAS-22(2), pp. 109-115, February 1975.

[Chawla75] B. R. Chawla, H. K. Gummel, and P. Kozak. MOTIS - An MOS timing simulator. *IEEE Transactions on Circuits and Systems*, vol. CAS-22(12), pp. 901-910, December 1975.

[Bryant80] R. E. Bryant. An algorithm for MOS logic simulation. *Lambda*, vol. 1(3), pp. 46-53, 1980.

[Bryant81] R. E. Bryant. MOSSIM: a switch level simulator for MOS LSI. In *Proceedings of the 18th Design Automation Conference*, June 1981.

[Hitchcock82] R. B. Hitchcock, G. L. Smith, and D. D. Cheng. Timing analysis of computer hardware. *IBM Journal of Research and Development*, vol. 26(1), pp. 100-105, January 1982.

[de Geus82] A. J. de Geus and R. A. Rohrer. A new circuit simulation program based on a tree-link approach. In *Proceedings of the International Symposium on Circuits and Systems (ISCAS)*, pp. 702-704, May 1982.

[Lelarasmee82] E. Lelarasmee, A. E. Ruehli, and A. L. Sangiovanni-Vincentelli. The waveform relaxation method for the time-domain analysis of large scale integrated circuits. *IEEE Transactions on Computer-Aided Design of ICs and Systems*, vol. CAD-1(3), pp. 131-145, July 1982.

[White83] J. White and A. L. Sangiovanni-Vincentelli. RELAX2: a new waveform relaxation approach for the analysis of LSI MOS circuits. In *Proceedings of the International Symposium on Circuits and Systems (ISCAS)*, May 1983.

[Newton84] A. R. Newton and A. L. Sangiovanni-Vincentelli. Relaxation based electrical simulation. *IEEE Transactions on Computer-Aided Design of ICs and Systems,* vol. CAD-3(4), pp. 308-330, October 1984.

[Chen84] C-F. Chen, C-Y. Lo, H. N. Nham, and P. Subramaniam. The second generation MOTIS mixed-mode simulator. In *Proceedings of the 21st Design Automation Conference*, pp. 10-16, 1984.

[de Geus84] A. J. de Geus. SPECS: simulation program for electronic circuits and systems. In *Proceedings of the International Symposium on Circuits and Systems*, pp. 534-537, May 1984.

[Sze85] S. M. Sze. *Semiconductor devices, physics and technology.* John Wiley and Sons, 1985.

[Hennion85] B. Hennion and P. Senn. A new algorithm for third generation circuit simulators: the one-step relaxation method. In *Proceedings of the 22nd Design Automation Conference*, pp. 137-143, June 1985.

[Tsao85] D. Tsao and C-F. Chen. A fast timing simulator for digital MOS circuits. *IEEE International Conference on Computer-Aided Design (ICCAD)*, pp. 185-187, November 1985.

[Sakallah85] K. A. Sakallah and S. W. Director. SAMSON2: an event driven VLSI circuit simulator. *IEEE Transactions on Computer-Aided Design of ICs and Systems*, vol. 4(4), pp. 668-684, October 1985.

[Mokari-Bolhassan85] M. E. Mokari-Bolhassan, D. Smart, and T. N. Trick. A new robust relaxation technique for VLSI circuit simulation. *IEEE International Conference on Computer-Aided Design*, pp. 26-28, November 1985.

[Vidigal86] L. M. Vidigal, S. R. Nassif, and S. W. Director. CINNAMON: coupled integration and nodal analysis of MOS networks. In *Proceedings of the 23rd Design Automation Conference*, pp. 179-185, June 1986.

[Muller86] R. S. Muller and T. I. Kamins. *Device electronics for integrated circuits*. John Wiley and Sons, 1986.

[Dumlugol86] D. Dumlugol, P. Odent, J. Cockx, and H. de Man. The segmented waveform relaxation method for mixed-mode switch electrical simulation of digital MOS VLSI and its hardware acceleration on parallel computers. *IEEE International Conference on Computer-Aided Design (ICCAD)*, pp. 84-87, November 1986.

[Bryant87] R. E. Bryant, *et al.* COSMOS: A COmpiled Simulator for MOS circuits. In *Proc. 24th Design Automation Conference*, pp. 9-16, 1987.

[Ruan88] G. Ruan, J. Vlach, and J. A. Barby. Current-limited switch-level timing simulator for MOS logic networks. *IEEE Transactions on Computer-Aided Design of ICs and Systems*, vol. CAD-7(6), pp. 659-667, June 1988.

[Visweswariah89] C. Visweswariah and R. A. Rohrer. Piecewise approximate circuit simulation. *IEEE International Conference on Computer-Aided Design (ICCAD)*, November 1989.

[Nguyen89] T. V. Nguyen, P. Feldmann, S. W. Director, and R. A. Rohrer. SPECS simulation validation with efficient transient sensitivity computation. *IEEE International Conference on Computer-Aided Design (ICCAD)*, pp. 252-255, November 1989.

[Kim89] Y. H. Kim, S. H. Hwang, and A. R. Newton. Electrical-logic simulation and its applications. *IEEE Transactions on Computer-Aided Design of ICs and Systems*, vol. CAD-8(1), pp. 8-22, January 1989.

[Roulston90] D. J. Roulston. *Bipolar Semiconductor Devices*. McGraw Hill, 1990.

[Ruan90] G. Ruan, J. Vlach, and J. A. Barby. Logic simulation with current-limited switches. *IEEE Transactions on Computer-Aided Design of ICs and Systems*, vol. CAD-9(2), pp. 133-141, February 1990.

[Visweswariah90] C. Visweswariah, P. Feldmann, and R. A. Rohrer. Incorporation of inductors in piecewise approximate circuit simulation. *IEEE International Conference on Computer-Aided Design (ICCAD)*, pp. 162-165, November 1990.

[Feldmann91] P. Feldmann, T. V. Nguyen, S. W. Director, and R. A. Rohrer. Sensitivity computation in piecewise approximate circuit simulation. *IEEE Transactions on Computer-Aided Design of ICs and Systems*, vol. CAD-10(2), pp. 171-183, February 1991.

[Visweswariah91a] C. Visweswariah and R. A. Rohrer. Piecewise approximate circuit simulation. *IEEE Transactions on Computer-Aided Design of ICs and Systems*, vol. CAD-10(7), pp. 861-870, July 1991.

[Visweswariah91b] C. Visweswariah and R. A. Rohrer. Efficient simulation of bipolar digital integrated circuits. In *Proceedings of the 28th Design Automation Conference*, pp. 32-37, June 1991.

[Visweswariah93] C. Visweswariah and J. A. Wehbeh. Incremental event-driven simulation of digital FET circuits. In *Proceedings of the 30th Design Automation Conference*, pp. 737-741, June 1993.

[Lin93] S. Lin, E. S. Kuh, and M. Marek-Sadowska. Stepwise equivalent conductance circuit simulation technique. *IEEE Transactions on Computer-Aided Design of ICs and Systems*, vol. CAD-12(5), pp. 672-683, May 1993.

[Devgan94] A. Devgan and R. A. Rohrer. Adaptively controlled explicit simulation. *IEEE Transactions on Computer-Aided Design of ICs and Systems*, vol. CAD-13(6), pp. 746-762, June 1994.

Appendix A Tree/Link Analysis

This appendix provides an introduction to the graph and circuit theory concepts related to Tree/Link Analysis (TLA), or Hybrid Analysis, and includes the derivation of the tree/link circuit equations. Sections A.1, A.2, and A.3 introduce graph concepts [Deo74, Bondy76] and Section A.4 applies these concepts to derive Sparse Tableau and Reduced Tableau circuit equations [Hachtel71, Weeks73]. Finally, Tree/Link Analysis and tree selection procedures are addressed in Sections A.5, A.6, and A.7 [Kuh65, Rohrer70, Russo71, Chua75]. Tree/Link Analysis is the basis of circuit equation formulation in the timing simulator described in Section 11.7.

A.1 Graphs

A (linear) graph $G = (V, E)$ consists of a set of objects $V = \{v_1, v_2, \ldots, v_n\}$ called *vertices* or *nodes*, and another set $E = \{e_1, e_2, \ldots, e_b\}$ of *edges* or *branches*, such that each edge e_k is identified with an unordered pair (v_i, v_j) of vertices.

Branches with ends that fall on a node are said to be *incident* at that node. If a branch is incident at the same node at both ends, it is called a *self-loop*. Self-loops are uncommon in electrical circuits, since those edges can be removed from the graph representing the circuit without affecting the operation of the circuit. The number of edges incident on a node i (with self-loops counted twice) is the *degree* of node i. A graph with branches that are oriented is called an *oriented* or a *directed* graph. Electrical circuits are mapped to oriented graphs.

A *subgraph* is a subset of the branches and nodes of a graph. The subgraph is said to be *proper* if it consists of strictly less than all the branches and nodes of the graph.

Figure A.2 shows examples of an undirected and directed graph. In Figure A.2(a), branch 1 is a self-loop. The incidence of node n4 is 3 and that of node n2 is 4. In Figure A.2(b), branch 3 is incident on nodes n1 and n3, and Figure A.1 shows a proper subgraph.

A *path* is defined as a finite, ordered, alternating sequence of nodes and branches, beginning and ending with nodes, such that each branch is incident at the nodes preceding and following it. No branch or node appears more than once in a path. Note the following:

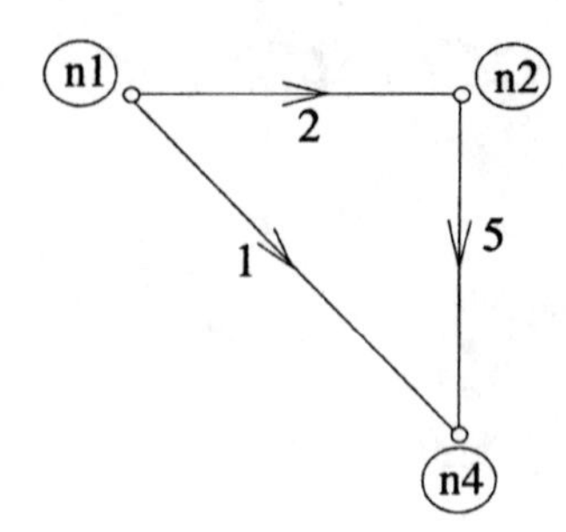

Figure A.1 Proper subgraph of the graph in Figure A.2(b).

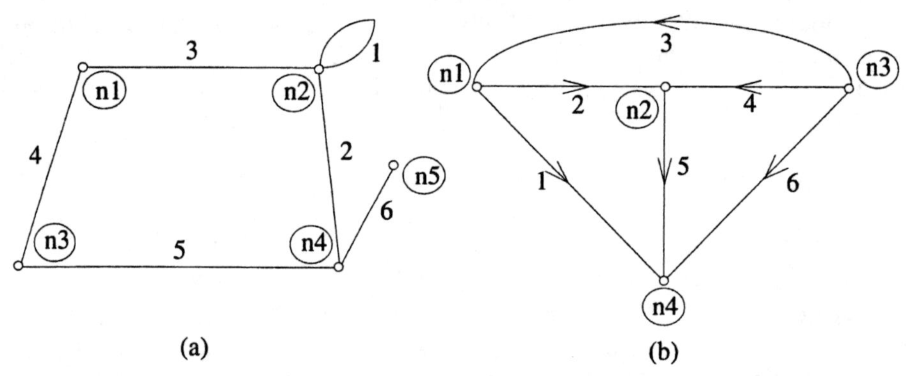

Figure A.2 Example of (a) an undirected graph and (b) a directed graph.

1. A path is a subgraph of the original graph.
2. The degree of all internal nodes of the subgraph is 2.
3. The degree of the terminal nodes of the subgraph is 1.
4. No proper subgraph of the path, having the same two terminal nodes, has properties (2) and (3) above.

In Figure A.2(b), $\{n1, 2, n2, 4, n3, 6, n4\}$ is a path, as is $\{n3, 3, n1, 2, n2, 5, n4\}$. A path is often designated by only a series of branches since the nodes are then obvious.

A graph is *connected* if there exists at least one path between any two nodes.

A *loop* is a connected subgraph of a graph at each node of which are incident exactly two branches of the subgraph. Clearly, the algebraic sum of branch voltages around any loop of a circuit is zero (as we will see later, this statement is an alternative way of expressing KVL). In Figure A.2(b), $\{3, 2, 4\}$ is a loop, as is $\{2, 4, 6, 1\}$.

A *tree* is a connected subgraph of a graph that contains all the nodes of the graph but no loops.

In Figure A.2(b), $\{1, 5, 6\}$ is a tree, as is $\{3, 2, 5\}$. $\{1, 3, 6\}$ is not a tree because it contains a loop, and $\{1, 5\}$ is not a tree because it does not contain node $n3$. Branches comprising the tree are called *tree branches.* The complement of the tree subgraph is called the *cotree.* Branches comprising the cotree are called *cotree links* or just *links.* Note the following:

- There is one and only one path between every pair of nodes in a tree.

  Proof: If there were two paths (or more), their union would form a loop and the tree would be invalid.

- If in a graph there is one and only one path between every pair of vertices, the graph is a tree.

  Proof: Clearly the graph is connected. Further, any loops would imply more than one path between two nodes; hence there are no loops.

- A graph with $(n+1)$ nodes has a tree with n branches.

  Proof: The proof is by induction:

  a) The proposition is true for $n = 1, 2$.

  b) Consider a graph with $(k+1)$ nodes. The part with k nodes has a tree that consists of $(k-1)$ branches by the induction hypothesis. The tree for the $(k+1)$ node graph has at least one node at which only one tree branch is incident. If that node and tree branch are removed, it leaves behind a k node graph with $(k-1)$ tree branches. Hence the tree of the $(k+1)$ node graph has $k-1+1 = k$ branches.

A.2 Fundamental Loops and Cutsets

Figure A.3 shows an example of a directed graph. The branches shown in solid lines have been selected to form the tree.

Given a graph G, choose a tree and remove the links. Replace one link l. By definition, it forms a loop, called a *fundamental loop.* A fundamental loop is a loop that consists of one link and all other tree branches. In Figure A.3, the fundamental loops are as follows:

- corresponding to link 5: $\{-2, +3\}$
- corresponding to link 4: $\{+2, +1\}$
- corresponding to link 6: $\{-1, +2, +3\}$

The *orientation* of the fundamental loop is determined by the orientation of the link that defines the loop. Tree branches that share the same orientation have been designated with a + sign above, and tree branches with the opposite orientation as the loop with a − sign.

In a connected graph G, a *cutset* is a set of edges whose removal from G leaves G dis-

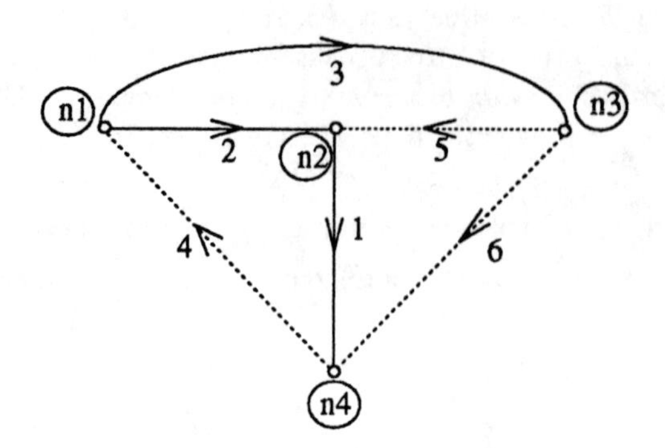

Figure A.3 Graph with tree shown in solid lines.

connected, provided removal of no proper subset of these edges disconnects G. In Figure A.3, the cutsets are $\{2, 3, 4\}$, $\{3, 5, 6\}$, $\{1, 2, 5\}$, $\{4, 1, 6\}$, $\{3, 2, 1, 6\}$, $\{3, 5, 1, 4\}$, and $\{2, 4, 5, 6\}$. A cutset can be thought of as a "Gaussian surface." Clearly, the algebraic sum of the branch currents of any cutset of a circuit is zero (this statement is an alternate way of expressing KCL). Note the following:

- Every edge of a graph that is a tree is a cutset of that graph.
- Every cutset in a connected graph G must contain at least one branch of every tree of G.
- In a connected graph G, every loop has an even number of branches in common with any cutset. As shown in Figure A.4, a loop can either be exclusively in part 1 (in which case it shares 0 branches with the cutset) or exclusively in part 2 (in which case it shares 0 branches with the cutset) or include branches from both part 1 and part 2. In the latter case, it must cross over from one part to another an even number of times.

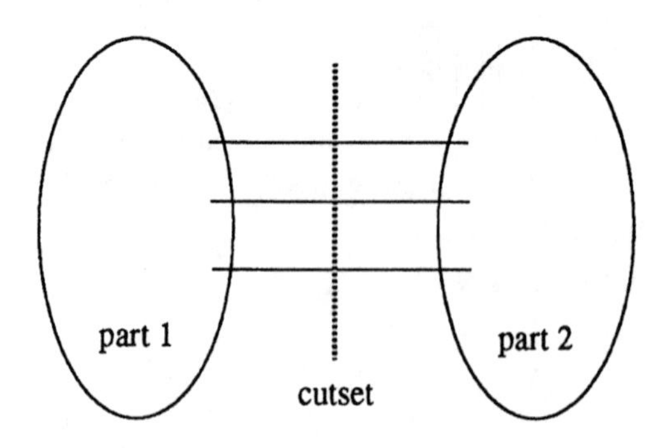

Figure A.4 Branches in common between loops and cutsets.

Consider a connected graph G as shown in Figure A.5 with a tree T. Consider one branch of T, t. Branch t is a cutset of T. Now add all the links of G back. It is clear that the cutset of G including t is unique, and may contain other links.

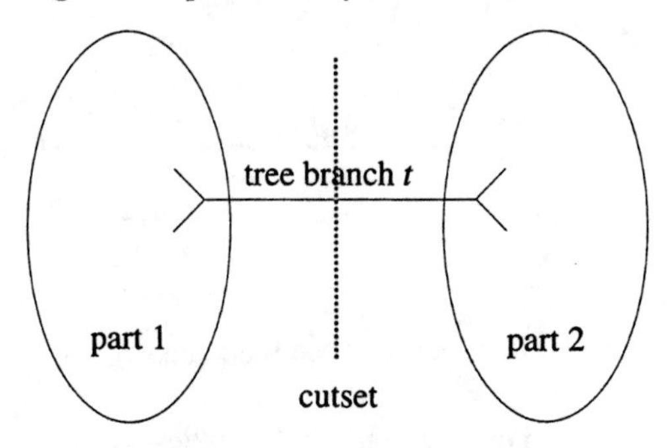

Figure A.5 Fundamental cutsets.

A *fundamental cutset* is a cutset that consists of one tree branch and all other links. In Figure A.3, the fundamental cutsets are:

- corresponding to tree branch 1: $\{-4, +6\}$
- corresponding to tree branch 2: $\{+5, +6, -4\}$
- corresponding to tree branch 3: $\{-5, -6\}$

The *orientation* of the cutset is defined by the tree branch (as being "into" or "out of" the Gaussian surface). A link that shares the orientation of the cutset is indicated with a + sign, while a link that has a direction opposite to that of the cutset is indicated with a - sign. Note the following:

- Consider a fundamental cutset including a tree branch t and all other links $\{l_1, l_2, \ldots, l_c\}$ as shown in Figure A.6. The fundamental loop of each of those links includes t. Further, the orientation of each link in the fundamental cutset and the tree branch in the corresponding fundamental loop are opposite.
- Consider a fundamental loop including a link l and all other tree branches. The fundamental cutsets of each of these tree branches includes l. Further, the orientation of each tree branch in the fundamental loop and the link in the corresponding fundamental cutset are opposite.

The above two intuitive graph theoretic concepts will be revisited in a more rigorous fashion in terms of the F matrix in Section A.5.

A circuit can be mapped to a graph and the concepts of loops and cutsets can be used to restate Kirchhoff's laws in an alternative but equivalent form:

Kirchhoff's Current Law (KCL): The algebraic sum of the currents in any cutset of a circuit is zero at any instant of time. The set of branches incident at each node form the sim-

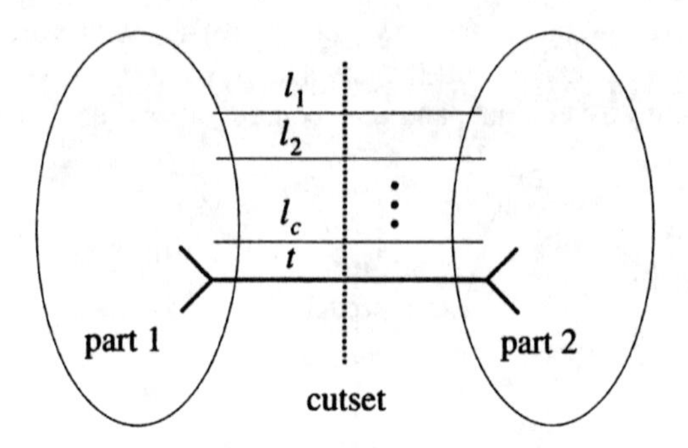

Figure A.6 Common branches between fundamental loops and cutsets.

plest cutsets, and MNA is based upon writing KCL for these cutsets.
Kirchhoff's Voltage Law (KVL): The algebraic sum of branch voltages in any loop of a circuit is zero at any given instant of time.

A.3 The Incidence Matrix

For an $(n+1)$ node, b branch graph without self-loops, the *complete incidence matrix* $A_a = [a_{ij}]$ is an $(n+1) \times b$ rectangular matrix whose elements have the following values:

- $a_{ij} = 1$ if branch j is incident at node i and oriented away from it.
- $a_{ij} = -1$ if branch j is incident at node i and oriented towards it.
- $a_{ij} = 0$ if branch j is not incident at node i.

For the graph in Figure A.3,

$$A_a = \begin{array}{c} \\ n1 \\ n2 \\ n3 \\ n4 \end{array} \begin{array}{c} \begin{array}{cccccc} 1 & 2 & 3 & 4 & 5 & 6 \leftarrow \text{branches} \end{array} \\ \begin{bmatrix} 0 & 1 & 1 & -1 & 0 & 0 \\ 1 & -1 & 0 & 0 & -1 & 0 \\ 0 & 0 & -1 & 0 & 1 & 1 \\ -1 & 0 & 0 & 1 & 0 & -1 \end{bmatrix} \end{array} \qquad \text{(A.3.1)}$$

$\uparrow$ nodes

Note the following:

- Every column has a +1 and a −1.
- The degree of a node i is the number of nonzeros in row i.
- The sum of all the rows is a row with all zeros, implying that the rows are linearly dependent.

For an $(n+1)$ node b branch connected graph, the rank of A_a is n.

Proof: Let us postulate that the rank of A_a is less than n. Then some nontrivial combination of n rows must result in a row of all zeros. Let k of the coefficients of this combination be nonzero. Let G_1 be the set of nodes corresponding to those k rows and G_2 the remaining nodes. Since the graph is connected, there is at least one branch connecting G_1 and G_2, causing a ± 1 in that column among the G_1 rows and a ∓ 1 in that column among the G_2 nodes and zeros in the rest of the column. Then the postulated linear combination yields a nonzero result in that column. Hence the rank cannot be $k < n$. However, the rank is at most n. Therefore the rank of A_a is exactly n.

Alternative (matrix-style) proof: Order the columns of A_a to have tree branches first. Eliminate any one row of A_a. Pick one tree branch that was incident at the eliminated row (there has to be at least one) and consider the other end of that branch. The determinant of the $(n \times n)$ submatrix of A_a is ± 1 times the cofactor of that element. The matrix associated with this cofactor corresponds to a connected subgraph with $(n-1)$ nodes; this matrix contains neither of the rows in which the eliminated column has nonzero elements. Since this subgraph is connected, it must have at least one tree branch incident at the eliminated row, and that column has only one other nonzero. Hence the determinant of the submatrix is ± 1 times the cofactor of the associated $(n-2 \times n-2)$ submatrix. Continue this procedure until one node is remaining, and the cofactor is again ± 1 because the graph is connected. Hence, an $(n \times n)$ submatrix of A_a whose columns correspond to tree branches and rows to all nodes but one is nonsingular. Therefore, the rank of A_a is at least n. But we know that the rank of A_a is less than $(n+1)$. Hence the rank of A_a is exactly n.

The *reduced incidence matrix* $\boldsymbol{A}$ of a graph is obtained by eliminating any one row i from its incidence matrix. The rank of $\boldsymbol{A}$ is n and node i is called the *datum* (or ground) node. In electrical circuits, one node needs to be designated as a reference node (ground) and the voltage of all other nodes is expressed in terms of this reference potential.

KCL can be conveniently written in terms of the reduced incidence matrix as follows:

$$\boldsymbol{A}\boldsymbol{i}_b = \boldsymbol{0} \tag{A.3.2}$$

Equation (A.3.2) assumes that the order of the branches in the columns of $\boldsymbol{A}$ is the same as the order of branches in the column vector of branch currents, $\boldsymbol{i}_b$. Clearly, each row represents the branches incident on that node, so when combined with the branch currents, equation (A.3.2) simply states that the sum of currents leaving (entering) every node is zero.

KVL can also be conveniently written in terms of the reduced incidence matrix as follows:

$$v_b = A^T v_n \tag{A.3.3}$$

Again, the order of nodes in the vector of node voltages in v_n is the same as the order of rows in A and the order of branch voltages in v_b is the same as the columns of A. The ith equation of (A.3.3) basically asserts that

$$v_{b_i} = v_{from-node_i} - v_{to-node_i} \tag{A.3.4}$$

A.4 Sparse Tableau and Reduced Tableau Analysis

Equations (A.3.2) and (A.3.3) comprise $(n+b)$ equations in $(n+2b)$ unknowns. The other b equations are the branch constitutive relations (BCRs) relating branch voltages and currents:

$$\alpha \mathbf{v}_b + \beta i_b = \gamma \tag{A.4.1}$$

where α and β are $(b \times b)$ matrices. These three sets of $(n+2b)$ unknowns in $(n+2b)$ unknowns constitute a complete set of sparse equations called the *Sparse Tableau* equations. The rationale behind *Sparse Tableau Analysis (STA)* is that even though the number of equations is much larger than MNA, the added sparsity of the equations render them efficient to solve. As with MNA, matrix stampings for various types of elements can be derived to formulate the equations efficiently. The STA equations can be represented in matrix form as follows:

$$\begin{array}{c} \\ b \\ n \\ b \end{array}\begin{array}{c} \begin{array}{ccc} b & b & n \end{array} \\ \begin{bmatrix} 1 & 0 & -A^T \\ 0 & A & 0 \\ \alpha & \beta & 0 \end{bmatrix} \end{array} \begin{bmatrix} v_b \\ i_b \\ v_n \end{bmatrix} \begin{array}{c} b \\ b \\ n \end{array} = \begin{bmatrix} 0 \\ 0 \\ \gamma \end{bmatrix} \begin{array}{c} b \\ n \\ b \end{array} \tag{A.4.2}$$

The dimensions of the various submatrices in (A.4.2) have been indicated along the margins of the matrices.

In *Reduced Tableau Analysis*, the KVL equations are substituted into the BCRs to get $(b+n)$ equations in $(b+n)$ unknowns (n node voltages and b branch currents) as follows:

$$\begin{array}{c} \\ n \\ b \end{array}\begin{array}{c} \begin{array}{cc} b & n \end{array} \\ \begin{bmatrix} A & 0 \\ \beta & \alpha A^T \end{bmatrix} \end{array} \begin{bmatrix} i_b \\ v_n \end{bmatrix} \begin{array}{c} b \\ n \end{array} = \begin{bmatrix} 0 \\ \gamma \end{bmatrix} \begin{array}{c} n \\ b \end{array} \tag{A.4.3}$$

A.5 Loop and Cutset Matrices

Find a tree in a graph. Partition the following vectors and matrices:

$$\boldsymbol{A} = \left[\boldsymbol{A}_t \; \vdots \; \boldsymbol{A}_l\right] \tag{A.5.1}$$

$$\boldsymbol{v}_b = \begin{bmatrix} \boldsymbol{v}_t \\ \cdots \\ \boldsymbol{v}_l \end{bmatrix} \tag{A.5.2}$$

$$\boldsymbol{i}_b = \begin{bmatrix} \boldsymbol{i}_t \\ \cdots \\ \boldsymbol{i}_l \end{bmatrix} \tag{A.5.3}$$

where the subscript t indicates tree branches and l indicates links.

Construct a $(b-n) \times n$ matrix $\boldsymbol{F}_{loop}$ such that each row represents a fundamental cutset. For example, considering the graph in Figure A.3,

$$\boldsymbol{F}_{loop} = \begin{matrix} & 1 & 2 & 3 \leftarrow \text{Tree branches} \\ 4 & 1 & 1 & 0 \\ 5 & 0 & -1 & 1 \\ 6 & -1 & -1 & 1 \\ \uparrow & & & \\ \text{Links} & & & \end{matrix} \tag{A.5.4}$$

The last row of (A.5.4) indicates that the fundamental loop corresponding to link 6 consists of tree branch 3 in the same direction as the loop direction defined by link 6 and of tree branches 1 and 2 in the opposite direction.

KVL can be written in terms of the loop matrix as follows:

$$\boldsymbol{v}_l = -\boldsymbol{F}_{loop}\boldsymbol{v}_t \tag{A.5.5}$$

or equivalently,

$$\boldsymbol{v}_b = \begin{bmatrix} \boldsymbol{v}_t \\ \cdots \\ \boldsymbol{v}_l \end{bmatrix} = \begin{bmatrix} \boldsymbol{1} \\ \cdots\cdots \\ -\boldsymbol{F}_{loop} \end{bmatrix} \boldsymbol{v}_t \tag{A.5.6}$$

Construct an $n \times (b-n)$ matrix $\boldsymbol{F}_{cutset}$ such that each row represents a fundamental cutset. For the example in Figure A.3,

$$F_{cutset} = \begin{array}{c} \\ 1 \\ 2 \\ 3 \end{array} \begin{array}{c} \begin{array}{ccc} 4 & 5 & 6 \leftarrow \text{Links} \end{array} \\ \begin{bmatrix} -1 & 0 & 1 \\ -1 & 1 & 1 \\ 0 & -1 & -1 \end{bmatrix} \end{array} \tag{A.5.7}$$

$\uparrow$

Tree branches

The first row of (A.5.7) implies that the fundamental cutset corresponding to tree branch 1 consists of link 4 in the opposite direction as the cutset and link 6 in the same direction.

KCL can be written in terms of the cutset matrix as follows:

$$\boldsymbol{i}_t = -\boldsymbol{F}_{cutset}\boldsymbol{i}_l \tag{A.5.8}$$

or

$$\boldsymbol{i}_b = \begin{bmatrix} \boldsymbol{i}_t \\ \cdots \\ \boldsymbol{i}_l \end{bmatrix} = \begin{bmatrix} -\boldsymbol{F}_{cutset} \\ \cdots\cdots \\ \mathbf{1} \end{bmatrix} \boldsymbol{i}_l \tag{A.5.9}$$

$$\begin{bmatrix} \mathbf{1} & \vdots & \boldsymbol{F}_{cutset} \end{bmatrix} \boldsymbol{i}_b = \mathbf{0} \tag{A.5.10}$$

Notice the similarity between (A.5.10) and (A.3.2). This similarity will be exploited in devising a method to find a tree in Section A.7.

We know that the instantaneous power in any circuit is zero -- the power delivered to the circuit by the sources is dissipated by the circuit in the resistive elements, and energy is neither created nor destroyed. The conservation of energy in a circuit is expressed as

$$\sum v_b i_b = v_b^T i_b = 0 \tag{A.5.11}$$

(A.5.11) is actually one way of expressing a more general theorem called Tellegen's theorem. Using this formula, let us try to relate $\boldsymbol{F}_{loop}$ and $\boldsymbol{F}_{cutset}$. Substituting for $\boldsymbol{v}_b$ and $\boldsymbol{i}_b$ in (A.5.11), we get

$$\left(\begin{bmatrix} \mathbf{1} \\ \cdots\cdots \\ -\boldsymbol{F}_{loop} \end{bmatrix} \boldsymbol{v}_t\right)^T \begin{bmatrix} -\boldsymbol{F}_{cutset} \\ \cdots\cdots \\ \mathbf{1} \end{bmatrix} \boldsymbol{i}_l = \boldsymbol{v}_t^T \begin{bmatrix} \mathbf{1} & \vdots & -\boldsymbol{F}_{loop}^T \end{bmatrix} \begin{bmatrix} -\boldsymbol{F}_{cutset} \\ \cdots\cdots \\ \mathbf{1} \end{bmatrix} \boldsymbol{i}_l = 0 \tag{A.5.12}$$

which leads to

$$-v_t^T\left[F_{cutset} + F_{loop}^T\right]i_l = 0 \tag{A.5.13}$$

or

$$F_{cutset} + F_{loop}^T = \mathbf{0} \tag{A.5.14}$$

Define a $(b-n) \times n$ matrix F such that

$$-F_{loop}^T = F_{cutset} \equiv F \tag{A.5.15}$$

Fundamental cutsets can be determined by the rows of F and fundamental loops by the columns, as shown in (A.5.16). Thus the relation between the fundamental loops and cutsets as alluded to in Section A.2 has been mathematically shown.

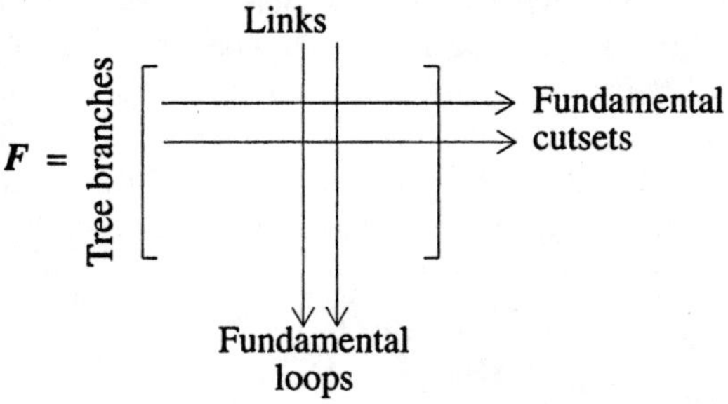

$$F = \left[\ \ \right] \tag{A.5.16}$$

Links

Tree branches

Fundamental cutsets

Fundamental loops

A.6 Circuit Equations in Terms of Loops and Cutsets

The n KCL equations can be written in any of the three forms below:

$$i_t = -Fi_l$$

$$i_b = \begin{bmatrix} -F \\ \cdots\cdots \\ 1 \end{bmatrix} i_l \tag{A.6.1}$$

$$\begin{bmatrix} 1 & \vdots & F \end{bmatrix} i_b = \mathbf{0}$$

The $(b-n)$ KVL equations can be written in any of the three forms below:

$$\begin{gathered} \boldsymbol{v}_l = \boldsymbol{F}^T \boldsymbol{v}_t \\ \boldsymbol{v}_b = \begin{bmatrix} \mathbf{1} \\ \cdots \\ \boldsymbol{F}^T \end{bmatrix} \boldsymbol{v}_t \\ \begin{bmatrix} -\boldsymbol{F}^T & \vdots & \mathbf{1} \end{bmatrix} \boldsymbol{v}_b = \mathbf{0} \end{gathered} \tag{A.6.2}$$

Tree/Link Analysis treats the circuit in terms of branch quantities rather than node voltages (node voltages can be derived after the analysis is complete using a matrix called the *path matrix* which specifies a path to ground from each node consisting of only tree branches). It turns out that there are numerical advantages to treating nonlinear elements on a branch basis rather than trying to solve for node voltages that are dictated by the nonlinearities of multiple elements. Given the n tree branch voltages, all branch voltages can easily be computed using the fundamental loops. Given the $(b-n)$ link currents, all branch currents can easily be computed using the fundamental cutsets. So Tree/Link Analysis chooses tree branch voltages and link currents as a set of basis variables. Hence Tree/Link Analysis is also called *Hybrid Analysis* because it solves for some currents and some voltages. There are b branch currents and b branch voltages that are unknown. We have b KCL and KVL equations. The other b equations are derived from BCRs. Let us assume that it is possible to write BCRs in the following form for trees and links:

$$\begin{gathered} \boldsymbol{v}_t = \boldsymbol{R}\boldsymbol{i}_t + \boldsymbol{V}_t \\ \boldsymbol{i}_l = \boldsymbol{G}\boldsymbol{v}_l + \boldsymbol{I}_l \end{gathered} \tag{A.6.3}$$

The first of these implies that tree branches are modeled as a generalized set of composite branches in Thevenin equivalent form as shown in Figure A.7. The second set of equations in (A.6.3) implies a generalized set of composite branches in Norton equivalent form for the links, as shown in Figure A.8. The set of tree branch voltages is sufficient to determine all branch voltages in the circuit, and likewise with link currents. Hence it helps to write the BCRs in the above form, with tree branch voltages and link currents being expressed in terms of other quantities.

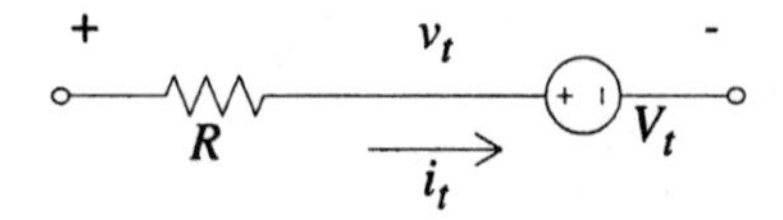

Figure A.7 Tree branches are modeled as Thevenin equivalents.

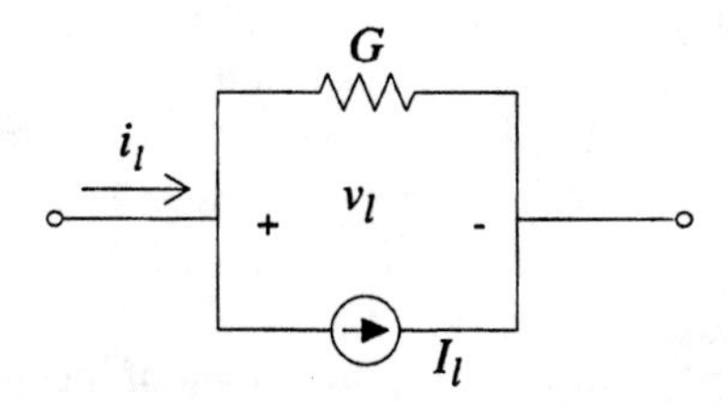

Figure A.8 Links are modeled as Norton equivalents.

In the extreme, a tree branch may be purely an independent voltage source, with $R = 0$. A link could be purely an independent current source, with $G = 0$. Of course, either branch type may be source free and represent a pure resistance or conductance element. Finally, off-diagonal entries in the $\boldsymbol{R}$ or $\boldsymbol{G}$ matrices can represent current-controlled voltage sources or voltage-controlled current sources, respectively. The other two kinds of controlled sources will be considered later. We should recognize the desirability of treating such composite branches as arising from the companion models that must be used for nonlinear iteration and time domain integration.

Substituting for $\boldsymbol{i}_t$ and $\boldsymbol{v}_l$ in terms of $\boldsymbol{i}_l$ and $\boldsymbol{v}_t$ from the KCL and KVL equations, we get

$$\boldsymbol{v}_t = -\boldsymbol{RFi}_l + \boldsymbol{V}_t \tag{A.6.4}$$

and

$$\boldsymbol{i}_l = \boldsymbol{GF}^T\boldsymbol{v}_t + \boldsymbol{I}_l \tag{A.6.5}$$

or

$$\begin{aligned} \boldsymbol{v}_t + \boldsymbol{RFi}_l &= \boldsymbol{V}_t \\ -\boldsymbol{GF}^T\boldsymbol{v}_t + \boldsymbol{i}_l &= \boldsymbol{I}_l \end{aligned} \tag{A.6.6}$$

or

$$\begin{bmatrix} \boldsymbol{1} & \boldsymbol{RF} \\ -\boldsymbol{GF}^T & \boldsymbol{1} \end{bmatrix} \begin{bmatrix} \boldsymbol{v}_t \\ \boldsymbol{i}_l \end{bmatrix} = \begin{bmatrix} \boldsymbol{V}_t \\ \boldsymbol{I}_l \end{bmatrix} \tag{A.6.7}$$

We could solve the hybrid set of b equations in b unknowns directly, or we could consider reducing it further. Note that the ones on the diagonal could be good choices for pivots provided that the nonzero-valued elements in the $\boldsymbol{R}$ and $\boldsymbol{G}$ matrices were small. We thus have a hint that it would be numerically advantageous to include large resistors (small Gs) in the cotree links and small resistances in the tree branches. Note too that pure independent voltage or current sources ($R = 0$ or $G = 0$) are easily accounted for in this hybrid formulation of the circuit equations, and that they produce trivial equations. Unlike MNA, this formulation treats independent voltage and current sources on an equal footing, which is an advantage. It does require that all independent voltage sources be assigned to the tree and all independent current sources to the cotree. However, this requirement does not pose any real restrictions. The only time independent voltage sources could not all be assigned to a tree is when they form a loop. But a loop of independent voltage sources is nonphysical, and the tree selection process should flag such a situation as an error. The same is true for a cutset of independent current sources.

We can substitute (A.6.5) into (A.6.4) to obtain

$$\boldsymbol{v}_t = -\boldsymbol{RF}(\boldsymbol{GF}^T\boldsymbol{v}_t + \boldsymbol{I}_l) + \boldsymbol{V}_t \qquad \text{(A.6.8)}$$

or

$$(1 + \boldsymbol{RFGF}^T)\boldsymbol{v}_t = \boldsymbol{V}_t - \boldsymbol{RFI}_l \qquad \text{(A.6.9)}$$

This set of n equations in n unknowns is called the set of *generalized cutset equations* for the circuit. Some texts prefer the form

$$(\boldsymbol{R}^{-1} + \boldsymbol{FGF}^T)\boldsymbol{v}_t = \boldsymbol{R}^{-1}\boldsymbol{V}_t - \boldsymbol{FI}_l \qquad \text{(A.6.10)}$$

but we don't, because it precludes the possibility of a singular $\boldsymbol{R}$ matrix.

Returning once again to equations (A.6.4) and (A.6.5) but this time substituting the first of these into the second we obtain

$$\boldsymbol{i}_l = \boldsymbol{GF}^T(-\boldsymbol{RFi}_l + \boldsymbol{V}_t) + \boldsymbol{I}_l \qquad \text{(A.6.11)}$$

or

$$(1 + \boldsymbol{GF}^T\boldsymbol{RF})\boldsymbol{i}_l = \boldsymbol{GF}^T\boldsymbol{V}_t + \boldsymbol{I}_l \qquad \text{(A.6.12)}$$

This set of $(b-n)$ equations in $(b-n)$ unknowns is called the set of *generalized loop equations* for the circuit. Again, some texts prefer the form

$$(\boldsymbol{G}^{-1} + \boldsymbol{F}^T\boldsymbol{RF})\boldsymbol{i}_l = \boldsymbol{F}^T\boldsymbol{V}_t + \boldsymbol{G}^{-1}\boldsymbol{I}_l \qquad \text{(A.6.13)}$$

but we do not so as to entertain the possibility of a singular $\boldsymbol{G}$ matrix.

The cutset equations are like nodal equations, only they employ n independent tree branch voltages as basis variables. The loop equations are like mesh equations, but they do not require circuit (graph) planarity. Each of the $(b-n)$ link currents defines a topologi-

cally independent loop current. Usually $n < (b-n)$, so we might prefer to work with the smaller set of cutset equations. As in the case of MNA, matrix stamps can be derived to populate the appropriate matrices efficiently. Two examples are shown below.
Example 1: Stamp of a conductance G to the matrix $\boldsymbol{FGF^T}$:

(i, i) term is incremented by G if tree branch i is in G's fundamental loop, else the diagonal term is unchanged;

(i, j) term is incremented by G if G's fundamental loop includes the ith and j^{th} tree branches in the same orientation;

(i, j) term is decremented by G if G's fundamental loop includes the ith and j^{th} tree branches in the opposite orientation;

(i, j) term is unchanged if G's fundamental loop does not include *both* the ith and j^{th} tree branches.

Example 2: Stamp of a resistance R to the matrix $\boldsymbol{F^TRF}$:

(i, i) term is incremented by R if link i is in R's fundamental cutset, else the diagonal term is unchanged;

(i, j) term is incremented by R if R's fundamental cutset includes the ith and j^{th} links in the same direction;

(i, j) term is decremented by R if R's fundamental cutset includes the ith and j^{th} links in the opposite direction;

(i, j) term is unchanged if R's fundamental cutset does not include *both* the ith and j^{th} links.

In order to be able to accommodate all four kinds of controlled sources, we generalize the BCRs of (A.6.3):

$$\begin{aligned} \boldsymbol{v}_t &= \boldsymbol{R}\boldsymbol{i}_t + \boldsymbol{V}_t + \alpha \boldsymbol{v}_l \\ \boldsymbol{i}_l &= \boldsymbol{G}\boldsymbol{v}_l + \boldsymbol{I}_l + \beta \boldsymbol{i}_t \end{aligned} \qquad \text{(A.6.14)}$$

Substituting for $\boldsymbol{i}_t$ and $\boldsymbol{v}_l$ in terms of $\boldsymbol{i}_l$ and $\boldsymbol{v}_t$, we get

$$\begin{aligned} \boldsymbol{v}_t &= -\boldsymbol{R}\boldsymbol{F}\boldsymbol{i}_l + \boldsymbol{V}_t + \alpha \boldsymbol{F}^T \boldsymbol{v}_t \\ \boldsymbol{i}_l &= \boldsymbol{G}\boldsymbol{F}^T\boldsymbol{v}_t + \boldsymbol{I}_l - \beta \boldsymbol{F}\boldsymbol{i}_l \end{aligned} \qquad \text{(A.6.15)}$$

or

$$\begin{bmatrix} 1-\alpha F^T & RF \\ -GF^T & 1+\beta F \end{bmatrix} \begin{bmatrix} v_t \\ i_l \end{bmatrix} = \begin{bmatrix} V_t \\ I_l \end{bmatrix} \quad \text{(A.6.16)}$$

Substituting for i_l, we get generalized cutset equations:

$$\left[1-\alpha F^T + RF(1+\beta F)^{-1}GF^T\right] v_t = V_t - RF(1+\beta F)^{-1} I_l \quad \text{(A.6.17)}$$

Substituting instead for v_t in (A.6.16), we get generalized loop equations:

$$[1+\beta F + GF^T(1-\alpha F^T)^{-1}RF]\, i_l = I_l + GF^T(1-\alpha F^T)^{-1} V_t \quad \text{(A.6.18)}$$

A.7 Tree Selection Procedure

This section describes how to select a tree, given a prioritized order of the elements we would like to have in the tree. The order of priority for inclusion in the tree is:

1. Independent voltage sources
2. Capacitances
3. Resistances or conductances (including dependent sources)
4. Inductances
5. Independent current sources.

Of course, all independent voltage sources must be in the tree and all independent current sources must be in the cotree or we could not proceed with the analysis in any case. To the extent possible, capacitances should be in the tree and inductances in the cotree. As noted in Chapter 4, companion models for capacitors in Thevenin form are preferable for situations with specified initial conditions or when dealing with small time steps. Similarly, companion models for inductors in Norton form are preferable, so we try to include as many inductances as possible in the cotree. With small time steps, note that both the resistance in the Thevenin companion model of a capacitor and the conductance in the Norton companion model of an inductor become smaller, leading to better numerical conditioning of the circuit equations. At the outset we allow resistances, conductances, and controlled sources to be in the tree or the cotree as dictated by circuit topology. Again, smaller resistances are preferred in the tree for numerical reasons.

Consider a fundamental cutset ξ. Let it divide a graph G into two sets of nodes, G_1 and G_2. Add up the nodal KCL equations for all the nodes in G_1. Clearly, the current of any branch completely within G_1 will cancel out because its "from" node and "to" node are both within G_1. Hence the sum of the nodal KCL equations for G_1 will result in the fundamental cutset equation corresponding to ξ. Likewise for G_2. This concept is demonstrated on an example in Figure A.9.

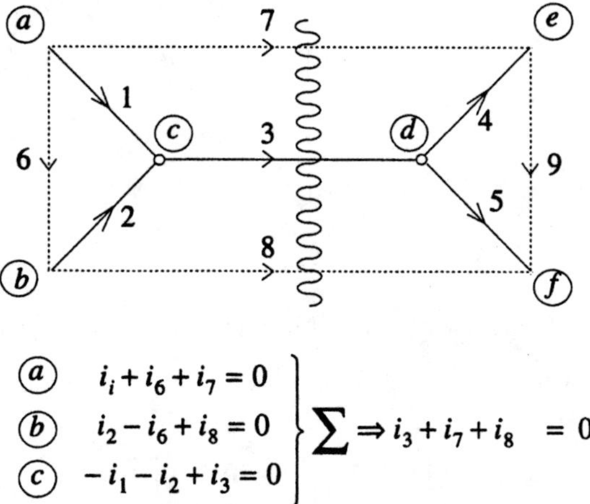

Figure A.9 Relation of nodal KCL equations to fundamental cutset equations.

Thus we recognize that if we can manipulate the nodal KCL equations

$$\boldsymbol{A}\boldsymbol{i}_b = \boldsymbol{0} \tag{A.7.1}$$

into the tree/link form

$$\left[\boldsymbol{1} \,\vdots\, \boldsymbol{F}\right]\hat{\boldsymbol{i}}_b = \boldsymbol{0} \tag{A.7.2}$$

then we will have determined the tree and also all the fundamental cutsets and loops in terms of the $\boldsymbol{F}$ matrix. We use $\hat{\boldsymbol{i}}_b$ to indicate the permutation of the original $\boldsymbol{i}_b$ vector that may be necessary to obtain the specific form $\left[\boldsymbol{1} \,\vdots\, \boldsymbol{F}\right]$. The manipulation should be possible by using only the following simple operations on the $\boldsymbol{A}$ matrix:

- adding or subtracting one row from another
- negating a row
- exchanging the order of the elements in $\boldsymbol{i}_b$, which is equivalent to swapping columns of $\boldsymbol{A}$

Figure A.10 shows a sample graph with the branches ordered in terms of tree priority. This graph has four nodes and therefore three tree branches. But we cannot use branches 1, 2, and 3 to form the tree, since they form a loop. Start with the reduced incidence matrix in tree branch priority order:

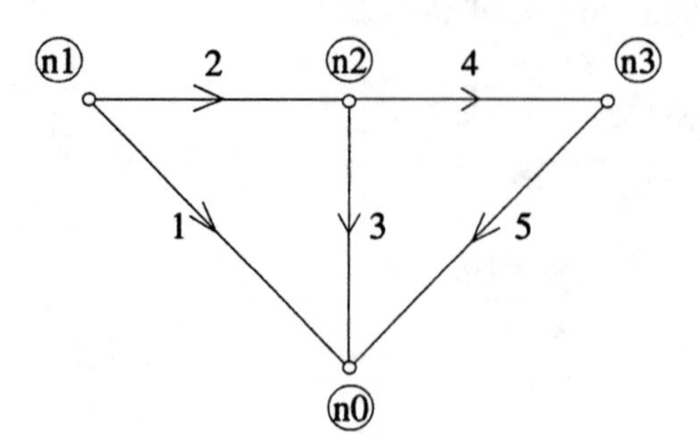

Figure A.10 Sample graph to illustrate tree-finding procedure.

$$A = \begin{array}{c} \\ n1 \\ n2 \\ n3 \end{array} \begin{array}{c} \begin{matrix} 1 & 2 & 3 & 4 & 5 \end{matrix} \\ \begin{bmatrix} 1 & 1 & 0 & 0 & 0 \\ 0 & -1 & 1 & 1 & 0 \\ 0 & 0 & 0 & -1 & 1 \end{bmatrix} \end{array} \tag{A.7.3}$$

The first column already matches the required identity matrix. Moving on to the second column, add row 2 to row 1 and then multiply row 2 by -1:

$$A^{(2)} = \begin{bmatrix} 1 & 0 & 1 & 1 & 0 \\ 0 & 1 & -1 & -1 & 0 \\ 0 & 0 & 0 & -1 & 1 \end{bmatrix} \tag{A.7.4}$$

We cannot work with column 3 because it has a zero in row 3, indicating that branch 3 cannot be a tree branch if branches 1 and 2 are tree branches as well. So, we move on to column 4 and add row 3 to row 1, subtract row 3 from row 2, and finally multiply row 3 by -1:

tree branches (columns 1, 2, 4)

$$A^{(3)} = \begin{bmatrix} 1 & 0 & 1 & 0 & 1 \\ 0 & 1 & -1 & 0 & -1 \\ 0 & 0 & 0 & 1 & -1 \end{bmatrix} \tag{A.7.5}$$

possible column interchange (columns 3 and 4)

Columns 3 and 4 could possibly be interchanged to make the first three columns of $A^{(3)}$ into an identity matrix.

What we have now are the KCL cutset equations

$$\begin{bmatrix} 1 & 0 & 1 & 0 & 1 \\ 0 & 1 & -1 & 0 & -1 \\ 0 & 0 & 0 & 1 & -1 \end{bmatrix} \begin{bmatrix} i_1 \\ i_2 \\ i_3 \\ i_4 \\ i_5 \end{bmatrix} = \begin{bmatrix} 0 \\ 0 \\ 0 \end{bmatrix} \tag{A.7.6}$$

corresponding to the tree shown in Figure A.11.

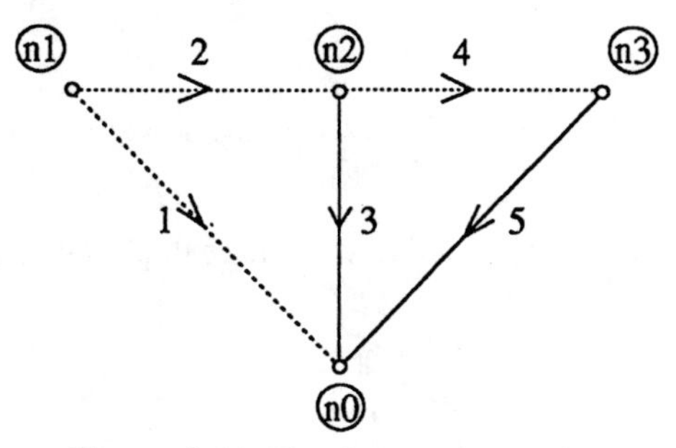

Figure A.11 Tree in sample graph.

As mentioned earlier, we can exchange columns 3 and 4 to obtain

$$A^{(4)} = \begin{bmatrix} 1 & 0 & 0 & 1 & 1 \\ 0 & 1 & 0 & -1 & -1 \\ 0 & 0 & 1 & 0 & -1 \end{bmatrix} \tag{A.7.7}$$

What we have now are the KCL cutset equations

$$\underbrace{\begin{bmatrix} 1 & 0 & 0 \\ 0 & 1 & 0 \\ 0 & 0 & 1 \end{bmatrix}}_{\boxed{1}} \vdots \underbrace{\begin{bmatrix} 1 & 1 \\ -1 & -1 \\ 0 & -1 \end{bmatrix}}_{\boxed{F}} \underbrace{\begin{bmatrix} i_1 \\ i_2 \\ i_4 \\ i_3 \\ i_5 \end{bmatrix}}_{\uparrow\ \hat{i}_b} = \begin{bmatrix} 0 \\ 0 \\ 0 \end{bmatrix} \tag{A.7.8}$$

Since we need only work with 0 and ± 1 valued elements and there are no carries, we can perform the transformation from A to $\begin{bmatrix} 1 \vdots F \end{bmatrix}$ very efficiently with integer -- or even logical -- operations on packed data.

The elementary operations that we performed above can be represented as the application to the reduced incidence matrix **A** of a sequence of *elementary matrices* as shown below.

Add row 2 to row 1 and multiply row 2 by -1:

$$\varepsilon_1 = \begin{bmatrix} 1 & 1 & 0 \\ 0 & 1 & 0 \\ 0 & 0 & 1 \end{bmatrix} \qquad \varepsilon_2 = \begin{bmatrix} 1 & 0 & 0 \\ 0 & -1 & 0 \\ 0 & 0 & 1 \end{bmatrix} \tag{A.7.9}$$

so that

$$A^{(2)} = \varepsilon_2 \varepsilon_1 A \tag{A.7.10}$$

Add row 3 to row 1, subtract row 3 from row 2, and multiply row 3 by -1:

$$\varepsilon_3 = \begin{bmatrix} 1 & 0 & 1 \\ 0 & 1 & 0 \\ 0 & 0 & 1 \end{bmatrix} \qquad \varepsilon_4 = \begin{bmatrix} 1 & 0 & 0 \\ 0 & 1 & -1 \\ 0 & 0 & 1 \end{bmatrix} \qquad \varepsilon_5 = \begin{bmatrix} 1 & 0 & 0 \\ 0 & 1 & 0 \\ 0 & 0 & -1 \end{bmatrix} \tag{A.7.11}$$

so that

$$A^{(3)} = \varepsilon_5 \varepsilon_4 \varepsilon_3 A^{(2)} = \varepsilon_5 \varepsilon_4 \varepsilon_3 \varepsilon_2 \varepsilon_1 A \tag{A.7.12}$$

Then exchange columns 3 and 4

$$\hat{\varepsilon}_1 = \begin{bmatrix} 1 & 0 & 0 & 0 & 0 \\ 0 & 1 & 0 & 0 & 0 \\ 0 & 0 & 0 & 1 & 0 \\ 0 & 0 & 1 & 0 & 0 \\ 0 & 0 & 0 & 0 & 1 \end{bmatrix} \tag{A.7.13}$$

so that

$$A^{(4)} = A^{(3)} \hat{\varepsilon}_1 = \varepsilon_5 \varepsilon_4 \varepsilon_3 \varepsilon_2 \varepsilon_1 A \hat{\varepsilon}_1 = \begin{bmatrix} \mathbf{1} & \vdots & F \end{bmatrix} \tag{A.7.14}$$

Note that if we omit the column (3 and 4) interchange, we get equally valid cutset equations without the branch current permutation

$$\varepsilon_5\varepsilon_4\varepsilon_3\varepsilon_2\varepsilon_1\boldsymbol{A}\boldsymbol{i}_b = \begin{array}{c} \text{tree branches} \\ \begin{array}{ccccc} \downarrow & \downarrow & & \downarrow & \end{array} \\ \begin{bmatrix} 1 & 0 & 1 & 0 & 1 \\ 0 & 1 & -1 & 0 & -1 \\ 0 & 0 & 0 & 1 & -1 \end{bmatrix} \\ \begin{array}{ccccc} & & \uparrow & & \uparrow \end{array} \\ \text{columns of } \boldsymbol{F} \end{array} \begin{bmatrix} i_1 \\ i_2 \\ i_3 \\ i_4 \\ i_5 \end{bmatrix} = \begin{bmatrix} 0 \\ 0 \\ 0 \end{bmatrix} \qquad \textbf{(A.7.15)}$$

We can look at the column permutation as follows.

$$\begin{bmatrix} \mathbf{1} & \vdots & \boldsymbol{F} \end{bmatrix} \hat{\boldsymbol{i}}_b = \boldsymbol{A}\boldsymbol{i}_b = 0$$

$$\varepsilon_5\varepsilon_4\varepsilon_3\varepsilon_2\varepsilon_1\boldsymbol{A}\hat{\varepsilon}_1\hat{\boldsymbol{i}}_b = \varepsilon_5\varepsilon_4\varepsilon_3\varepsilon_2\varepsilon_1\boldsymbol{A}\boldsymbol{i}_b = \mathbf{0} \qquad \textbf{(A.7.16)}$$

$$\hat{\varepsilon}_1\hat{\boldsymbol{i}}_b = \boldsymbol{i}_b$$

Therefore, $\hat{\varepsilon}_1$ represents the permutation of $\boldsymbol{i}_b$ that is necessary to manipulate A so that the first n columns are tree branches. Given the above, we can find the permuted branch voltages from

$$\hat{\boldsymbol{v}}_b = \begin{bmatrix} \mathbf{1} \\ \cdots \\ \boldsymbol{F}^T \end{bmatrix} \boldsymbol{v}_t = \hat{\varepsilon}_1^T\boldsymbol{A}^T\varepsilon_1^T\varepsilon_2^T\varepsilon_3^T\varepsilon_4^T\varepsilon_5^T\boldsymbol{v}_t \qquad \textbf{(A.7.17)}$$

Again, if we do not insist on permuting the branch voltage vector so that the tree branches are numbered first, we can ignore $\hat{\varepsilon}_1^T$:

$$\boldsymbol{v}_b = \boldsymbol{A}^T\varepsilon_1^T\varepsilon_2^T\varepsilon_3^T\varepsilon_4^T\varepsilon_5^T\boldsymbol{v}_t \qquad \textbf{(A.7.18)}$$

We recall that

$$\boldsymbol{v}_b = \boldsymbol{A}^T\boldsymbol{v}_n \qquad \textbf{(A.7.19)}$$

so

$$\boldsymbol{v}_n = \varepsilon_1^T\varepsilon_2^T\varepsilon_3^T\varepsilon_4^T\varepsilon_5^T\boldsymbol{v}_t \qquad \textbf{(A.7.20)}$$

If we want ultimately to find the node voltages, we can push the premultiplying elementary matrices onto a LIFO stack in the order that we find them. Then to obtain $\boldsymbol{v}_n$ from $\boldsymbol{v}_t$ we pop their transposes from that stack in inverse order to apply sequentially to $\boldsymbol{v}_t$ once it has been computed. Note that the elementary matrices are sparse, simple, and only come in three types, so we can code, store, and manipulate them efficiently.

Other than to identify the tree branches, we need not worry about the identity partition of $\left[\mathbf{1} \vdots \boldsymbol{F}\right]$. We can obtain the $\boldsymbol{F}$ partition merely by applying the elementary matrix operations to the potential link-related columns of $\boldsymbol{A}$ at any stage of the tree-finding process.

As a digression, we consider the possibility of having ideal switch elements in the circuit. A switch is a simple and useful generalization of many nonlinear elements. When the switch is open, it is equivalent to a zero-valued independent current source, and must be a link. When it is closed, it acts like a zero-valued independent voltage source and must be a tree branch, as shown in Figure A.12.

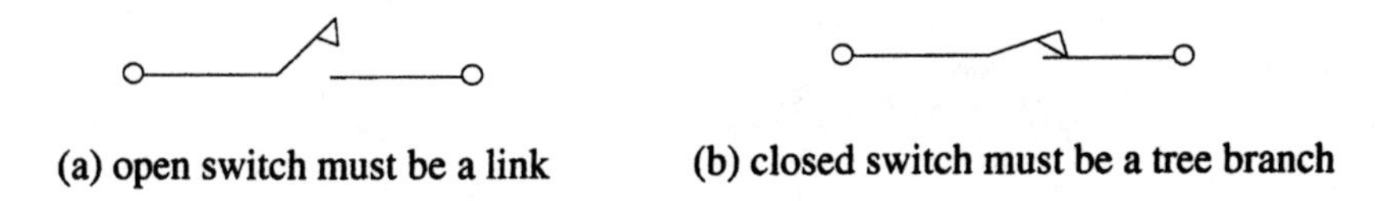

(a) open switch must be a link (b) closed switch must be a tree branch

Figure A.12 Ideal switches in Tree/Link Analysis.

As switches open and close, we must dynamically change tree/link partitioning priorities. Even though finding a tree for the first time may entail some effort, new trees can dynamically be found on subsequent passes in an incremental manner. The procedure is as follows:

- Make initially closed switches lowest priority among the independent voltage sources of the circuit.
- Make initially open switches highest priority among the independent current sources of the circuit.
- Store elementary operations or matrices during the original tree determination procedure on a stack.
- Any time that a switch changes state, change the position of that column of the $\boldsymbol{A}$ matrix as dictated by the new priorities. Then pop elementary matrices off the stack until the one(s) that dealt with the column corresponding to that particular switch. These operations are now useless and must be redone. However, operations that are left on the stack are still useful and need not be redone. Then the elementary matrices required to find the new tree are computed and pushed back on the stack to help find the new tree next time one of the switches changes state.

For example, in the graph of Figure A.10, any of the new tree priority orders {1,2,3,5,4}, {1,2,5,3,4}, {1,5,2,3,4} and {5,1,2,3,4} results merely in branch 5 becoming a tree branch and branch 4 a link.

Note that there are other methods to determine a tree given the priority ordering of the branches. The matrix method presented in this section has an asymptotic order of complexity that is superlinear in the size of the graph. It also requires the incidence matrix to be stored, which can take up a lot of memory for large circuits. However, the technique can be used to find the tree as well as the $\boldsymbol{F}$ matrix. The elementary matrix method is equivalent to the matrix method, and would only be used in a situation where the tree had to be dynamically reformulated frequently. If we had to implement elementary matrices in a computer program, we would not actually store entire matrices and carry out full matrix multiplication -- there are implementation techniques that are more efficient and compact. There are also graph searching methods [Deo74] of finding trees that have linear computational complexity. Once the tree has been determined, fundamental loops and cutsets can also be identified by graph-searching methods. Depending on the application, circuit equations may be formulated directly by means of stamps. Hence the implementation of tree-determination and the construction of the $\boldsymbol{F}$ matrix is often application-dependent.

A.8 References

[Kuh65] E. S. Kuh and R. A. Rohrer. The State Variable Approach to Network Analysis. *Proceedings of the IEEE*, vol. 53, pp. 672-686, July 1965.

[Rohrer70] R. A. Rohrer. *Circuit Theory: An Introduction to the State Variable Approach.* McGraw-Hill, 1970.

[Hachtel71] G. D. Hachtel, R. K. Brayton, and F. G. Gustavson. The Sparse Tableau Approach to Network Analysis and Design. *IEEE Transactions on Circuit Theory*, vol. CT(18), pp. 101-118. January 1971.

[Russo71] P. M. Russo and R. A. Rohrer. The Tree-link Analysis Approach to Network Analysis. *IEEE Transactions on Circuit Theory*, vol. CT-18(3), pp. 400-403, May 1971.

[Weeks73] W. T. Weeks, A. J. Jimenez, G. W. Mahoney, D. Mehta, H. Quassemzaheh, and T. R. Scott. Algorithms for ASTAP - a Network Analysis Program. *IEEE Transactions on Circuit Theory*, vol. CT(20), pp. 628-634, November 1973.

[Deo74] N. Deo. *Graph Theory with Applications to Engineering and Computer Science.* Prentice-Hall, 1974.

[Chua75] L. O. Chua and P. M. Lin. *Computer-aided Analysis of Electronic Circuits: Algorithms and Computational Techniques.* Prentice-Hall, 1975.

[Bondy76] J. A. Bondy and U. S. R. Murty. *Graph Theory with Applications.* North Holland, 1976.

Index

H

I

K

L

M

O

P

R

S

T

V

W

Z

About the Authors

LAWRENCE T. PILLAGE is an associate professor of Electrical and Computer Engineering at the University of Texas, Austin, and a Temple Endowed Faculty Fellow in Engineering. He was an integrated circuit and system designer at Westinghouse Research and Development from 1984 to 1986, where he was recognized with the corporation's highest engineering achievement award.

RONALD A. ROHRER is the Wilkoff University Professor of Electrical and Computer Engineering at Carnegie-Mellon University. In addition to prior academic positions, he has been involved in a variety of research and development, marketing, and management positions in the electronic design automation industry.

CHANDRAMOULI VISWESWARIAH is a research staff member at IBM's T. J. Watson Research Center in Yorktown Heights, New York. His research interests include circuit and timing simulation, modeling, circuit optimization, and mixed mode simulation.